Statistics for Marketing and Consumer Research

Statistics for Marketing and Consumer Research

Mario Mazzocchi

SAGE

Los Angeles • London • New Delhi • Singapore

Statistics for marketing and consumer research 2008.

2009 02 09

© Mario Mazzocchi 2008

First published 2008

Apart from any fair dealing for the purposes of research or
private study, or criticism or review, as permitted under the
Copyright, Designs and Patents Act, 1988, this publication
may be reproduced, stored or transmitted in any form, or by
any means, only with the prior permission in writing of the
publishers, or in the case of reprographic reproduction, in
accordance with the terms of licences issued by the Copyright
Licensing Agency. Enquiries concerning reproduction
outside those terms should be sent to the publishers.

SAGE Publications Ltd
1 Oliver's Yard
55 City Road
London EC1Y 1SP

SAGE Publications Inc.
2455 Teller Road
Thousand Oaks, California 91320

SAGE Publications India Pvt Ltd
B 1/I 1 Mohan Cooperative Industrial Area
Mathura Road
New Delhi 110 044

SAGE Publications Asia-Pacific Pte Ltd
33 Pekin Street #02-01
Far East Square
Singapore 048763

Library of Congress Control Number: 2007936615

British Library Cataloguing in Publication data

A catalogue record for this book is available from
the British Library

ISBN 978-1-4129-1121-4
ISBN 978-1-4129-1122-1 (pbk)

Typeset by CEPHA Imaging Pvt. Ltd., Bangalore, India
Printed in India at Replika Press Pvt Ltd
Printed on paper from sustainable resources

To Lucie

Contents

Preface   xii

List of Acronyms   xvii

Part I   Collecting, Preparing and Checking the Data   **1**

1   Measurement, Errors and Data for Consumer Research   **2**
   1.1   Measuring the world (of consumers): the problem of
          measurement   3
   1.2   Measurement scales and latent dimensions   6
   1.3   Two sample data-sets   11
   1.4   Statistical software   18
          Summing up   23
          Exercises   23
          Further readings and web-links   25
          Hints for more advanced studies   26
          Notes   26

2   Secondary Consumer Data   **27**
   2.1   Primary and secondary data   28
   2.2   Secondary data sources   32
   2.3   Household budget surveys   36
   2.4   Household panels   40
   2.5   Commercial and scan data   41
          Summing up   43
          Exercises   44
          Further readings and web-links   45
          Hints for more advanced studies   45
          Notes   45

3   Primary Data Collection   **46**
   3.1   Primary data collection: surveys errors and the research design   47
   3.2   Administration methods   53
   3.3   Questionnaire   58
   3.4   Four types of surveys   63
          Summing up   74
          Exercises   75
          Further readings and web-links   76
          Hints for more advanced studies   76
          Notes   76

4 **Data Preparation and Descriptive Statistics** 77
4.1 Data preparation 78
4.2 Data exploration 85
4.3 Missing values and outliers detection 93
Summing up 100
Exercises 101
Further readings and web-links 102
Hints for more advanced studies 102
Notes 102

Part II Sampling, Probability and Inference 103

5 **Sampling** 104
5.1 To sample or not to sample 105
5.2 Probability sampling 108
5.3 Non-probability sampling 123
Summing up 125
Appendix 126
Exercises 128
Further readings and web-links 128
Hints for more advanced studies 129
Notes 129

6 **Hypothesis Testing** 130
6.1 Confidence intervals and the principles of hypothesis testing 130
6.2 Test on one mean 137
6.3 Test on two means 139
6.4 Qualitative variables and non-parametric tests 142
6.5 Tests on proportions and variances 145
Summing up 146
Exercises 147
Further readings and web-links 147
Hints for more advanced studies 148
Notes 148

7 **Analysis of Variance** 150
7.1 Comparing more than two means: analysis of variance 151
7.2 Further testing issues in one-way ANOVA 154
7.3 Multi-way ANOVA, regression and the general linear model (GLM) 160
7.4 Starting hints for more complex ANOVA designs 162
Summing up 167
Exercises 167
Further readings and web-links 168
Hints for more advanced studies 169
Notes 169

Part III   Relationships Among Variables   **171**

8   Correlation and Regression   **172**
   8.1   Covariance and correlation measures   173
   8.2   Linear regression   179
   8.3   Multiple regression   185
   8.4   Stepwise regression   187
   8.5   Extending the regression model   189
      Summing up   193
      Exercises   193
      Further readings and web-links   194
      Hints for more advanced studies   195
      Notes   195

9   Association, Log-linear Analysis and Canonical Correlation
Analysis   **196**
   9.1   Contingency tables and association statistics   196
   9.2   Log-linear analysis   199
   9.3   Canonical correlation analysis   208
      Summing up   214
      Exercises   215
      Further readings and web-links   216
      Hints for more advanced studies   216
      Notes   217

10   Factor Analysis and Principal Component Analysis   **218**
   10.1   Principles and applications of data reduction techniques 219
   10.2   Factor analysis   221
   10.3   Principal component analysis   229
   10.4   Theory into practice   235
      Summing up   244
      Exercises   245
      Further readings and web-links   246
      Hints for more advanced studies   246
      Notes   246

Part IV   Classification and Segmentation Techniques   **247**

11   Discriminant Analysis   **248**
   11.1   Discriminant analysis and its application to consumer and
       marketing data   248
   11.2   Running discriminant analysis   250
   11.3   Multiple discriminant analysis   254
      Summing up   260
      Exercises   260
      Further readings and web-links   261
      Hints for more advanced studies   261
      Notes   262

12   Cluster Analysis   263
   12.1   Cluster analysis and its application to consumer and
           marketing data   263
   12.2   Steps in conducting cluster analysis   265
   12.3   The application of cluster analysis in SAS and SPSS – empirical
           issues and solutions   271
        Summing up   280
        Exercises   280
        Further readings and web-links   281
        Hints for more advanced studies   281
        Notes   282

13   Multidimensional Scaling   283
   13.1   Preferences, perceptions and multidimensional scaling   283
   13.2   Running multidimensional scaling   286
   13.3   Multidimensional scaling and unfolding using SPSS and SAS   292
        Summing up   297
        Exercises   297
        Further readings and web-links   298
        Hints for more advanced studies   299
        Notes   299

14   Correspondence Analysis   300
   14.1   Principles and applications of correspondence analysis   300
   14.2   Theory and techniques of correspondence analysis   303
   14.3   Running correspondence analysis   305
        Summing up   311
        Exercises   312
        Further readings and web-links   312
        Hints for more advanced studies   313
        Notes   313

Part V   Further Methods in Multivariate Analysis   315

15   Structural Equation Models   316
   15.1   From exploration to confirmation: structural equation
           models   316
   15.2   Structural equation modeling: key concepts and
           estimation   318
   15.3   Theory at work: SEM and the Theory of Planned
           Behavior   322
        Summing up   333
        Exercises   333
        Further readings and web-links   335
        Hints for more advanced studies   335
        Notes   336

16   Discrete Choice Models   337
   16.1   From linear regression to discrete choice models   337
   16.2   Discrete choice models   339

16.3	Discrete choice models in SPSS	342
16.4	Choice modeling and conjoint analysis	347
	Summing up	349
	Exercises	350
	Further readings and web-links	351
	Hints for more advanced studies	351

17 The End (and Beyond) | **352**
17.1	Conclusions	353
17.2	Data mining	353
17.3	The Bayesian comeback	354
	Summing up	358
	Notes	358

Appendix Fundamentals of Matrix Algebra and Statistics | **359**
A.1	Getting to know x and y	359
A.2	First steps into statistical grounds	367
	Notes	374

Glossary | **375**

References | **387**

Index | **399**

Preface

I read somewhere (probably my Editor told me...) that sales of a marketing research textbook are inversely proportional to the number of equations. I wonder whether there is an equation to quantify the above relationship? This is a good example on how marketing experts and statisticians relate to each other. However, it is without doubt that a good marketing strategy needs good numbers. Nowadays, getting numbers and statistics out of a software is rarely a problem, and the big step is making good use of the increasingly powerful computing and statistical tools.

Thus, complex methods that once were exclusive to highly trained statisticians are now physically accessible to most students and practitioners through relatively user-friendly software like SPSS. Still, physical access does not really mean that the methods are entirely accessible. Attempts to understand the functioning of complex methods out of the help windows of statistical packages, web-sites or forums may be a frustrating and misleading experience.

This is where the idea of this book stems from. The main objective is to provide a soft acquaintance with the main multivariate statistical methods employed in consumer and marketing research. Most students in the final undergraduate year or in a postgraduate course in consumer and marketing disciplines have already had an introduction to the founding techniques in marketing research, with an emphasis on data collection and preliminary data analysis. After that, usually the focus is on hypothesis testing, correlation and regression, as in the brilliant textbook by Andy Field (2006), frequently referenced in this book. With no pretension of replicating the success and completeness of Field's textbook, this one shares some of its objectives (a quest for simplification and the use of SPSS) and extends the discussion to multivariate methods not covered there, namely cluster analysis, multidimensional scaling, correspondence analysis, structural equation modeling, discrete choice models and sampling techniques. To ensure a fair coverage, the first four chapters summarize measurement and data issues, chapter 5 explores sampling while chapters 6 to 9 discuss hypothesis testing, ANOVA, correlation, regression and log-linear analysis, albeit not so extensively as in Field's book.

The step between introductory statistical methodologies and more advanced multivariate methods can be a difficult one, because it usually requires a deeper familiarity with equations, matrix algebra and theoretical statistics. Actually, it is almost impossible to get rid of equations and matrix algebra, which would make (most) students and (most) editors happy, but a text which explains multivariate research methods only requires a minimal formalization to provide the basic understanding. Furthermore, the physical accessibility to powerful software suggests that explanations may become easier if accompanied with practical examples. Thus, as discussed later in this preface, all applied examples in this textbook are based on data-sets which are available to readers, so that examples can be replicated and extended. For a better grasp of the remaining technicalities, this book tries to go back to the basics of matrix algebra in a final appendix, which can be ignored by those who are already familiar with matrices and probability distributions.

In many instances, attempts to simplify the treatment do not give justice to the power and flexibility of methods. Being short and simple is not fair to the myriad of researchers who have dissected and are currently dissecting the statistical techniques covered in the following 17 chapters, with fantastic progress in the usefulness of statistics for the 'real' world. Thus, this book responds to two hopes. First, that it may be a relatively easy introduction to sometimes complex statistical techniques and their application to consumer research. Second, that the relatively light mathematical treatment might invite more advanced readings. To this purpose, each chapter is concluded by a selection of further reading and suggestions for becoming more knowledgeable in our preferred technique(s).

Intended audience

From the above considerations, it follows that this book is targeted at undergraduate students in their final year, MSc or postgraduate students, preferably with some preliminary knowledge of the essential marketing research, mathematics and statistics, although all concepts and terminology relevant to full comprehension are quickly explained in the final appendix.

The book can be read at different levels. All chapters emphasize an intuitive explanation of the techniques and their application to marketing and consumer research, which aims to be useful to all readers. Where necessary or useful, some chapters do get into technical details (for example the chapters on factor analysis or particular aspects of sampling, linear regression, structural equation models and discrete choice models), but these parts are either confined in non-essential paragraphs or boxes (see discussion of the book's structure below) or equations are accompanied by hopefully adequate explanations. All of the techniques are discussed here in a chapter or less, but have been the subjects of entire book. Thus, there will be readers who wish to go beyond the basic understanding and applications to gain more detailed knowledge about the methods. For this reason, special care has been devoted to provide advanced references for those issues that seemed more relevant and each chapter ends with suggestion for further readings and study.

Structure of the book

This book is divided in five parts. The **first part** includes four chapters about measurement and data, with emphasis on collection, preparation, description and quality checks. It is not intended to be a detailed discussion of the multitude of aspects, but as an introduction to multivariate techniques it is important to recall where the data come from and how they are produced. Although the first few chapters briefly cover the founding steps of social research and data collection, they are probably not sufficiently exhaustive as compared to the range of issues which emerge in measurement theory, data collection and access to secondary data. Luckily, there is a myriad of good textbooks in the market which provide a more comprehensive discussion of these topics.

The book then moves to the **second part**, which takes a step forward and makes use of probability concepts to introduce sampling techniques and hypothesis testing. It starts with a sampling chapter, since the techniques explored in subsequent chapters are grounded on probabilistic extraction of samples. Probability designs open the way to inference and hypothesis testing techniques, like mean comparison tests and analysis of variance. Again, treatment is not comparable with Field's book, but it is intended to work as a quicker guide to the application of these techniques.

The **third part** starts looking into multivariate relationships among variables, first with a chapter on correlation and regression, then extending the discussion to the broader concept of association of qualitative variables, with an eye to more advanced techniques like log-linear analysis and canonical correlation analysis. The concepts used for correlation and regression should help the reader through the chapter on data reduction techniques, factor and principal component analysis.

The **fourth part** dives into deeper waters of multivariate analysis with four chapters reviewing multivariate classification and segmentation techniques, discriminant analysis, cluster analysis, multidimensional scaling and correspondence analysis.

The book ends with a **fifth part** which explores two more techniques looking into relationships among variables, but more advanced and complex than those of the third section. These techniques are those of structural equation modeling and discrete choice modeling, while the final chapter is a rather personal summary of the (intended) achievements of the book, plus a rather short and rough guide to new directions in marketing research, like data mining and the application of Bayesian techniques.

A quick guide to reading the book

One of the hopes of this book is avoiding being 'another textbook in marketing research.' It is not an easy task, but one feature that could help in that direction is the attempt to explain theory in a concise way and with applied examples, while covering more techniques than the typical marketing research book. A second aim of this book is to serve as a quick reference, which means that one should be able to grasp the key aspects of a single method without too much reading. Again, this can be accomplished by allowing readers to distinguish what is essential to know before starting their SPSS or SAS analysis from more advanced technicalities. The book tries to guide readers across the pages by following a structure enabling different levels of reading and study.

For example, readers will find four types of boxes scattered throughout the book. Boxes generally contain independent and more specific information that is not necessarily relevant to all readers, like the description of a specific secondary data-set, the explanation of features in packages other than SPSS, and the mathematical equations for sampling methods or some examples from marketing research.

> Boxes like this add some guidelines and further explanation to a topic – useful for quick reference

> Boxes like this highlight examples – readers may skip them if they have already understood

> Boxes like this gets a bit deeper into methods and technicalities – for starting more advanced studies

> Boxes like this provide information on computer packages, mainly SPSS – helpful in trying techniques out

After a content summary, each chapter is concluded by a set of three exercises (largely based on SPSS applications to the sample data-sets), by a selection of interesting points for further study, accompanied by a few highly selected references and web-links. Some hints for further (and usually quite advanced) study are also provided in the form of questions.

Supplementary features

A companion web-site (www.sagepub.co.uk/mazzocchi) accompanies this textbook to direct readers to a range of additional features for students and lecturers:

- Three SPSS data-sets, including the EFS and Trust data-sets used throughout the book (see chapter 1) and the MDS data-set for the application of multidimensional scaling techniques; plus, for each chapter;
- PowerPoint lecture slides for each chapter;
- A set of multiple choice questions per chapter

The web-site also provides more interactive evolving features, including a FAQ section and SPSS syntax for running the techniques, plus an opportunity to provide comments, feedback and (quite likely) submit corrections to an *errata corrige*, as interaction with readers is greatly appreciated.

In summary the features of this book are:
- Concise and relatively light treatment of multivariate techniques for consumer and marketing research
- Exemplification through SPSS, using the same two data-sets throughout the book
- Access to an evolving web-site with further information.

Acknowledgments

So many people helped me (knowingly or unknowingly) in this adventure, that I am really worried about forgetting someone. First the editorial team at Sage, Patrick and Claire. Patrick must be a trained psychologist to understand when to encourage me and when to be firm… (thanks for the excellent London meal, too, I will return the invitation if I sell more than 20 copies). The long list of people who contributed to turn a set of messy pages into a proper *looking* book includes Rosemary, Rathi, Nic and Jais plus many others whose names I don't even know, but my gratitude is for all of them, for their patience and e-mailing skills.

Second, this whole project was born a few years back at the Department of Agricultural and Food Economics of the University of Reading, where I found a fantastic work environment, thanks to a brilliant group of colleagues and friends. This must be why at first the book project did not seem so complicated as it turned out to be.

Thanks also to my undergraduate and postgraduate students, excellent guinea pigs for this material and especially the PhD and post-doctoral students at the University of Bologna. Among colleagues who helped me on practical grounds, a special mention for Bhavani Shankar and also many thanks for their help to Silvia Cagnone, Luca Fanelli, Maddalena Ragona, Garth Holloway (he's my Bayesian friend of chapter 17 – although I did not show him that chapter), Alexandra Lobb, Stephanie Chambers, Sara Capacci, Francesca Bruno, Sergio Brasini, Bruce Traill, and all my current colleagues-friends, especially my office-mates Rosella, Cristina and Roberto. All these people, plus others I have forgotten and (especially) those anonymous reviewers, helped me to complete and improve this effort. All remaining errors and confusing bits are solely mine, so the least I can do is to maintain an errata corrige section in the above mentioned web-site (by the way, thanks Orazio!) and reply to readers' (if any…) questions and comments (…if any).

Dedication

This book is dedicated to Lucie, who has patiently and lovingly supported me in climbing this mountain (first draft), then crossing the ocean (revision and indexing), coping with my night shifts in the 'Altana tower' and monothematic conversations.

My gratitude also goes to my parents and family (with a special mention for Luca and Clara) and to my friends, especially (alphabetical order) Barbara, Elena, Giorgio, Jessica, Jumpi, Meie, Monica, Nadia, Paolo and Valentina, who showed their enthusiasm to the project although they will never read the book apart from the cover and dedication section.

List of Acronyms

AGFI	Adjusted Goodness-of-Fit Index
AIC	Akaike Information Criterion
AMOS	Analysis of Moment Structures
ANCOVA	Analysis of Covariance
ANOVA	Analysis of Variance
APS	Annual Population Surveys
BHPS	British Household Panel Survey
BIC	Schwartz (Bayes) Information Criterion
CAPI	Computer-Assisted Personal Interview
CATI	Computer-Assisted Telephone Interviews
CAWI	Computer-Assisted Web Interviews
CA	Cluster Analysis
CCA	Canonical Correlation Analysis
CCC	Cubic Clustering Criterion
CES	US Consumer Expenditure Survey
CFI	Comparative Fit Index
CMDS	Classical Multidimensional Scaling
CMIN	Minimum Sample Discrepancy
COICOP	Classification of Individual Consumption by purpose
DA	Discriminant Analysis
DF	Degrees of Freedom
DM	Data Mining
DEFRA	Department for Environment, Food and Rural Affairs etc
ECHP	European Community Household Panel
EFS	Expenditure and Food Survey
EPM	External Preference Mapping
EPOS	Electronic Point of Sale Scanners
EU-SILC	(European Union) Statistics on Income and Living Conditions
GFI	Goodness-of-Fit Index
GLLM	General Log-Linear Model
GLM	General Linear Model
HBS	Household Budget Surveys
HHFCE	Household Final Consumption Expenditure
HMSO	UK Statistical Office
HSD	Honestly Significant Difference
IFI	Incremental Fit Index
IPM	Internal Preference Mapping
ISER	Institute for Social and Economic Research
ITP	Intentions to Purchase
KDD	Knowledge Discovery in Databases
MANOVA	Multivariate Anova

MAR	Missing at Random
MANCOVA	Multivariate Analysis of Covariance
MCA	Multiple Correspondence Analysis
MDA	Muliple Discriminant Analysis
MDS	Multidimensional Scaling
MLE	Maximum Likelihood Estimator
NCP	Non-Centrality Parameter
NFI	Normed Fit Index
OLAP	On-Line Analytical Processing
ONS	Office for National Statistics
PBC	Perceived Behavioral Control
PCA	Principal Component Analysis
PCLOSE	Test of Close Fit
REGWF	Ryan-Einot-Gabriel-Welsch (F distribution)
REGWQ	Ryan-Einot-Gabriel-Welsch (Studentized-range distribution)
RFI	Relative Fit Index
RMR	Root Mean Square Residual
RMSE	Root Mean Standard Error
RMSEA	Root Mean Square Error of Approximation
RNI	(Shepard's) Rough Non-degeneracy Index (MDS)
RNP	Relative Non-centrality Parameter
SAS	Statistical Analysis Software
SEM	Structural Equation Model
SN	Subjective Norm
SNK	Studentized Newman-Keuls
SPSS	Statistical Analysis Software
SRS	Simple Random Sampling
SSR	Regression deviance (Sum of Squares)
SST	Total deviance (Sum of Squares)
SSE	Sum of Squared Errors
TLI	Tucker-Lewis Coefficient
TNS	Taylor-Nelson Sofres
TPB	Theory of Planned Behavior
WLS	Weighted Least Squares

PART I

Collecting, Preparing and Checking the Data

This first part of the book synthesizes the founding steps of the statistical analysis, those leading to the construction of a data-set meeting the necessary quality requirements.

Chapter 1 reviews the basic measurement issues and measurement scales, which are the elemental tools for measuring constructs whose quantification is not straightforward. It also introduces the distinction between metric (scale) and non-metric (nominal or ordinal) variables and the concept of error, central to statistical theory. Finally, it presents two sample data-sets which are used for examples throughout the book, together with a short review of the most popular statistical packages for data analysis. The distinction between secondary and primary data is explored in **chapter 2** and **chapter 3** respectively. **Chapter 2** deals with the use of secondary data, that is, information not explicitly collected for the purpose of the research, and explores the main available sources for secondary data relevant to consumer and marketing research, bringing examples of official consumer surveys in the UK, Europe and the US. **Chapter 3** gets into the process of primary data collection and provides a synthesis of the main steps of the data collection process, with an overview of planning and field work issues. This chapter emphasizes the role of non-random errors in collecting data as compared to random errors. **Chapter 4** moves a step forward and begins with the applied work. First, an overview of data quality issues, diagnostics and some solutions is provided, with a special emphasis on missing data and outlier problems. Then, with the aid of the SPSS examples based on the two sample data-sets, it illustrates some graphical and tabular techniques for an initial description of the data-set.

CHAPTER 1

Measurement, Errors and Data for Consumer Research

T HIS CHAPTER reviews the key measurement and data collection issues to lay the foundation for the statistical analysis of consumer and marketing data. Measurement is inevitably affected by errors and difficulties, especially when the objects of measurement are persons and their psychological traits, relevant to consumer behavior. The chapter looks at the distinction between systematic and random sources of error, then presents the main measurement scales and discusses the concept of latent dimension for objects – like quality – that cannot be defined or measured in an unequivocal way. Finally, this chapter introduces the two data-sets exploited for applied examples throughout the book and briefly reviews the main statistical packages commonly used in consumer and marketing research.

Section 1.1 introduces the problem of measurement in relation to consumer research

Section 1.2 introduces the fundamental measurement scales and data types

Section 1.3 presents the two data-set used in applied examples throughout the book

Section 1.4 provides a brief overview of commercial software for data analysis

THREE LEARNING OUTCOMES

This chapter enables the reader to:

➡ Focus on the main problems in measuring objects and collecting data
➡ Review the key data types and measurement scales
➡ Be introduced to the data-sets and statistical packages used later in the book

PRELIMINARY KNOWLEDGE: It is advisable to review the basic probabilistic concepts described in the appendix, with particular attention to frequency and probability distributions and the normal curve.

1.1 Measuring the world (of consumers): the problem of measurement

Measuring the World is the title of an enjoyable novel built upon the lives of the most famous of statisticians, Carl Friedrich Gauss, and Alexander von Humboldt, the naturalist and explorer who founded biogeography.[1] The two scientists have something in common besides their German roots: they both devoted their lives to the problem of measuring objects. Interestingly, the novel portraits Gauss as a quite unsociable mathematician who preferred working at home with numbers, while von Humboldt was traveling around Latin America and Europe with his instruments to measure scientifically lands and animal species. Consumer research needs good measurement and a bit of both personalities, that is, some 'unsociable' math and hard field work. These ingredients can create good marketing intelligence. The good news is that both scientific celebrities were successful with money ... in the sense that their faces ended up on German banknotes.

This chapter lays the foundations for reading the rest of the textbook. While the math is kept to a minimum, those readers willing to refresh or familiarize with the key mathematical notation and the basic concepts of matrix algebra, probability and statistics are invited to read the appendix, which should help them get a better grasp of the more complex concepts discussed in the rest of the textbook.

However, before getting into quantification and data analysis, it is probably useful to go back to the origin of data and think about the object of measurement and the final purpose of consumer and marketing research. There are several definitions of marketing. A widely used one is that adopted by the Chartered Institute of Marketing (CIM), which states that 'Marketing is the management process which identifies, anticipates, and supplies customer requirements efficiently and profitably.'

Little more is required to show how consumer research is the foundation of any marketing activity. While in the CIM definition the target is the customer, the focus here is rather on consumers in general, as one might argue that customers are only a subset of consumers. Politically correct language is now even suggesting that the term citizen should be preferred, although few publishers would consider a title on citizen research.

As a matter of fact, consumer research should ideally be able to explore even those behaviors and attitudes which consumers are not even consciously considering at the time of the study. The basis of product innovation is to create something which will be welcome by those consumers who are not yet customers.

Let us take one of the controversial topics in current consumer research – genetically-modified foods. This is one of the biggest challenges for those who want to measure and predict behavior relying on a scientific approach. What is the value of asking someone an opinion, an attitude or a statement on purchasing intentions for something which they don't know or – in many circumstances – does not even exist? Are there procedures to ensure that the collected data is not completely useless? What is the point of running complex statistical analysis on huge samples if the single data in itself has little meaning?

A well-known example of consumer research failure is the New Coke case in 1985 (see Gelb and Gelb, 1986), when Coca Cola proposed a new formula based on blind sensory tests on 200,000 consumers, testing preferences versus Old Coke and Pepsi. Since tests were blind, research ignored the role of brand image and failed to predict consumer reaction, which was a wide rejection of the new product in favor of the old one, so that Coca Cola had to go back to a single product after 77 days, with a cost $35 billion. While some conspiracy theorists argue that the New Coke case was a successful marketing move to strengthen brand loyalty, this is an unlikely justification for the waste of resources in running the consumer research.

This sort of failure is one of the preferred arguments for those researchers who advocate qualitative research methods over quantitative and statistically-based consumer research.

While this whole textbook argues that only probability and rigorous statistics allow generalization of results from a sample to the target population, it is difficult to deny that 'ten thousand times nothing is still nothing' and big representative samples are useless if not misleading when the collected information is not appropriate.

1.1.1 Issues in measurement

Get a ruler and measure the width of the cover of this book. Repeat the exercise 20 times and write down the measurements. Then get a different instrument, a folding rule, and measure the book width again for 20 times. It would be surprising if you had *exactly* the same measurement 40 times, especially when your instruments are slightly different. More precise instruments and higher target precisions of the measure make it less likely that all observations return the same measure. There may be different reasons for this, for example a slightly different inclination of the ruler, an error in reading the measure or a minor difference between the ruler and the folding rule. Measuring objects is not that straightforward.

Things get much more complicated when the objects of measurement are persons or their tastes, attitudes, beliefs ... Needless to say, it might be a waste of time to run complex statistical methods on badly measured information.

The key point here is that empirical measurements and the characteristics being measured are different things, although they are related. *Measurement* is usually defined as the assignment of numerals to objects or events according to rules (Stevens, 1946). Thus, a transparent solution to the problem of measuring requires a clear and explicit definition of the measurement rule including its mathematical and statistical properties. The rule is usually defined as the *scale of measurement*.

Before getting into the discussion of measurement scales, it is worth looking at other aspects of measurement theory, those more directly related with the statistical foundation of this book. For example, before using any statistical methodologies one should carefully check whether it is appropriate (Stevens uses the word 'permissible') for a given measurement rule. Another issue which follows from the initial example of this section regards the relationship between *measurement* and *reality*, with an eye at the concepts of precision and accuracy. While at a first glance it might look like a purely philosophical issue, the question is whether 'numbers are real' is quite important to obtain meaningful results from data analysis. For example, take a measurement of attitude toward math taken in three different years on a class of first-year undergraduates. If the measured value increases over time, does it mean that first-year students increasingly like math?

This example draws a line between the contributions of measurement and statistical theory to the knowledge of reality. On the one hand, more accurate measurements lead to better research conclusions. On the other hand, statistics is needed to deal with measurement error (unavoidable, to some extent), to generalize measurements taken on samples and to explore relationship among different measurements.

1.1.2 Errors and randomness

Measuring objects is a big step toward learning about a phenomenon, but statistics can further increase the value of the collected data. Statistics helps through two main routes: dealing with *errors* and *inference*.

It should be clear from the first paragraphs of this chapter that measurements are subject to errors, for many reasons. Consider again the example of the width of this book's cover.

Even if the width is a constant, as it is not supposed to change unless something bad happens to the book, it is likely to get different measures over 40 attempts. In other words, our measures are an approximation of the true measure.

Errors in measurement are potentially made of two main components, a systematic error and a random error. In other words, each measure can be expressed as:

$$empirical\ measure = true\ value + systematic\ error + random\ error$$

A *systematic error* is a bias in measurement which makes each of the measures systematically too high or too low. For example, suppose the ruler used to measure the width of the book is made of a material which is sensitive to humidity. In very humid days it becomes slightly longer, so that it reads 10 cm when the actual measure is 10.1 cm. Measures taken in humid days are subject to a systematic downward bias. There are several potential sources of systematic errors in consumer research; some of them are discussed in relation to survey and data collection (see chapter 3). For example, if a lecturer asks students how much they enjoyed the lecture on a scale between 1 and 9, using a non-anonymous questionnaire, it is quite likely that responses will show a systematic upward bias. Ignoring systematic errors can have dramatic effects on the quality of research, but it is quite difficult to identify them after the data have been collected. Thus, countermeasures should be taken before running a survey, as discussed in chapter 3.

Instead, *random errors* are fluctuations which do not follow any systematic direction, but are due to factors that act in a random fashion. For example, small changes in the inclination of the ruler can lead to differences in measurements, in either direction. A consumer asked to rate the taste of a cheese, could give different ratings depending on how much time has elapsed since last meal. Luckily, randomness is never a problem for statisticians. The idea (further discussed in chapter 5) is that across a sufficiently large sample positive random errors compensate negative random errors, so that a sum of all random errors would be zero. This is a key point for dealing with errors – while with a single measurement it is not possible to quantify the amount of random error, over multiple measurements the 'average' (or total) random error would near zero. This means that the sample mean is the best possible measure for the true value, provided that there are no systematic errors.

This opens the way to the foundation of the statistical theory of errors and its key element: the normal (or Gaussian) curve. The normal curve is the probability distribution representing perfect randomness around a mean value and is bell-shaped. The larger are the random errors (less precise measurements), the flatter is the bell-shaped curve.

All of the methods discussed in this book deal with normal distributions in some way, but they play a major role especially in sampling theory (see chapter 5) and hypothesis testing (see chapter 6). At this point, before getting acquainted with the normal curve, those who feel little familiarity with the essential statistical and probability concepts are advised to have a look at the appendix.

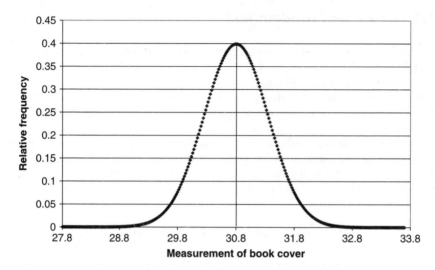

Figure 1.1 *Measurement error and the normal distribution*

Consider error theory again and the example on the width of the book cover. Suppose the 'true' width is 30.8 cm and that some 100,000 measurements were made. By plotting the relative frequencies on a graph, with the measurements on the horizontal axis and their relative frequencies on the vertical axis, according to error theory the resulting graph should look as the one in figure 1.1

This is the shape of a normal distribution. The average value (30.8 cm) is also the most likely one and – provided there are no systematic errors – corresponds to the true value.

As one moves away from the mean value, probability decreases symmetrically. Thus, the probability of committing a 1 cm error in excess is equal to the probability of a 1 cm error in defect. Besides error, the normal curve may well represent the influence of other random factors (see appendix).

1.2 Measurement scales and latent dimensions

The *Oxford English Dictionary* defines measurement as 'A dimension ascertained by measuring; a magnitude, quantity, or extent calculated by the application of an instrument or device marked in standard units.' It is an interesting definition, as it relates to a *latent dimension* and some sort of instrument which translates that dimension into some standardized unit. The latent dimension is real and unique, measurement is artificial and not unique. Thus, the researcher may choose the instrument for translating the latent dimension into units. A *latent* dimension or construct is something which cannot be defined and measured in an unequivocal and verifiable way. For example, any consumer would recognize in products and services a quality dimension, and some products are of better quality than others. However, the quality dimension of a product is latent, since it is impossible to quantify the quality content of a product with a single and objective measure. While quality is a clear example, one may argue that also the width of this book's cover could be a latent dimension. Without a perfect measurement instrument, if 40 measures are taken and there is some variability, how can one decide which one is the true and real one?

Thus, measurement is first needed to ascertain this dimension. Then – as it is shown later in this book – statistical methods can help in guessing the true latent dimension.

The basic distinction is between *qualitative* or *non-metric* scales (not allowing for mathematical operations) and *quantitative* or *metric* scales, but these two categories can be further split in the four main typologies of measurement scales. These were developed in 1940 by a committee of nineteen scientists who worked for almost eight years to report on measuring human sensations (Stevens, 1946).

Qualitative scales are further divided into:

- *Nominal*, which only requires nominal assignment to a specific class. The basic empirical operation is checking whether two elements are equal (they belong to the same class) or different (they belong to different classes), but no ranking or distance measurement is possible. For example, the job type is usually measured through a nominal scale; and
- *Ordinal*, where classes can be ranked according to some criterion and it becomes possible to determine which class is greater and which is smaller, albeit no distance measurement is possible. This also includes variables which are ordered by nature. For example, consumer perception of quality is usually measured through ordinal scales and consumer may be asked to rate the quality of a product with an integer between 1 (very poor quality) and 9 (top quality). This is the case of Likert scales, discussed in section 1.2.2. Otherwise, the respondent can rank-order several products in terms of perceived quality. In terms of basic operations, besides equality, it is possible to perform ranking operations (smaller than, greater than).

Although this distinction is rarely relevant to consumer research, quantitative scales can be split into:

- *Interval scales*, where the distance between two points is meaningful, but there is no absolute reference point which can be set conventionally and the measurement units likewise. Temperature is the typical example. Whatever the measurement unit (Fahrenheit or Celsius), it is possible to say that the temperature change (distance) between today and yesterday is twice the temperature change of the day before. But if one says that today's temperature is twice yesterday's, this is not independent from the measurement unit. Besides equality and ranking,

---

BOX 1.1 *Types of variables in SPSS*

To avoid some confusion that might arise from the interchangeable use of terms like qualitative, categorical and non-metric variables, it is useful to review the classification employed in SPSS. The key distinction is on the MEASURE of the variables. Three types exist:

(a) SCALE variables (which correspond to *quantitative* and *metric*);
(b) NOMINAL variables (a sub-group of *qualitative* or *non-metric* variables also called *categorical* which includes those variables whose categories cannot be ordered); and
(c) ORDINAL variables (which completes the set of *qualitative* or *non-metric* variables for those categories that can be ranked).

it is possible to compute relative differences, hence calculating the variability measures discussed later in this chapter; and

- *Ratio scale,* where an absolute and natural reference point (zero) exists. This applies to most quantitative variables (height, weight, monetary, etc.). Different measurement units (for example currencies) are still possible, but if I say that today my wallet contains half of the money I had yesterday, this sentence is meaningful whatever the currency used. It is called ratio scale, because it is possible to perform ratios between two measures and this ratio will be the same independently by the measurement unit. Hence, besides equality and ranking relative differences it is also possible to compute ratios.

SPSS exploits a slightly simpler classification, which is explained in box 1.1. This primary classification does not solve the many problems in quantifying latent constructs. These are very common in consumer and marketing research, especially when dealing with psychographics, lifestyle or attitude measurements (see chapter 3). The scaling techniques are usually classified as *comparative* or *non-comparative*, where the former require relative comparison between objects and the latter rely on individual assessment for each of the objects.

1.2.1 Comparative scaling

The basic comparative scaling technique is based on *pair-wise comparison*, where the respondent simply chooses between two alternatives according to some ordinal criterion. If transitivity exists, by means of paired comparisons one may obtain the ranking of three or more objects.[2] Another approach to measuring attitudes is given by *Guttman scaling*, where the same concept is expressed at different and ranked level of intensity in several questions. For example, one might elicit attitudes toward chocolate with three questions – (a) 'I adore chocolate,' (b) 'Chocolate is good,' (c) 'I don't mind eating chocolate,' asking whether they agree or not with the sentence. Those who agree with (a) are expected to agree with (b) and (c), those who disagree with (a) but agree with (b) are expected to agree with (c) too. The design can be more complex with more intensity levels and provides to measure the intensity of an attitude through a set of binary questions.

If the respondent is asked to rank more than two items simultaneously, the measurement scale is called *rank order scale* and returns an ordinal variable. This scale is especially useful when one wants to rank products according to perceptions of their single attributes. *Constant sum scaling* is based on the allocation of a constant sum to each object within the set of alternatives. For example, a respondent may be asked to allocate 100 points into the relevance of each of a set of product attributes when choosing whether to buy it or not. In other words, the question requires giving a relative weight to each of the attributes. It is not advisable to use the resulting variable as a quantitative one, because of the many bias involved in allocating sums, but the outcome can be exploited as an ordinal variable. When there is a large amount of objects to be evaluated (usually around 100 and up to 140), a suitable method is based on *Q-sorting* where the respondent is given a set of cards, each corresponding to an item, and is asked to stack them into different piles, ranked according to same criterion. This might be a suitable method for the Brunsø and Grunert (1998) approach to measuring food lifestyles (see chapter 3), where respondents are asked to express the level of agreement with 69 sentences.

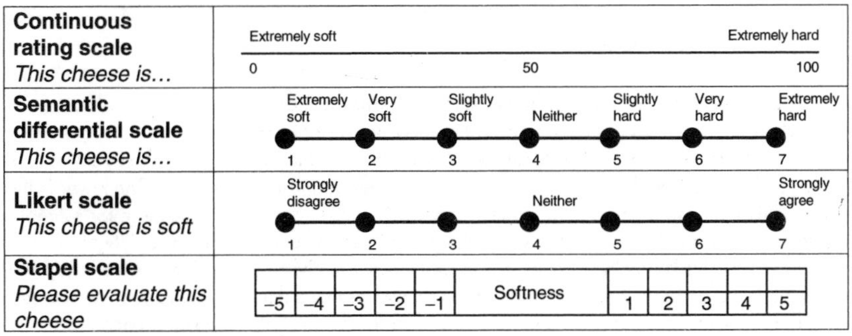

Continuous rating scale *This cheese is...*	Extremely soft 0			50			Extremely hard 100
Semantic differential scale *This cheese is...*	Extremely soft ● 1	Very soft ● 2	Slightly soft ● 3	Neither ● 4	Slightly hard ● 5	Very hard ● 6	Extremely hard ● 7
Likert scale *This cheese is soft*	Strongly disagree ● 1	● 2	● 3	Neither ● 4	● 5	● 6	Strongly agree ● 7
Stapel scale *Please evaluate this cheese*	−5 −4 −3 −2 −1		Softness		1 2 3 4 5		

Figure 1.2 *Rating scales*

1.2.2 Non-comparative scaling

With non-comparative scales, the respondent provides a measure for each item without referring to an alternative or any other benchmark. The basic rating scales are shown graphically in figure 1.2.

In *continuous rating* the respondent must tick in a continuous line, running between two extremes of the attribute being investigated. There is no restriction on where the mark should be placed and the line may be accompanied by numbers so that evaluation can be translated into scores at a later stage, although this might be a demanding procedure. The *semantic differential scale* is very similar to the above rating but is itemized in categories, each associated with a number and/or a description, so that the respondent is bound to choose between 5 or 7 points, with a neutral point in the middle. The end points are two bipolar attributes. Typically, the overall attitude is measured through a set of bipolar adjectives, as this scale is exploited for comparisons across products and brand images, in order to detect strengths and weaknesses. An issue is whether the two adjectives are actually bipolar and equidistant from the neutrality point. The *Likert scale* is the most popular measurement scale in marketing, because it is simple and powerful. It measures the intensity of a single attribute, usually by asking the level of agreement with a given statement. As for the Semantic differential scale, it is made by 5 or 7 points (sometimes 9) with a middle point for neutrality. In some situations the middle point is not included, so that the respondent is forced to agree or disagree. There may be problems in Likert scales – respondents that tend to avoid the first of last category as they are perceived as extreme, or the tendency to agree with the sentence, so that if the statement is reversed responses could lose consistency. Likert scales generate ordinal variables.

The *Stapel scale* is used less frequently and is a unipolar scale with 10 points, from −5 to 5 with no neutrality. The respondent ticks the number that best reflects the level of accuracy of a statement, where a value of 5 means maximum accuracy and negative values indicate inaccuracy. Stapel scales work around the bipolarity problem, but some bias is potentially induced by the phrasing of the question, whether it is negative or positive.

1.2.3 Choice and evaluation of measurement scales

Some properties to be considered in relation to any measurement are accuracy, precision, reliability, and validity. Note the distinction between *accuracy* and *precision*.

BOX 1.2 *Cronbach's Alpha*

The Cronbach's Alpha is a widely used measure of internal reliability for measurement scales. Its rationale is that if the set of items supposed to measure a single latent construct, the total variability of these items should approximate the variability of the true score for the latent variable. However, since we do not know the true score, we commit a measurement error. The lower is this measurement error, the higher the reliability. Cronbach (1951) devised a procedure for estimating reliability by looking at the sum of the variabilities within each item and the variability of the summated score. This basically corresponds to looking at the bivariate correlations (see chapter 8) between the items across the sample. In high reliability situations it is possible to sum the scores obtained in the single items to estimate the score of the latent variable. The computation of the Cronbach's Alpha is based on the average of all bivariate correlations between items of the same construct set, with an adjustment for the number of items. The Alpha reliability coefficient has a maximum value of 1, while it can be negative in presence of negative correlations for some of the items. A value below 0.70 indicate scarce reliability of the item, although the coefficient is sensitive to the number of items and can be artificially inflated. A meta-analysis is found in Peterson (1994). The Cronbach's Alpha can be computed in SPSS through the ANALYZE/SCALE/RELIABILITY menu, while in SAS it can be computed as an option (ALPHA) of the CORR procedure in SAS BASE.

The former refers to the degree to which the measurement reflects the true value (how close is the measure of the book's cover to the real cover's width?), the latter to the degree of detail in measurement (for example, whether the ruler is able to catch differences of millimeters) and it is usually related to the variability around the average measurement when carrying out repeated measurements. *Reliability* is the property referring to the consistency of the measurement across several questionnaire items measuring the same latent construct or over time. For example, one might measure the same concept (attitude toward watching television) asking different questions like 'Do you like watching television?,' 'Do you find television programs interesting?,' 'Do you think watching television is good?.' These are attempts of measuring the 'true' attitude toward watching television. If a battery of possible questions is used, responses to those more reliable will show a higher degree of consistency among themselves and across respondents. A typical measure of reliability is the Cronbach's Alpha (see box 1.2) or, if one tries to measure the same thing several times (like the width of the book cover), the more measures are consistent, the more measurement is reliable. Finally, *validity* is the extent to which measurement reflects the 'true' phenomenon under study. For example, suppose that a measure of product quality is obtained by combining proxies like price, packaging and other technical characteristics. If this measure has a high correlation (see chapter 8) with the quality actually perceived by consumers, then it shows a high validity and it can be used to predict product quality in absence of the perceived quality information.

There is a vast literature exploring the empirical performance of alternative non-comparative scales. Evaluations are based on the reliability of scales, their validity and also *generalizability*. As explained above, reliability measures the consistency across various items of the scale which are supposed to be part of the same construct (*internal consistency*), but other forms of reliability look at the stability of results when the survey is repeated in different circumstances (*test-retest*) or when the scale is used in different formats (*alternative forms*). Validity checks look at the observed differences in scale scores and purge them from the influence of random errors or systematic biases.[3] Finally, with generalizability, the researcher tries to generalize the results obtained to a broader situation which goes beyond the sampling process and includes different administration methods, different timings, etc. (Rentz, 1987). Churchill and Peter (1984) had a close look to the performance of different scales by performing a meta-analysis (see box 2.3) on 89 marketing research studies. They concluded that the choice of the measurement scale did not generally affect the result of the studies, apart from the number of items in the final scale and the number of scale points. If a construct is measured through a large number of items, the reliability measure tends to be higher (and the questionnaire longer). Higher numbers of points in the scales also increase reliability, while other choices including the type of non-comparative scale may actually affect the result, but not in a single direction. One should accurately weigh the pros and cons of each of the measurement scales described in this chapter according to the research question and the environment of the study, especially the administration method and the socio-cultural characteristics of the surveyed population.

1.3 Two sample data-sets

The previous sections have summarized the main issues in measuring objects, with an emphasis on the potential impact of measurement problems and methods on the research output. Over the next three chapters, the focus will move to data collection, organization and preparation prior to more advanced statistical analysis, discussed in the following chapters. To facilitate understanding and replication of examples, all methods discussed in this book with a very few exceptions are based on two sample data-sets:[4]

- The **Expenditure and Food Survey (EFS)** data-set; and
- The **Trust** data-set.

The EFS data-set is a subset from the 2004–05 UK Expenditure and Food Survey (which is described in some detail in chapter 2). The file includes a simple random sample of 500 households and a selection of 420 variables out of the 1952 available in the officially released data-set.[5] The complete list of variables is listed in box 1.3. This sample data-set is a good example of secondary data (see chapter 2) with household expenditure figures. Classification follows the *European classification of individual consumption by purpose* scheme (COICOP, see table 2.1).

A second data-set is interesting on a different perspective. In this case, the aim of the original survey was to collect attitudinal and psychographic data to explain chicken purchasing behavior in five European countries. Again, the simplified sample data-set is a selection of the original one and contains 500 cases (100 per country selected randomly) out of the original 2,725 household surveyed and 138 variables out of 226 are in the sample data-set.

The Trust survey is a good example of the outcome of primary data collection (see chapter 3), which means that the data were explicitly collected for the purpose of the consumer research. The questionnaire for the Trust survey was developed within a European Research Project[6] and its core section was based on an extension of the Theory of the Planned Behavior (see chapter 15 for further details). Further questions covered socio-demographic characteristics of the respondent, lifestyle questions, questions related to chicken purchasing behavior and a set of question on risk perceptions and trust in information. The list of variables included in the data-set is shown in box 1.4. A nationally representative survey based on probabilistic area sampling was conducted in five countries (UK, Italy, Germany, the Netherlands and France) in May 2004 on

BOX 1.3 *Variables in the EFS data-set*

Variable	Description	Variable	Description
case	Case number	c11321	Seafood (fresh, chilled or frozen)
weighta	Annual weight	c11331	Dried, smoked or salted fish and
weightq	Quarterly weight		seafood
a040	Number of children children-age	c11341	Other preserved/processed
	under 2		fish/seafood
a041	Number of children – age 2 and	c11411	Whole milk
	under 5	c11421	Low fat milk
a042	Number of children – age 5 and	c11431	Preserved milk
	under 18	c11441	Yoghurt
a043	Number of adults – age under 45	c11451	Cheese and curd
a044	Number of adults – age 45 but	c11461	Other milk products
	under 60	c11471	Eggs
a045	Number of adults – age 60 but	c11511	Butter
	under 65	c11521	Margarine and other vegetable
a046	Number of adults – age 65 but		fats
	under 70	c11522	Peanut butter
a047	Number of adults – age 70 and over	c11531	Olive oil
a049	Household size	c11541	Edible oils
a062	Composition of household	c11551	Other edible animal fats
a071	Sex of oldest person in household	c11611	Citrus fruits (fresh)
a121	Tenure – type	c11621	Bananas (fresh)
a124	Cars and vans in household	c11631	Apples (fresh)
a1661	Home computer in household	c11641	Pears (fresh)
a172	Internet connection in household	c11651	Stone fruits (fresh)
b010	Rent rates – last net payment	c11661	Berries (fresh)
b1661	Mobile telephone account	c11671	Other fresh, chilled or frozen
b216	Bus tube and/or rail season ticket		fruits
b260	School meals-total amount paid last	c11681	Dried fruit and nuts
	week	c11691	Preserved fruit and fruit-based
g018	Number of adults		products
g019	Number of children	c11711	Leaf and stem vegetables (fresh
c11111	Rice		or chilled)
c11121	Bread	c11721	Cabbages (fresh or chilled)
c11122	Buns, crispbread and biscuits	c11731	Vegetables grown for their fruit
c11131	Pasta products		(fresh, chilled or frozen)
c11141	Cakes and puddings	c11741	Root crops, non-starchy bulbs,
c11142	Pastry (savory)		mushrooms (fresh/chilled/
c11151	Other breads and cereals		frozen)
c11211	Beef (fresh, chilled or frozen)	c11751	Dried vegetables
c11221	Pork (fresh, chilled or frozen)	c11761	Other preserved or processed
c11231	Lamb (fresh, chilled or frozen)		vegetables
c11241	Poultry (fresh, chilled or frozen)	c11771	Potatoes
c11251	Sausages	c11781	Other tubers and products of
c11252	Bacon and ham		tuber vegetables
c11253	Offal, pâté, etc.	c11811	Sugar
c11311	Fish (fresh, chilled or frozen)	c11821	Jams, marmalades

BOX 1.3 *Cont'd*

Variable	Description	Variable	Description
c11831	Chocolate	cb111i	Beer and lager (away from home)
c11841	Confectionery products	cb111j	Round of drinks (away from home)
c11851	Edible ices and ice cream		
c11861	Other sugar products	cb1121	Food non-alcoholic drinks eaten drunk on premises
c11911	Sauces, condiments		
c11921	Salt, spices and culinary herbs	cb1122	Confectionery
c11931	Bakers yeast, dessert preparations, soups	cb1123	Ice cream
		cb1124	Soft drinks
c11941	Other food products	cb1125	Hot food
c12111	Coffee	cb1126	Cold food
c12121	Tea	cb1127	Hot take-away meal eaten at home
c12131	Cocoa and powdered chocolate		
c12211	Mineral or spring waters	cb1128	Cold take-away meal eaten at home
c12221	Soft drinks		
c12231	Fruit juices	cb112b	Contract catering (food)
c12241	Vegetable juices	cb1213	Meals bought and eaten at the workplace
c21111	Spirits and liqueurs (brought home)		
		c31111	Clothing materials
c21211	Wine from grape or other fruit (brought home)	c31211	Men's outer garments
		c31212	Men's under garments
c21212	Fortified wine (brought home)	c31221	Women's outer garments
c21213	Ciders and Perry (brought home)	c31222	Women's under garments
c21214	Alcopops (brought home)	c31231	Boys outer garments (5-15)
c21221	Champagne and sparkling wines (brought home)	c31232	Girls outer garments (5-15)
		c31233	Infant's outer garments (under 5)
c21311	Beer and lager (brought home)	c31234	Children's under garments (under 16)
c22111	Cigarettes		
c22121	Cigars	c31311	Men's accessories
c22131	Other tobacco	c31312	Women's accessories
cb1111	Catered food non-alcoholic drink eaten/drunk on premises	c31313	Children's accessories
		c31314	Haberdashery
cb1112	Confectionery eaten off premises	c31315	Protective head gear (crash helmets)
cb1113	Ice cream eaten off premises		
cb1114	Soft drinks drunk off premises	c31411	Clothing hire
cb1115	Hot food eaten off premises	c31412	Dry cleaners and dyeing
cb1116	Cold food eaten off premises	c31413	Laundry, launderettes
cb111c	Catered food non-alcoholic drink eaten/drunk on premises	c32111	Footwear for men
		c32121	Footwear for women
cb111d	Wine from grape or other fruit (away from home)	c32131	Footwear for children (5-15) and infants
cb111e	Fortified wines (away from home)	c32211	Repair and hire of footwear
cb111f	Ciders and Perry (away from home)	c41211	Second dwelling – rent
		c43111	Paint, wallpaper, timber
cb111g	Alcopops (away from home)	c43112	Equipment hire, small materials
cb111h	Champagne and sparkling wines (away from home)	c44211	Refuse collection, including skip hire

(Continued)

BOX 1.3 *Cont'd*

Variable	Description	Variable	Description
c45112	Second dwelling: electricity account payment	c54121	Cutlery and silverware
		c54131	Kitchen utensils
c45114	Electricity slot meter payment	c54132	Storage and other durable household articles
c45212	Second dwelling: gas account payment	c54141	Repair of glassware, tableware and house
c45214	Gas slot meter payment		
c45222	Bottled gas – other	c55111	Electrical tools
c45312	Paraffin	c55112	Lawn mowers and related accessories
c45411	Coal and coke		
c45412	Wood and peat	c55211	Small tools
c45511	Hot water, steam and ice	c55212	Door, electrical and other fittings
c51113	Fancy decorative goods	c55213	Garden tools and equipment
c51114	Garden furniture	c55214	Electrical consumables
c51212	Hard floor coverings	c56111	Detergents, washing-up liquid, washing powder
c51311	Repair of furniture, furnishings and floor coverings	c56112	Disinfectants, polishes, other cleaning materials, etc.
c52111	Bedroom textiles, including duvets and pillows	c56121	Kitchen disposables
c52112	Other household textiles, including cushions, towels, curtains	c56122	Household hardware and appliances, matches
c53111	Refrigerators, freezers and fridge-freezers	c56123	Kitchen gloves, cloths, etc.
		c56124	Pins, needles and tape measures
c53121	Clothes washing machines and clothes drying machines	c56125	Nails, nuts, bolts, washers, tape and glue
c53122	Dish washing machines	c56211	Domestic services, including cleaners, gardeners, au pairs
c53131	Gas cookers		
c53132	Electric cookers, combined gas electric	c56221	Cleaning of carpets, curtains & household
c53133	Microwave ovens	c56222	Other household services
c53141	Heaters, air conditioners, shower units	c56223	Hire of household furniture and furnishings
c53151	Vacuum cleaners and steam cleaners	c61111	NHS prescription charges and payments
c53161	Sewing and knitting machines	c61112	Medicines and medical goods (not NHS)
c53171	Fire extinguisher, water softener, safes	c61211	Other medical products (e.g. plasters, condoms, etc.)
c53211	Small electric household appliances, excluding hairdryers	c61311	Purchase of spectacles, lenses, prescription sunglasses
c53311	Spare parts: gas and electric appliances	c61312	Accessories repairs to spectacles lenses
c53312	Electrical appliance repairs	c61313	Non-optical appliances and equipment
c53313	Gas appliance repairs		
c53314	Rental hire of major household appliance	c62111	NHS medical services
		c62112	Private medical services
c54111	Glassware, china, pottery	c62113	NHS optical services

BOX 1.3 *Cont'd*

Variable	Description	Variable	Description
c62114	Private optical services	c73311	Air fares (within UK)
c62211	NHS dental services	c73312	Air fares (international)
c62212	Private dental services	c73411	Water travel
c62311	Services of medical analysis laboratories and x-ray centers	c73512	Combined fares other than season tickets
c62321	Services of NHS medical auxiliaries	c73513	School travel
c62322	Services of private medical auxiliaries	c73611	Delivery charges and other transport services
c62331	Non-hospital ambulance services etc.	c81111	Postage and poundage
		c82111	Telephone purchase
c63111	Hospital services	c82112	Mobile phone purchase
c71112	Loan HP purchase of new car van	c82113	Answering machines, fax machines, modem
c71122	Loan HP purchase of second-hand car van	c83112	Telephone coin and other payments
c71212	Loan/HP purchase of new or second-hand motorcycle	c83114	Mobile phone – other payments
c71311	Purchase of bicycle	c83115	Second dwelling – telephone account payments
c71411	Animal drawn vehicles	c91111	Audio equipment, CD players
c72111	Car van accessories and fittings	c91112	Audio equipment – in car
c72112	Car van spare parts	c91113	Accessories for audio equipment – headphones etc.
c72113	Motor cycle accessories and spare parts	c91121	Television set purchase
c72114	Anti-freeze, battery water, cleaning materials	c91122	Satellite dish purchase
		c91123	Satellite dish installation
c72115	Bicycle accessories, repairs and other costs	c91124	Video recorder purchase
		c91125	Purchase of digital TV decoder
c72211	Petrol	c91126	Spare parts for TV, video, audio
c72212	Diesel oil		
c72213	Other motor oils	c91127	Cable TV connection
c72313	Motoring organization subscription (e.g. AA and RAC)	c91128	DVD purchase
		c91211	Photographic cinematographic equipment
c72314	Car washing and breakdown services	c91221	Optical instruments, binoculars, telescopes, microscopes
c72411	Parking fees, tolls and permits (excluding motoring fines)	c91311	Personal computers, printers and calculators
c72412	Garage rent, other costs (excluding fines)	c91411	Records, CDs, audio cassettes
c72413	Driving lessons	c91412	Blank and pre-recorded video cassettes
c72414	Hire of self-drive cars, vans, bicycles	c91413	Camera films
c73112	Railway and tube fares other than season	c91511	Repair of AV/photographic/ information processing equipment
c73212	Taxis and hired cars with drivers	c92111	Purchase of boats, trailers and horses
c73213	Taxis and hired cars with drivers		
c73214	Other personal travel		

(Continued)

BOX 1.3 *Cont'd*

Variable	Description	Variable	Description
c92112	Purchase of caravans, mobile homes (including decoration)	c94232	TV licence payments (second dwelling)
c92114	Purchase of motor caravan (new) – loan/HP	c94236	TV slot meter payments
c92116	Purchase of motor caravan (second-hand)	c94238	Video cassette rental
		c94239	Cassette hire (library), CD hire (library)
c92117	Accessories for boats, horses, caravans	c94241	Admissions to clubs, dances, discos, bingo
c92211	Musical instruments (purchase and hire)	c94242	Social events and gatherings
		c94243	Subscriptions for leisure activities
c92221	Major durables for indoor recreation (e.g. snooker table etc.)	c94244	Other subscriptions
		c94245	Internet subscription fees
		c94246	Development of film, photos, etc.
c92311	Maintenance/repair other major durables for recreation	c94311	Football pools stakes
		c94312	Bingo stakes excluding admission
c93111	Games, toys and hobbies (excluding artists materials)	c94313	Lottery (not National Irish Lottery) stakes
c93112	Computer software and game cartridges	c94314	Bookmaker, tote, other betting stakes
c93113	Console computer games	c94315	Irish Lottery Stakes
c93114	Games toys etc. (misc fancy, decorative)	c94316	National Lottery instants. Scratchcards
c93211	Equipment for sport, camping and open-air recreation	c94319	National Lottery stakes
		c9431a	Football pools winnings
c93212	BBQ and swings	c9431b	Bingo winnings
c93311	Plants, flowers, seeds, fertilizers, insecticides	c9431c	Lottery (not National Irish Lottery) winnings
c93312	Garden decorative	c9431d	Bookmaker, tote, other betting winnings
c93313	Artificial flowers, pot pourri		
c93411	Pet food	c9431e	Irish Lottery Winnings
c93412	Pet purchase and accessories	c9431f	National Lottery instants. Scratchcards
c93511	Veterinary and other services for pets identified separately	c9431i	National Lottery winnings
c94111	Spectator sports: admission charges	c95111	Books
		c95211	Newspapers
c94112	Participant sports (excluding subscriptions)	c95212	Magazines and periodicals
		c95311	Cards, calendars, posters and other prin
c94113	Subscriptions to sports and social clubs	c95411	Stationery, diaries, address books, art
c94115	Hire of equipment & accessories for sports	ca1113	Pre-primary/primary edn: (school trips, ad hoc school expenditure)
c94211	Cinemas		
c94212	Live entertainment: theater, concerts, shows	ca2113	Secondary edn: (school trips, other ad hoc school expenditure)
c94221	Museums, zoological gardens, theme parks		

BOX 1.3 *Cont'd*

Variable	Description	Variable	Description
ca3113	Further edn: (school trips, other ad hoc school expenditure)	cc6214	Commission travelers cheques and currency
ca4113	Higher edn: (school trips, other ad hoc school expenditure)	cc7111	Legal fees paid to banks
ca5113	Education not definable: (school trips, other ad hoc school expenditure)	cc7112	Legal fees paid to solicitors
		cc7113	Other payments for services e.g. photocopying
cb2114	Room hire	cc7114	Funeral expenses
cc1111	Hairdressing salons/personal grooming (exc health and slimming clubs)	cc7115	Other professional fees incl court fines
		cc7116	TU and professional organizations
cc1211	Electrical appliances for personal care	ck1211	Outright purchase of, or deposit on, main dwelling
cc1311	Toilet paper	ck1313	Central heating installation (DIY)
cc1312	Toiletries (disposable inc tampons, lipsyl, toothpaste, deodorant, etc.)	ck1314	Double Glazing, Kitchen Units, Sheds, etc.
		ck1315	Purchase of materials for Capital Improvements
cc1313	Bar of soap, liquid soap, shower gel, etc.	ck1316	Bathroom fittings
cc1314	Toilet requisites (durable inc razors, hairbrushes, toothbrushes, etc.)	ck1411	Purchase of second dwelling
		ck2111	Food stamps, other food related expenditure
cc1315	Hair products	ck3111	Stamp duty, licences and fines (excluding motoring fines)
cc1316	Cosmetics and related accessories		
cc1317	Baby toiletries and accessories (disposable)	ck3112	Motoring Fines
		ck4111	Money spent abroad
cc3111	Jewelry, clocks and watches	ck4112	Duty free goods bought in UK
cc3112	Repairs to personal goods	ck5111	Savings, investments (exc AVCs)
cc3211	Leather and travel goods (excluding baby items)	ck5113	Additional Voluntary Contributions
cc3221	Other personal effects n.e.c.	ck5115	Superannuation deductions – subsidiary employee job
cc3222	Baby equipment (excluding prams and pushchairs)	ck5116	Widows/Dependants/Orphans fund
cc3223	Prams, pram accessories and pushchairs	ck5212	Money given to children for specific purposes: pocket money
cc3224	Sunglasses (non-prescription)		
cc4111	Residential homes	ck5213	Money given to children for specific purposes: school dinner
cc4112	Home help		
cc4121	Nursery, creche, playschools	ck5214	Money given to children for specific purposes: school travel
cc4122	Child care payments		
cc5213	Insurance for household appliances	ck5215	Money given to children for specific purposes
cc5412	Boat insurance (not home)	ck5216	Cash gifts to children
cc5413	Non-package holiday, other travel insurance	ck5221	Money given to those outside the household
cc6212	Bank and Post Office counter charges	ck5222	Present – not specified

(Continued)

BOX 1.3 *Cont'd*

Variable	Description	Variable	Description
ck5223	Charitable donations and subscriptions	year	Survey Year
		p600	EFS: Total consumption expenditure
ck5316	Pay off loan to clear other debt		
a054	Number of workers in household	p601	EFS: Total Food & Non-alcoholic beverage
a056	Number of persons economically active		
		p602	EFS: Total Alcoholic Beverages, Tobacco
a060	Gross normal income of HRP by range		
		p603	EFS: Total Clothing and Footwear
a091	Socio-economic group – Household Reference Person	p604	EFS: Total Housing, Water, Electricity
a093	Economic position of Household Reference Person	p605	EFS: Total Furnishings, HH Equipment, Carpets
a094	NS-SEC 8 Class of Household Reference Person	p606	EFS: Total Health expenditure
		p607	EFS: Total Transport costs
p344	Gross normal weekly household income	p608	EFS: Total Communication
		p609	EFS: Total Recreation
p348	Social security benefits – household	p610	EFS: Total Education
		p611	EFS: Total Restaurants and Hotels
p352	Gross current income of household		
		p612	EFS: Total Miscellaneous Goods and Services
p396	Age of Household Reference Person		
		p620p	EFS: Total Non-Consumption Expenditure (anonymized)
sexhrp	Sex of Household Reference Person		
		p630p	EFS: Total Expenditure (anonymized)
a055	Sampling month		
a190	Internet access via Home Computer	incanon	Anonymized hhold inc + allowances
gor	Government Office Region		
gorx	Govt. Office Region modified	a070p	Age of oldest person in hhold – anonymized
p389	Normal weekly disposable hhld income		
		p396p	Age of HRP – anonymized

Source: Office for National Statistics and DEFRA (through www.esds.ac.uk)

a total of 2725 respondents via face-to-face, in-home interviews. The adopted sampling method was Random Location Sampling, a two-stage sampling method which provides a country-representative subdivision into locations; the locations are selected randomly across potential locations to ensure national representativeness. The sampling unit was the household and the respondent the person responsible for the actual purchase of food. The questionnaire took approximately 30 minutes to complete with 'prompts' on certain questions from the interviewer when required by the respondent.

1.4 Statistical software

Another aim of this book is to provide an applied view of the methodologies for consumer and marketing research. To this purpose, the discussion of statistical methods is accompanied by specific consumer research examples and some key rules for

BOX 1.4 *Variables in the TRUST data-set*

Variable	Description	Variable	Description
code	ID		*Please indicate the extent to*
q1	How many people do you regularly buy food for home consumption (including yourself)?		*which you agree or disagree with each of the statements you find below by circling the number that most closely*
	How frequently do you buy...		*describes your personal*
q2a	Food for your household's home consumption	q9	*view.* In my household we like chicken
q2b	Any type of chicken for your household's home consumption	q10	A good diet should include chicken
q2c	Fresh chicken		*My decision whether or not to*
q2d	Frozen chicken		*buy chicken next week is based*
q2e	Chicken as part of a prepared meal	q12a	*on the fact that:* Chicken tastes good
q2f	Cooked chicken	q12b	Chicken is good value for
q2g	Processed chicken		money
q2h	Chicken as a meal outside your home	q12c	Chicken is not easy to prepare
		q12d	Chicken is a safe food
		q12e	All the family likes chicken
q4kilos	In a typical week how much fresh or frozen chicken do you buy for your household consumption (kg)?	q12f	Chicken works well with lots of other ingredients
		q12g	Chicken is low in fat
		q12h	Chicken is low in cholesterol
q5	In a typical week how much do you spend on fresh or frozen chicken (Euro)?	q12i	Chicken lacks flavor
		q12j	Buying chicken helps the local farmers and economy
q6	In a typical week, what type of fresh or frozen chicken do you buy for your household's home consumption?	q12k	I do not like the idea of chickens being killed for food
		q12l	Chicken is not produced taking into account animal welfare
q7	How likely or unlikely is it that you will buy fresh or frozen chicken for your household's home consumption at least once in the next week?	q14	Others' opinions on chicken are important to me
		q15	I take others' opinions into account when making decisions on whether or not to buy chicken
	In a typical week where do you purchase your fresh or frozen chicken?	q16	Other people suggest chicken in the diet is?
q8a	Discount supermarket	q20a1	I typically store chicken in my freezer
q8b	Supermarket		
q8c	Local shop	q20b1	We eat too much chicken
q8d	Butcher	q20a2	Let's say you do have some chicken in your freezer. Is it likely you would buy more next week?
q8e	Farmer		
q8f	Market		
q8g	Online shopping/home delivery		
q8h	Other		

(Continued)

BOX 1.4 *Cont'd*

Variable	Description	Variable	Description
q20b2	Let's say last week you ate a lot of chicken. Is it likely you would not buy chicken at all next week?	q25f	I am constantly sampling new and different foods
		q25g	I don't trust new foods
	Safe chicken is...	q25h	I will eat almost anything
q21a	Packaged	q25i	If I don't know what is in a food, I won't try it
q21b	Clearly labeled		
q21c	Whole chicken	q25j	I am afraid to eat things I have never eaten before
q21d	From the butcher		
q21e	From the supermarket	q26a	I usually aim to eat natural foods
q21f	Produced in your own country		
q21g	Produced in the EU	q26b	I am willing to pay more for a better quality product
q21h	Produced in Asia		
q21i	Expensive	q26c	Quality is decisive for me when purchasing foods
q21j	Free range, organic or corn-fed		
q21k	Recognizable by color, taste or smell	q26d	I always aim for the best quality
		q26e	When choosing foods, I try to buy products that do not contain residues of pesticides or antibiotics
	In general, how important are each of the following to your household?		
q21l	Fresh	q26f	I am willing to pay more for foods containing natural ingredients
q24a	Tasty food		
q24b	Value for money	q26g	For me, wholesome nutrition begins with the purchase of high quality foods
q24c	Ease of preparation		
q24d	Food safety		
q24e	Food that everyone likes		*How would you rate these activities in terms of risk to health?*
q24f	Variety in our meals		
q24g	Fat content	q27a	Smoking cigarettes
q24h	Cholesterol content	q27b	Driving
q24i	Ethical food production methods	q27c	Eating beef
		q27d	Eating chicken
q24j	Local community livelihood	q27e	Taking illegal drugs
q24k	Animal welfare	q27f	Scuba diving
	Please indicate the extent to which you agree or disagree with each of the statements you find below by circling the number that most closely describes your personal view.	q27g	Swimming
		q28	Risk aversion
		q34	Have you actively searched for any information on food safety in the last two weeks?
q25a	I like foods from different countries	q35	How many hours per day do you watch TV?
q25b	Ethnic food looks too weird to eat	q36	How many hours per day do you listen to the radio?
q25c	I like to try new ethnic restaurants	q37	How many hours per day do you surf the Internet?
q25d	I like to purchase the best quality food I can afford	q38	How many different newspapers do you read in a typical week?
q25e	At parties, I will try a new food		

BOX 1.4 *Cont'd*

Variable	Description	Variable	Description
	Suppose that each of the following has provided information about potential risks associated with salmonella in food. Please indicate to what extent you would trust that information	q50	Gender
		q51	Age
		q52	Marital status
		q54	Job Status
		q55	If employed, what is your occupation?
q43a	Shopkeepers	q56	Number of people currently living in your household (including yourself)
q43b	Supermarkets		
q43c	Organic shop		
q43d	Specialty store		*Number and age of children*
q43e	Farmers/breeders	q57a	No children
q43f	Processors	q57b	Less than 3 years
q43g	Doctors/health authority	q57c	3–10 years
q43h	University scientists	q57d	11–16 years
q43i	National authority in charge of food safety	q57e	Greater than 16 years
q43j	Government	q58a	Are there other members of the household who are dependant on you (e.g. elderly or disabled)?
q43k	Political groups		
q43l	Environmental organizations		
q43m	Animal welfare organizations		
q43n	Consumer organizations	q58b	If yes, how many?
q43o	European Union authority in charge of food safety	q59	On average, how much does your household spend on food each week
q43p	Television documentary		
q43q	Television news/current affairs	q60	Please indicate your gross annual household income range
q43r	Television adverts	q61	How would you describe the financial situation of your household?
q43s	Newspapers		
q43t	Internet		
q43u	Radio	q62a	Do you belong to any consumer or environmental organizations?
q43v	Magazines		
q43w	Product label		
q49	How do you rate your ability to assess food quality and safety?	q63	Which size is the town where you live?
		q64	Country

applying these methods in SPSS and/or SAS are provided. SPSS and SAS are the most widely employed statistical packages in consumer and marketing research.[7] However, as it will be shown for specific cases, the performance, flexibility and output of these packages varies with different methodologies. Furthermore, for specific statistical methods and applications, other statistical packages like STATA, Limdep, LISREL or Econometric Views might provide better solutions. Most advanced consumer research could also gain maximum flexibility by exploiting statistical programming languages like Gauss or Matlab. While an exhaustive review of statistical software is not feasible within the aims of this book, not least because of the rapidly evolving characteristics of the aforementioned packages, within each chapter an indication of alternative software choices will be provided when relevant.

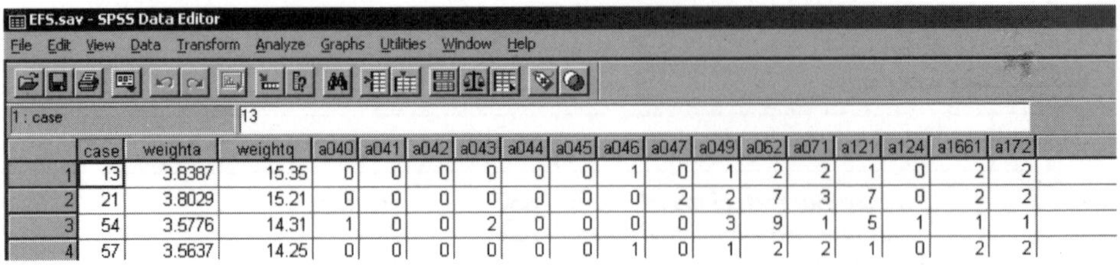
EFS.sav - SPSS Data Editor
File Edit View Data Transform Analyze Graphs Utilities Window Help
1 : case
13
case weighta weightq a040 a041 a042 a043 a044 a045 a046 a047 a049 a062 a071 a121 a124 a1661 a172
1 13 3.8387 15.35 0 0 0 0 0 0 1 0 1 2 2 1 0 2 2
2 21 3.8029 15.21 0 0 0 0 0 0 0 2 2 7 3 7 0 2 2
3 54 3.5776 14.31 1 0 0 2 0 0 0 0 3 9 1 5 1 1 1
4 57 3.5637 14.25 0 0 0 0 0 0 1 0 1 2 2 1 0 2 2

Figure 1.3 *The SPSS menus and data window*

Here it may be worth to explore the fundamental characteristics of the main packages.

SPSS has the advantage of combining a good range of methodologies with a relative ease of use. This is sometimes achieved at the cost of reduced control on options for some methodologies, which may result in a 'black box' problem. Other advantages of SPSS are the data management design (see figure 1.3), which allows one to manage large data-sets and is calibrated to fit very well with marketing research, especially for dealing with sampling weights, complex sampling designs, variable definitions, missing data treatment, etc. The user-friendly interface with dialog boxes allows quick implementation of standard methods, while the syntax editor allows one to save sequences of commands for repetitive tasks. The syntax commands can be easily pasted into the editor through the dialog boxes. Output is collected in a separate viewer and objects can be copied and pasted into word processing and spreadsheet software. SPSS is probably the most accessible software for those with a limited background in statistics. The software in its base version provides a quick and effective solution to most marketing research problems. However, more complex tasks like estimation of systems of simultaneous equations cannot be easily accomplished in SPSS.

SAS differs from SPSS in several respects. Essentially it trades off part of the user-friendly features for increased control on statistical methodologies and flexibility. While some dialog-boxes and pre-compiled routines are available in some SAS programs,[8] the software is particularly powerful when its languages are exploited. A strength of SAS is the well-documented on-line user's guide. The relative complexity of SAS derives from the fact that it is an integrated system of different software items rather than a single packages. Thus, SAS provides a very flexible programming language and a wide range of procedures as compared to SPSS, which is specifically calibrated to applications in social science and is particularly suitable for those with a marketing research background. SAS is a statistical software with a broader concept and requires a stronger background in statistics.

Other software products cited throughout this book may be useful to users with a specific interests in more advanced methodologies:

- *LISREL* is a software specifically developed for structural equation systems (chapter 15);
- *LIMDEP* is a general econometrics computer program with a specific versatility for discrete choice and limited dependent variable models (chapter 16);
- *Econometric Views (Eviews)* is also an econometric software, particularly helpful for dealing with advanced time series models and simultaneous equation system.

Summing up

'He who makes no mistakes, makes nothing' or – to translate – it is necessary to deal with error to measure objects. Measurement and error are closely related concepts, as measurements are potentially affected by systematic and random errors. Systematic errors make each of the repeated measures on the same object systematically too high or too low, while random errors are fluctuations without any fixed direction due to factors that act in a random fashion. Care in measurement can avoid systematic errors, while statistics can deal with random errors. For example, random measurement errors are expected to follow a bell-shaped probability distribution, the normal or Gaussian curve.

The measurement scales for recording consumer responses determine the content of the variables for subsequent analysis. These variables can be quantitative or qualitative and the latter can be either nominal or ordinal. There are various scaling rates which serve to the purpose. A first set of scales covers situations where a comparison between alternatives is requested to the respondent (comparative scales), while non-comparative scales are designed when each item is measured separately. For example, the Likert scale is the most frequently employed and is aimed at measuring the intensity of a single attribute, usually by asking the level of agreement with a sentence. The choice of the best structure and measurement scale depends on various aspects, especially how self-consistent is the measure, if it is a valid measure and whether it can be generalized.

The next chapters will explore the range of statistical techniques by referring to application on two sample data-set introduced in this chapter, EFS and TRUST. EFS is a secondary data-set which contains a selection of expenditure and household data from the 2004/05 UK Expenditure and Food Survey. The TRUST data-set is an example of primary data collection and contains behavioral and attitudinal data on chicken purchases in five different European countries.

To analyze this data and apply the techniques described over the next chapters, there are several commercial statistical packages; SPSS and SAS are those especially covered in this book.

EXERCISES

1. A survey has been conducted on 30 students to measure how much time they spend in the University canteen, between 9 am and 9 pm. There are 30 observations in the data-set, measured in minutes:

3	29	0	0	25	60	48	16	25	29
25	31	45	90	0	0	30	32	0	65
45	15	28	33	10	38	5	110	30	27

   a. What type of variable is this? And how should it be defined in the SPSS Variable view?

    b.   While the chronometer could record minutes and seconds, data were only registered in minutes. Two rounding procedures can be chosen: (a) rounding to the nearest minute (e.g. 10 minutes and 20 seconds = 10 minutes; 10 minutes and 40 seconds = 11 minutes); (b) registering only minutes whatever the number of seconds (e.g. 10 minutes and 1 second = 10 minutes, 10 minutes and 59 seconds = 10 minutes). Are the errors generated by rounding through (a) or (b) systematic or random? If systematic errors are generated, are they downward or upward biases?

2.   Classify the following variables according to their type and measure, following the scheme below:

METRIC VARIABLES	
Discrete	(MD)
Continuous	(MC)
NON-METRIC VARIABLES	
Nominal	(NN)
Ordinal	(NO)

And indicate the corresponding classification in SPSS (Scale, Nominal or Ordinal)

Variable	Example of measurements	Classification	SPSS classification
Price of a laptop	$1300.00; $1098.22; $2000.00	MC	Scale
Job	Bus driver; Marketing consultant; Singer	NN	Nominal
Education level	Higher education degree, High School Diploma	NO	Ordinal
Monthly income	$7000; $3000; $4112		
Monthly income bracket	1 = [0,$2000]; 2 = [$2001,$5000]; 3 = [more than $5000]		
Customer satisfaction	1 to 7 where 1 = completely unsatisfied; 7 = completely satisfied		
Perceived risk of skying	1 to 10 where 1 = not risky at all; 10 = extremely risky		
Ranking of preferred hobbies	1) Reading; 2) Playing baseball; 3) Gardening		
Amount of water drank in a day	2.3 liters; 1 liter; 5 liters		
Days off work over the last month	12 days; 6 days; 8 days		
Number of words in a document	1319 words; 5412 words; 1000 words		
Time spent walking yesterday	38 minutes; 12 minutes and 40 seconds; 15 minutes and 20 seconds		
Number of SMS sent last month	18; 12; 15; 31		

Variable	Example of measurements	Classification	SPSS classification
Phone bill	£41; £32; £55		
Weather today	Sunny; Rain; Snow		
Preferred mean of transport to work	Bike; Walking; Car;		
Telephone number	0044-123-456789;0039-987-654321		

3. Open the TRUST data-set in SPSS:

   • Go to VARIABLE view
   • Consider the TYPE of variable and the MEASURE column and compare them with the LABEL and VALUES column
   • Are the MEASURE types appropriate? Correct them as appropriate

Further readings and web-links

❖ **Measurement scales** – Ofir et al. (1987) compare the statistical performance of the Stapel, Semantic Differential and Likert scales and suggest that the three scales are interchangeable. A special issue of the review *Marketing Research* (Fall, 2003) contains a set of interesting papers on the use of measurement scales in consumer perception surveys. This special issue was published to revisit an article previously published on the same journal looking for a good scale to measure quality (Devlin et al., 1993) and a number of reactions sparked by that article. For those interested in measurement scales, a very comprehensive review of the key issues.

❖ **Measurement and error theory** – An excellent review of the problem of measurement, with a specific focus on the identification of error sources and the minimization of errors is provided in the book by Madhu Viswanathan (2005).

❖ **Software reviews** – It is not easy to compare the performances of statistical packages like SAS and SPSS in general terms, although for specific techniques the approaches and outcomes can be very different. When this is the case, details are provided in this book. Furthermore, statistical packages evolve quickly. See www.sas.com and www.spss.com to be updated on the latest versions and features of SAS and SPSS, respectively. However, for some specific computational aspect (including the key one of generating random numbers), different packages may perform very differently. McCullough (1998 and 1999) looks into the statistical and numerical accuracy of SPSS versus SAS and S-Plus, identifying some pitfalls.

❖ **The UK Expenditure and Food Survey** – The EFS is further discussed in chapter 2. Anything else you may want to know about questionnaires, variables, data-set, etc. can be found by referring to www.data-archive.ac.uk

(Continued)

and search the catalog using the key-word EFS plus the year of the survey. For example, with 'EFS 2004-2005' you will be referred to www.data-archive.ac.uk/findingData/snDescription.asp?sn=5375

❖ **The Trust project** – The web-site contains all of the working papers (including those on sampling and questionnaire preparation), plus contacts for more information: www.trust.unifi.it

Hints for more advanced studies

☞ Is it better to have an even or odd number of points in a Likert scale?
☞ What are skewness and kurtosis in a normal curve? How are they measured?
☞ How does the wording of multiple questions influence computation of the Cronbach's alpha?
☞ What are the methods to measure consumer attitudes?

Notes

1. Daniel Kehlmann (2007), *Measuring the World*. Quercus.
2. There are also procedures to combine pair-wise comparisons into interval scales as the Thurstone Scale (see Thurstone, 1927). It is also possible to consider individual personal effects in choosing between the alternatives through the Rasch model (see e.g. Wright, 1977).
3. For a review of techniques for testing validity, see Peter (1981).
4. The data-sets EFS and TRUST can be downloaded from www.sagepub.co.uk/Mazzocchi
5. The EFS data collection includes several derived and raw files, accessibile from www.data-archive.ac.uk. The data-set used in this book is derived from the household file 'dvhh.'
6. www.trust.unifi.it. Data were produced within the project 'Food Risk Communication and Consumer's Trust in the Food Supply Chain – TRUST,' funded by the European Commission (QLK1-CT-2002-02343).
7. In 2005 revenues for SPSS and SAS were $236 millions and $1,680 millions, respectively.
8. See the latest addition SAS Enterprise Guide.

CHAPTER 2

Secondary Consumer Data

T HIS CHAPTER introduces the distinction between primary and secondary data and explores the main official and commercial sources of secondary consumer and marketing data. Data may be gathered through ad-hoc surveys specifically built for the purposes of the study (primary data) or assembled by accessing available information like previous studies, official statistics, etc. (secondary data). Usually, the trade-off is apparent; good primary data are generally more informative than secondary data, but also more expensive. This chapter further discusses the above trade-off. The availability and detail of secondary data have greatly improved over the last decade in most countries. This chapter provides details on consumer surveys in the UK, with a look at US and European data.

Section 2.1   explains the differences and trade-offs between primary and secondary data
Section 2.2   reviews the main types of secondary data, with a focus on UK official surveys
Section 2.3   looks into the availability of household budget data in the UK, US and Europe
Section 2.4   describes the main characteristics and advantages of consumer panel data
Section 2.5   lists the main commercial providers of secondary marketing and consumer data

THREE LEARNING OUTCOMES

This chapter enables the reader to:

➡ Appreciate the difference and trade-offs between primary and secondary data
➡ Understand the different types of secondary data available to marketing researchers
➡ Become familiar with the official and commercial data sources in the UK and other countries

PRELIMINARY KNOWLEDGE: None in particular

2.1 Primary and secondary data

Measurement and dealing with its potential and unavoidable errors is at the core of any sound statistical analysis, as emphasized in chapter 1. However, measurement is not necessarily under the direct control of the researcher processing the data, who nevertheless needs to deal with its errors. This is especially true when the statistical analysis is dealing with already available data, which someone else has produced, like official statistics or published data. This requires a careful evaluation of data quality (more details are provided in chapter 4). In other situations, the research process explicitly includes an initial phase of field work for data collection and gives control and responsibility to the researcher.

The basic distinction introduced above is the one between *primary data* and *secondary data*. The former indicates data that are directly collected for the purposes of the research under the researcher control, the latter refers to the exploitation of already available data, which are useful to the research aims although not being explicitly built for its scopes. Although the primary aim of this book is to deal with the statistical analysis of data rather than their collection and evaluation, it is important to recall the key characteristics of primary and secondary data to ensure that more complex analyses take into account possible data problems. This chapter and the next review the main characteristics and issues for secondary and primary data, but here it may be helpful to provide a quick comparison between these two types.

Since access to secondary data is generally much cheaper than primary data collection through a survey, the key question is whether and when it is acceptable to settle for existing data without running an ad-hoc survey. There is no recipe for such a decision, but a series of schematic considerations on available secondary data might be useful to decide whether primary data collection is necessary (see box 2.1).

In general, any research plan should schedule a phase of exploration of available secondary sources. In the worst case scenario, this could be a useful preliminary step toward primary analysis.

With respect to the checklist indicated in box 2.1, the items are sorted according to their relevance. With a ten out of ten score, the advice is to give up expensive primary data collection and focus on the available published data. However, even if only the first 5 or 6 points are met, it may be convenient to rely on statistical analysis of secondary data to attain the research objectives, given some cautions listed below.

BOX 2.1   *A ten-item checklist before using secondary data*

1. The difference between the costs of secondary data and primary data collection is relevant (or timing is not enough for primary data collection)
2. Detailed information on data quality exists (survey method, questionnaire if relevant, etc.)
3. The sample is probabilistic and the sampling method is known
4. The target population of the secondary data sources coincides with the research target population
5. Secondary data are not outdated
6. The source of secondary data is independent and free from biases (critical view is advisable)
7. The purpose of secondary data fits with those of the research
8. There is no or little risk that some key variable is missing from secondary data
9. The secondary data source allows for time comparisons
10. Micro-data are available

Difference in costs and time. Sometimes, even when secondary data is available, costs might be prohibitive. When the researchers need to have up-to-date or sensitive market information at a great detail, market research companies can charge very high fees, comparable to the costs of running a primary data collection. In such a situation, given that primary data give more control (and responsibility) to the researcher, marginally higher costs should not prevent the running of an ad-hoc survey. In other circumstances, the client needs a very quick response and the timing does not allow to set up a proper data collection step. In these circumstances, the choice is between giving up the research or exploiting the available information and the circumstances in which the former option is chosen are very rare. When research findings are based on secondary data with limitations, a transparent discussion of these limitations and a quantification of the potential errors and biases accompanying the results may be the preferable alternative.

Detailed information on the quality of secondary data. If secondary data are accompanied by a detailed description of the sampling plan, questionnaire, interview methods, non-response treatments, etc., it becomes possible to have a clear idea of their quality. This is a necessary condition to avoid meaningless (or misleading) research conclusions. A critical review of secondary data construction may lead to two outcomes: (a) data are of an acceptable quality, hence statistical processing will be meaningful or (b) quality is inadequate because of serious shortcomings. Even in case (b) the review proves very useful; the necessary primary research can now be set up exploiting strengths and fixing weaknesses of the secondary source.

Probabilistic sampling. Chapter 5 warns of the risks of attempting inference (generalizations of sampling data to the target population) with non-representative samples emphasizing the role of probabilistic sampling methods. Non-probabilistic samples (such as quota sampling or haphazard samples) might be biased and misleading. When using secondary data, a necessary condition is to be informed of the sampling method and of the steps taken to ensure sample representativeness.

Comparison of the target populations and representativeness. Even with probabilistic methods, representativeness is a relative concept, usually based on one or a few variables. If one relies on secondary data, besides checking for an accurate sampling strategy, it is necessary to ensure that the representativeness criteria fit with ours. For example, if we are interested in a given geographical region and we use micro-data from a nationally representative survey, we have to make sure that the sampling size and the sampling strategy also ensure representativeness at regional level. Alternatively, if representativeness is for the whole population and we are concerned with a sub-group of the population (for example a specific age group or income bracket), we should make sure that we can actually rely on the sub-sample. An example of misuse of secondary data is a consumer survey which explores purchases of a wide basket of goods. Suppose that one is interested in fizzy drinks consumption and these are part of the basket. Even if the sample is geographically representative and data quality is adequate, but the data collection step has been carried out in July, one will probably get an upward biased estimate of consumption, because the target population of the secondary data (consumers in July) does not fit adequately with the target population (fizzy drinks consumers throughout the year).

Time lag between collection and publication of secondary data. When looking for published data, it is not infrequent to find a potentially interesting data-set, but too far in time with respect to the research objective. For examples, household budget survey data produced by national statistical institutes are generally made available one or two years after their collection. The relevance of the time lag depends on the research objectives. If the aim is to explore whether there is room for a novel product, it may be very risky to rely on outdated information. If the research purpose is to explore a relationship between two or more variables which can be safely assumed to be stable in the medium term (like income elasticities), then a two-year-old data-set may be acceptable.

Source reliability. It is often the case that there are two very different pieces of information for the same variable, especially when measurement of the target variable is a sensitive issue, for instance in electoral surveys (see box 2.2). Sometimes it is possible to evaluate source reliability in a relatively objective way by employing meta-analysis (see box 2.3). In the given example, a simple averages of the electoral surveys returns a 51.2% for the left-wing coalition and a 47.6% for the right-wing coalition. Bearing in mind that in this case the size of the meta-analysis is far from being sufficient (one should rely on at least 40–50 measures), this result would indicate that those surveys that show relevant discrepancies from the average might be affected by some bias.

Purpose of the research. Besides the cautions on source reliability, similar care should be taken in considering the purpose driving data collection for the secondary sources. For example, if household food expenditure data is collected by an official source as part of a budget survey within the national accounting system, it could be misleading to use that data source for evaluating food consumption and nutrition intakes. Expenditures depend on price levels and the sample could be representative in terms of the range of price levels faced by the populations, while food consumption may depend on a number of lifestyle variables not necessarily accounted for in designing the expenditure survey. While this source of bias is likely to exist and should be explicitly evaluated, it is not necessarily a reason for giving up the analysis of secondary data.

Completeness of the survey. When the objective of the research is broad, for example consumption of fruit and vegetables of British people, it may be relatively easy to identify a number of secondary sources providing that sort of information; food consumption surveys, scan data from supermarkets, disappearance data from food balance sheets are examples of secondary sources available for this purpose. However, the researcher must take care to ensure that all the necessary variables are considered in these data-sets. Scan data will exclude all those shoppers who still shop at local stores rather than supermarkets; disappearance data does not take into account household waste; food consumption surveys might only provide household consumption data without disaggregation across household components.

Secondary surveys repeated over time. Sometimes secondary data, especially those provided by national statistical offices, are collected at regular intervals, sometimes even on the same individuals. Since repeated surveys are extremely expensive and consistency across surveys is not easy to be guaranteed, when the research purpose is to explore trends or monitor a population over time, secondary sources could have a key advantage on running an expensive panel survey.

Availability of individual data. Another situation when the use of secondary data could be encouraged is the availability of micro-data, that is individual information on the sample unit. This could help to correct sampling biases or target population mismatches

BOX 2.2 *Electoral surveys and source reliability*

Take the electoral surveys published in Italy before the 2006 parliamentary elections. Within four days six different surveys have been published. One of them (commissioned by a political party) reversed the results presented by most sources giving rise to a hot debate in the media on the statistical reliability of electoral surveys. Given that all sources claimed to use nationally representative probabilistic sampling, and that all the interviews methods were computer-assisted telephone interviews (CATI), similar results should be expected. All sources recognized that about one quarter of the respondents were undecided about their voting preference. This already poses a big question mark on the utility of these surveys, but it is recognized that they are often used as an instrument to influence voters rather than monitoring them (see for instance Alves et al., 2002). Thus, knowing the client who commissioned the survey becomes a key step in order to exploit the published data. In these circumstances, a useful technique to explore secondary data sources is *meta-analysis* (box 2.3). Professor Sam Wang at Princetown University shows how the use of electoral surveys in the week before the elections led to an exact prediction of the final electoral results of the 2004 US elections (election.princeton.edu). Predictions were available on-line before the elections on the Princeton web-site, which received up to 100,000 visitors per day and was covered in the Wall Street Journal.

Italian elections 2006 – Electoral surveys two months before elections

Source	Date of survey	Left-wing coalition	Right-wing coalition	Sample size	Survey information
SGW for magazine 'L'Espresso'	10 February	52%	47.5%	1,000	Stratified random sampling, CATI interviews
TSN Abacus for 'Sky television'	13–14 February	51%	47%	1,000	Stratified random sampling, CATI interviews
IPR Marketing for newspaper 'La Repubblica'	13–14 February	52%	47.5%	1,000	Simple random sampling with phonebook sampling frame, CATI interviews
PSB for political party 'Forza Italia'	7–9 February	48.2%	48.4%	1,920	Simple random sampling, CATI interviews, sampling error of 2.2%
Ekma for press agency 'News'	13 February	51.5%	47.5%	1,000	Stratified random sampling, CATI interviews
ISPO Ltd. for newspaper 'Il Corriere della Sera'	9–10 February	51.5%	47.5%	1,602	Stratified random sampling, CATI interviews

Note: Results and survey information from the government web-site
www.sondaggipoliticoelettorali.it

BOX 2.3 *Meta-analysis*

In a situation where there are several surveys which share the research purpose (or simply some of the measured variables), a useful tool increasingly used in marketing research is *meta-analysis*, which is the statistical analysis of several research findings. The aim of meta-analysis is to generalize specific researches to gain broader knowledge of a phenomenon, by finding consistencies across studies and isolating inconsistencies (see for instance Sawyer and Peter, 1983; Farley and Lehmann, 1986).

A simple meta-analysis could consist in averaging estimates for a given variable obtained in different studies. More complex meta-analyses relate the target *effect* (the target/dependent variable) on the characteristic of the research environment of the selected studies (the explanatory variables). A regression analysis of the target variable on a set of variables derived from the individual studies can (1) provide useful information on the direction of the relationship and (2) identify potential biases in research designs and sources. The contribution of meta-analysis depends on the number of studies taken into account and is based on a rigorous and systematic selection and classification of the *meta-data*, that is 'data about data.'

There are several approaches to combining the research findings from different studies. For example, an issue is whether the weight of different surveys should be proportional to their size. Or – as in the *inverse variance method* – when an *average effect* (which is an average of the target variable) is computed, a weighted average could be based on weights equal to the inverse variance of the estimator for each study, so that more precise studies have more relevance in computing the average. Meta-analysis helps correcting measurement errors or other methodological weaknesses.

A drawback of meta-analysis is that sources of bias are difficult to be controlled. If one can include some indicator of the methodological quality of the study, then the effect of improper methods can be evaluated. Otherwise, improper studies may reflect on the final estimates of meta-analysis. The method is especially popular in medicine and education research, but is increasingly used in marketing and consumer research. Eisend (2005) provide a vast bibliography on meta-analysis in consumer research.

(for example through post-stratification), besides allowing to take into account as many control factors as the number of variables available for each statistical unit.

2.2 Secondary data sources

It is not big news that there is a huge proliferation of secondary data sources on consumer behaviors and attitudes. As it happens, supply increases in response to a growing demand. Furthermore, data collection methods are becoming cheaper and cheaper. Internet-based communication systems made it easier to reach a wide number of individuals at low costs and home phones are increasingly targeted by marketing calls and surveys. While this has an unquestionable impact on the quality of collected data (and the quality of life at home ...), it is also increasingly difficult to choose among the potential sources.

This section tries to provide a 'rough guide' to secondary sources in consumer research according to the type of sources, listing their strength and weaknesses and some examples of real surveys . Practical examples focus on UK sources,[1] but surveys of the same types exist in most countries. Some detail on non-UK surveys are provided in the next section.

Although it is difficult to compile a complete list of sources and despite the international differences in surveys and collection strategies, most of the existing data can be classified into one of the five following categories:

1. Household surveys
2. Lifestyle and social trend surveys
3. Consumer panels
4. Attitude surveys
5. Retail level surveys

2.2.1 Household surveys in the UK

Almost all countries run official household budget surveys (or expenditure surveys), since they allow computation of consumption aggregates in national accounting and provide the information basis for fiscal policies and studies on income distribution. The main purpose of these studies is to gather data on household expenditures, although in some countries consumed quantities are also recorded. Samples are nationally representative, but in most cases also the sampling design and size allow for representativeness of population sub-groups (regions, age groups, etc.). Data collection is usually based on weekly or bi-weekly diaries where the household representative is expected to record any expenditure (in some cases receipts are collected as well). At the end of the period an interviewer submits a questionnaire (which includes socio-demographic and other information) and collects the diary. Household budget surveys are a valuable source of consumer information for several reasons:

a. they are based on large and representative samples;
b. they allow to explore relation between expenditures and a large set of household information;
c. they are carried out at regular intervals;
d. they are generally accessible at a very low cost, often for free;
e. it is usually possible to access individual household data (treated to guarantee anonymity), hence allowing for a high degree of personalization in cross-tabulation and processing.

Drawbacks usually lie in some respondent biases (especially in diary keeping), in the frequent lack of information on prices and quantities and in the delay between data collection and data publication (in most of the cases at least 1 or 2 years). Also, they do not allow evaluation of non-economic factors such as attitudes and they rarely record information on the point of purchase.

The main household budget survey in the UK is the *Expenditure and Food Survey*, which was initiated in 2001 following the merger of the National Food Survey and the Household Expenditure Survey. The survey covers about 7,000 British households, interviewed through face-to-face interviews and diaries. Europe is working toward the harmonization of expenditure surveys (although the target is still far away), through the COICOP classification system (table 2.1). More details are provided in section 2.3.1.

Among other surveys, the *Family Resource Survey* gathers information on tenure and housing costs including council tax, mortgages, insurance, water and sewage rates, consumer durables, vehicles, use of health services, welfare/school milk and meals, educational grants and loans, children in education, informal care (given and received), occupation and employment, health restrictions on work, children's health, wage details, self-employed earnings, personal and occupational pension schemes and more.

The *National Diet and Nutrition Survey* which is run irregularly every four to five years, is especially useful to monitor food consumption and dietary habits. Information is gathered through a seven-day dietary record.

2.2.2 Lifestyle and social trend surveys

A second type of large household surveys is dedicated to record a wide range of variables, focusing on social dynamics, habits and lifestyle trends. These surveys are multi-purpose, with some sections that remain unchanged over time to allow for comparisons purposes, other sections that change cyclically, so that time comparisons are still possible although not on a yearly basis, and some sections change every year to look into specific issues of current relevance.

In the UK, this survey is the *General Household Survey*, a multi-purpose sample survey where the household is the sample unit, but information is collected on individuals. Approximately 9,000 households are involved and about 16,000 adults aged 16 and over. Data are collected on five core topics, namely education, employment, health, housing, and population and family information. Other areas such as leisure, household burglary, smoking and drinking are covered periodically. The interview methods are based on face-to-face computer-assisted personal interview (CAPI), with some self-completion sections.

A second survey is the Office for National Statistics (ONS) *Omnibus Survey*, where about 1,800 adults are interviewed by computer-aided personal interviewing on a multi-purpose questionnaire, relatively short and made of simple sets of questions on a varying and large range of topics, such as contraception, unused medicines, tobacco consumption, changes to family income, Internet access, arts participation, transport, fire safety and time use.

A survey on *Smoking, Drinking and Drug Use among Young People* is carried out every year to monitor smoking behaviors, family and individual attitudes toward smoking, drinking behaviors, methods of obtaining alcohol, 'binge' drinking, drug use, and drug education. Classificatory, school and demographic information is also included in the data file.

The *Time Use Survey* is a one-time source of information on lifestyles. About 6,400 respondents were asked to provide information through a household questionnaire, individual questionnaire and self-completion 24 hour diaries, broken down into 10-minute slots. Topics covered in the questionnaires include employment, qualifications, care of dependants and children, leisure time activities and demographic details, such as age, gender, marital status, ethnicity and housing. Occasionally (for example in 2001), the survey on *National Surveys of Sexual Attitudes and Lifestyles* monitors sexual habits of British people.

2.2.3 Consumer panels

When the objective is to monitor consumption dynamics or response to changing factors (like advertising), a key source of information is given by consumer panels. These are repeated surveys where the same individuals are interviewed over time. Panels have higher response rates, but are very expensive, so access to secondary data is generally the preferred option.

The Office for National Statistics *British Household Panel Survey* monitors social and economic change at the individual and household level in Britain through a range of socio-economic variables. The panel structure of the survey implies that the same individuals are re-interviewed in successive waves. More details on the BHPS are provided in section 2.4.

Private market research companies also run regular household panel surveys, like the ACNielsen *HomeScan Consumer Panel*™ described in section 2.5.

2.2.4 Attitudes and values

Attitudinal data are a valuable source of information on consumer behavior, but given the range of attitudes that could be measured and the specificity to target behaviors, this is an area where regular and official surveys do not usually provide sufficient detail.

The main official source for attitudinal data in Britain is the *British Social Attitudes Survey*, a series of annual surveys to monitor dynamics in British social, economic, political and moral values. Around 3,600 individuals are contacted each year to answer to a set of questions, some of which are the same throughout the years, while others have a lower frequency. Attitudes are related to a large range of issues, such as newspaper readership, political parties and trust, public spending, welfare benefits, health care, childcare, poverty, the labor market and the workplace, education, charitable giving, the countryside, transport and the environment, Europe, economic prospects, race, religion, civil liberties, immigration, sentencing and prisons, fear of crime and the portrayal of sex and violence in the media.

2.2.5 Retail level surveys

A continuous monitoring of market dynamics can be based on data collection at the point of purchase. Retail level surveys generally exploit scan tracking at the retail level, although this excludes those stores with no scanning facilities.

The official survey at the retailing level is the *Retail Sales Inquiry*, which monthly collects sales data on a sample of 5,000 retailers, including all retailers with more than 100 employees and a stratified sample of smaller retailers. The survey records total retail turnover (including sales from stores, e-commerce, Internet sales, etc.), which is the value of sales of goods to the general public for personal or household use. Sales volume is also recorded following the COICOP classification. Retail tracking services based on scan data are also provided by Gfk, ACNielsen and IRI.

2.2.6 Consumer trends

The UK Statistical office (HMSO) provides free-of-charge time series data,[2] which includes:

a. quarterly data on household final consumption expenditure (from National Accounts) according to the COICOP international classification;

b. monthly consumer and retail price indices;
c. monthly data on consumer credit and other households borrowing;
d. monthly retail sales indices, for different types of retailers and goods.

Consumer trends contain details on Household Final Consumption Expenditure (HHFCE) at aggregate level for the whole UK on a quarterly basis. Final consumption includes also goods consumed by non-profit institutions serving households.

For each item of the COICOP classification it is possible to download the quarterly time series from 1963 to the latest available information, choosing between *national expenditure* (by UK citizens) and *domestic expenditure* (on the UK territory), between *raw* (not-seasonally adjusted) and *seasonally adjusted* data (where seasonal factors have been removed through statistical methods) and between *current prices* and *constant prices*, where the latter implies that expenditures have been deflated so that the whole time series refers to the prices of a given base year.

2.3 Household budget surveys

In all countries, Household Budget Surveys (HBS) constitute a valuable and official source of data on household expenditures. While Europe is slowly going toward a harmonization of classifications, contents and methods of HBS, (the UK survey, now named Expenditure and Food Survey, EFS), is certainly one of the most complete existing household surveys. As introduced in chapter 1, this book exploits a sample data-set extracted from the EFS to provide applied examples of statistical analysis.

2.3.1 The UK Expenditure and Food Survey

The UK Expenditure and Food Survey (EFS) has existed in its current version since 2001 (see also section 2.2.1). Household surveys in the UK date back to the 1940s.

The EFS is also based on the COICOP classification and micro-data are made available for research purposes. The procedure for collecting data is based strongly on consumer-assisted methods. Each individual above 16 in the household sampled keeps a diary recording all expenditure, with a simplified version for children between 7 and 15. This information is integrated by a detailed household questionnaire collecting socio-demographic information and data on regular household bills, large items of expenditure such as cars and holidays, and ownership of consumer durables. Members above 16 also fill a questionnaire about income. Paper questionnaires are read by a data inputting software, which also drives the computer-assisted personal interviews. The target sample includes about 12,000 households, although the actual data reduces to about 7,000 because of coverage errors and missing responses. Sampling is based on post codes (see chapter 5). Micro (household level) data of the EFS are available as SPSS files through the UK Data Archive.[3]

The household questionnaire covers both one-time purchases and regular payments, including expenditure on rent, rates, mortgages, insurance, council tax, pensions, consumer durables, telephone, electricity and gas, central heating, television and satellite/digital services, vehicles, season tickets, credit cards, loans, hire purchase,

Table 2.1   *COICOP classification*

1 FOOD AND NON-ALCOHOLIC BEVERAGES
1.1 Food
1.1.1 Bread and cereals
1.1.1.1 Bread
1.1.1.2 Cereals
1.1.1.3 Biscuits and cakes
1.1.1.4 Pasta and pastry-cooked products
1.1.2 Meat
1.1.2.1 Beef
1.1.2.2 Lamb
1.1.2.3 Pork and bacon
1.1.2.4 Poultry
1.1.2.5 Other meat
1.1.3 Fish
1.1.3.1 Fresh fish
1.1.3.2 Processed fish
1.1.4 Milk, cheese and eggs
1.1.5 Oils and fats
1.1.6 Fruit
1.1.7 Vegetables
1.1.7.1 Fresh vegetables (including potatoes)
1.1.7.2 Processed vegetables and vegetables preparations
1.1.8 Sugar, jam, honey, chocolate and confectionery
1.1.9 Food products n.e.c.
1.2 Non-alcoholic beverages
1.2.1 Coffee, tea, cocoa and other hot drinks
1.2.2 Mineral waters, soft drinks, fruit and vegetable juices

2 ALCOHOLIC BEVERAGES AND TOBACCO
2.1 Alcoholic beverages ('off' sales)
2.1.1 Spirits
2.1.2 Wine
2.1.3 Beer
2.2 Tobacco
2.2.0 Tobacco

3 CLOTHING AND FOOTWEAR
3.1 Clothing
3.1.2 Garments

3.1.2.1 Men's outerwear
3.1.2.2 Women's outerwear
3.1.2.3 Children's outerwear
3.1.2.4 Other garments
3.1.3/4 Other articles of clothing and clothing accessories & Cleaning, repair and hire of clothing
3.2 Footwear
3.2.1/2 Shoes and other footwear & Repair and hire of footwear

4 HOUSING, WATER, ELECTRICITY, GAS AND OTHER FUELS
4.1 Actual rentals for housing
4.1.1/2 Actual rentals for housing
4.3 Maintenance and repair of the dwelling
4.3.1 Materials for the maintenance and repair of the dwelling
4.3.2 Services for the maintenance and repair of the dwelling
4.4 Water supply and miscellaneous services relating to the dwelling
4.4.1 Water supply
4.4.3 Sewerage collection
4.5 Electricity, gas and other fuels
4.5.1 Electricity
4.5.2 Gas
4.5.3/4 Liquid and solid fuels
4.6 Owner-occupier housing costs
4.6.1 Mortgage Interest Payments
4.6.2 Council Tax
4.6.3 Depreciation and ground rent

5 FURNISHINGS, HOUSEHOLD EQUIPMENT AND ROUTINE MAINTENANCE OF THE HOUSE
5.1 Furniture and furnishings, carpets and other floor coverings
5.1.1 Furniture and furnishings
5.1.2 Carpets and other floor coverings
5.2 Household textiles
5.2.0 Household textiles
5.3 Household appliances
5.3.1/2 Major and small household appliances whether electric or not

(Continued)

Table 2.1 *Cont'd*

5.3.3 Repair of household appliances

5.4 Glassware, tableware and household utensils

5.4.0 Glassware, tableware and household utensils

5.5 Tools and equipment for house and garden

5.5.1/2 Tools and equipment for house and garden

5.6 Goods and services for routine household maintenance

5.6.1 Non-durable household goods

5.6.2 Domestic services and household services

6 HEALTH

6.1 Medical products, appliances and equipment

6.1.0 Medical products, appliances and equipment

6.2 Out-patient services

6.2.0 Out-patient services

7 TRANSPORT

7.1 Purchase of vehicles

7.1.1 Motor cars

7.1.1.1 New cars (proxy)

7.1.1.2 Second-hand cars

7.1.2/3 Motor cycles and bicycles

7.2 Operation of personal transport equipment

7.2.1 Spare parts and accessories for personal transport equipment

7.2.2 Fuels and lubricants for personal transport equipment

7.2.3 Maintenance and repair of personal transport equipment

7.2.4 Other services in respect of personal transport equipment

7.3 Transport services

7.3.1 Passenger transport by railway

7.3.2 Passenger transport by road

7.3.2.1 Bus/coach fares

7.3.2.2 Taxi and minicab

7.3.3/4 Passenger transport by air, sea and inland waterway

7.4 Annual Road Licence

7.4.0 Annual Road Licence

8 COMMUNICATION

8.1 Postal services

8.1.0 Postal services

8.2/3 Telephone and telefax equipment and services

8.2/3.0 Telephone and telefax equipment and services

9 RECREATION AND CULTURE

9.1 Audio-visual, photographic and information processing equipment

9.1.1/2 Equipment for the reception, recording and reproduction of sound and pictures & Photographic and cinematographic equipment and optical instruments

9.1.3 Information processing equipment

9.1.4 Recording media

9.1.5 Repair of audio-visual, photographic, and information processing equipment

9.2 Other major durables for recreation and culture

9.2.0 Other major durables for recreation and culture

9.3 Other recreational items and equipment, gardens and pets

9.3.1/2 Games, toys and hobbies and equipment for sport, camping and open-air recreation

9.3.3 Gardens, plants and flowers

9.3.4 Pets and related products

9.3.5 Veterinary and other services for pets

9.4 Recreational and cultural services

9.4.1 Recreational and sporting services

9.4.2 Cultural services

9.4.2.1 TV licences/rent

9.4.2.2 Other cultural services

9.5 Newspapers, books and stationery

9.5.1 Books

9.5.2 Newspapers and periodicals

9.5.4 Stationery and drawing materials

9.6 Package holidays

9.6.0 Package holidays

10 EDUCATION

10.0 Education

10.0.0 Education

Table 2.1 *Cont'd*

11 RESTAURANTS AND HOTELS	12.3 Personal effects n.e.c.
11.1 Catering services	12.3.1 Jewellery, clocks and watches
11.1.1 Restaurants, cafés and the like	12.3.2 Other personal effects
11.1.1.1 Restaurant meals	12.4 Social protection
11.1.1.2 Take-aways and snacks	12.4.0 Social protection
11.1.1.3 Beer 'on' sales	12.5 Insurance
11.1.1.4 Wines and spirits 'on' sales	12.5.1/3/5 Life, health and other insurance
11.1.2 Canteens	12.5.2 Insurance connected with the dwelling
11.2 Accommodation services	
11.2.0 Accommodation services	12.5.4 Insurance connected with transport
	12.6 Financial services n.e.c.
12 MISCELLANEOUS GOODS AND SERVICES	12.6.2 Financial services n.e.c.
12.1 Personal care	12.7 Other services n.e.c.
12.1.1 Hairdressing salons and personal grooming establishments	12.7.0 Other services n.e.c.
	12.8 House-buying services
12.1.2/3 Appliances, articles and products for personal care	12.8.0 House-buying services

club credit, second dwellings, capital improvements and house maintenance, house moves, furniture, carpets, holidays and flights, banking, welfare and free school milk and meals, education, separation/maintenance allowances, employer refunds, items paid from outside the household and money given to household and Internet purchasing. Demographic information on household composition, age, gender, benefit units, accommodation and tenure is also collected. The income questionnaire collects detailed information on employment status, job description, pay details, income from self-employment and subsidiary employment, national insurance contributions, redundancy payments, concessionary bus passes, social security benefits, pensions, allowances, money sent abroad, investments and children's income. Personal information such as age, gender and marital status is also gathered.

2.3.2 *The US Consumer Survey*

Similarly to the UK EFS, the United States Consumer Expenditure Survey (CES) also collects information through CAPI interviews and a diary filled by the household members. However, the two collection strategies are based on two separate and independent samples which have in common data on household income and socio-economic characteristics. The Bureau of Labor Statistics holds responsibility for the CES,[4] data are publicly available and date back to 1980. About 7,500 households enter each of the surveys. The interview survey includes monthly out-of-pocket expenditures such as housing, apparel, transportation, health care, insurance, and entertainment, while the diary survey includes weekly expenditures of frequently purchased items such as food and beverages, tobacco, personal care products and non-prescription drugs and supplies.

BOX 2.4 *Household Budget Surveys in the European Union (EU-15)*

Belgium:	Enquête sur les Budgets des Ménages
Denmark:	Forbrugerundersøgelsen
Germany:	Einkommens- und Verbrauchsstichprobe
Greece:	Family Budget Survey
Spain:	Encuesta Continua de Presupuestos Familiares
	Encuesta Basica de Presupuestos Familiares
France:	Enquête Budgets des Familles
Ireland:	Household Budget Survey
Italy:	Rilevazione sui consumi delle famiglie italiane
Luxembourg:	Enquête Budgets Familiaux
Netherlands:	Budgetonderzoek
Austria:	Konsumerhebung
Portugal:	Inquérito aos orçamentos familiares
Finland:	Kulutustukimus
Sweden:	Hushållens utgifter
UK:	Expenditure and Food Survey

Source: Household Budget Survey Working Group (1997). *Household Budget Surveys in the EU: Methodology and recommendations for harmonization.* Eurostat, February 1997

2.3.3 Surveys in Europe

Each EU member state conducts its own household budget survey (see box 2.4), but since 1988 the European Statistical Institute (Eurostat) has started a harmonization effort which has also led to an EU household budget survey. The latter consists in the ex-post harmonization of National micro-data, an operation which is carried out every five or six years.

These aggregate data are available on the Eurostat New-Cronos Data Bank, which contains the average expenditure by country and COICOP category.

2.4 Household panels

In household panels, the same households take part in the survey in successive waves. This allows to monitor changes and dynamics in consumption patterns. Since 1991, the *British Household Panel Survey* (BHPS) follows 5,500 households every year. Its coverage includes income, labor market behavior, social and political values, health, education, housing and household organization. The BHPS is run by the ESRC UK Longitudinal Studies Centre (ULSC), together with the Institute for Social and Economic Research (ISER) at the University of Essex.[5] Although the household is the sampling unit, the target population is made by individuals above 15, and the survey consists of about 10,000 interviews per year. When individuals leave their original households, the sample is extended to include the members of their new households. While individual household members enter the sample when they are 16, since 1995 there is a special survey for children aged between 11 and 15. The survey is based on a mixed computer-assisted administration method, which currently includes a CAPI face-to-face interview, a CATI interview and a self-administered questionnaire for attitudinal and sensitive questions. The original sampling technique was a two stage stratified systematic sampling. The survey collects information on household composition and demographics, data on accommodation and tenure and some household-level measures of consumption, residential mobility, health and caring, current employment and earnings, employment changes, lifetime childbirth, marital and relationship history, household finances and organization. The BHPS has an interesting section on values and opinions and the self-administered questionnaire includes subjective or attitudinal questions that are perceived as sensitive to the influence of other household members or the interviewer (see box 2.5). The telephone interview is used as a substitute after failure of efforts for conducting the CAPI interview. Interestingly, the special questionnaire for children aged between 11 and 15 is administered through a self-completion grid guided by a walkman tape to cover sensitive issues.

Eurostat has recently launched a new EU-wide household panel survey, called EU-SILC (where SILC stands for Statistics on Income and Living Conditions). Every year (since 2005), the survey collects micro-data on income, poverty, social exclusion, labor, education, health information and living conditions in all 25 member States plus Norway and Iceland and is expected to include Turkey, Romania, Bulgaria and Switzerland in coming years. It provides cross-sectional data on income, poverty, social exclusion and other living conditions and longitudinal data to detect individual-level changes, referring to a four year period.[6] The EU-SILC survey is the evolution of a previous panel survey at European level, the ECHP (European Community Household Panel) which ran eight waves of surveys between 1994 and 2001.

2.5 Commercial and scan data

A vast amount of consumer data is sold at commercial rates by companies specialized in market research. Some of these companies operate at international level and follow standards and procedures allowing for cross-country comparisons. Furthermore, market research companies provide brand data, which is not available in official surveys. This allows it to carry out research on market shares, loyalty, impact of brand advertising, promotions and other brand-specific marketing activities. Another advantage of commercial data is the quick availability, while official data may take several months. The drawback is that commercial rates are usually extremely high when compared to cheap and often free secondary data provided by national statistical offices.

Among the main companies providing secondary consumer data for UK and European markets are ACNielsen, GfK, IRI International, Taylor-Nelson Sofres, NPD and Ipsos. Besides doing personalized research, these companies run continuous consumer and household surveys.

ACNielsen has run a continuous consumer panel survey since 1989 using the home scanning technology (HomeScan™), with consumers scanning the purchased products when they bring them home and all information on the barcode is memorized. Scan data is integrated with traditional diaries and with barcode scanners (MyScan™) for out-of-home purchases. The ACNielsen consumer panel is nationally representative and data can be disaggregated by retail outlet type and by demographic group. The ACNielsen consumer panel in the UK includes 10,000 households and is representative at national level. Besides the scanning technology, collected data includes variable weight and non-bar-coded fresh products. A major advantage of this survey is the comparability with surveys run with the same methodology in other 26 countries, for a total of about 210,000 households.

BOX 2.5 *Questions in the self-completed questionnaire of the BHPS*

1. Have you recently....
   a. been able to concentrate on whatever you're doing?
   b. lost much sleep over worry?
   c. felt that you were playing a useful part in things?
   d. felt capable of making decisions about things?
   e. felt constantly under strain?
   f. felt you couldn't overcome your difficulties?
   g. been able to enjoy your normal day-to-day activities?
   h. been able to face up to problems?
   i. been feeling unhappy or depressed?
   j. been losing confidence in yourself?
   k. been thinking of yourself as a worthless person?
   l. been feeling reasonably happy, all things considered?
2. Do you personally agree or disagree with the following statements?
   a. It is alright for people to live together even if they have no interest in considering marriage.
   b. It is better to divorce than continue an unhappy marriage.
   c. When there are children in the family, parents should stay together even if they don't get along.
   d. It makes no difference to children whether their parents are married to each other or just living together.
   e. Adult children have an obligation to look after their elderly parents.
   f. Homosexual relationships are always wrong.
3. Here are some questions about how you feel about your life. Please tick the number which you feel best describes how dissatisfied or satisfied you are with the following aspects of your current situation.
   a. Your health
   b. The income of your household
   c. Your house/flat
   d. Your husband/wife/partner
   e. Your job (if in employment)
   f. Your social life
   g. The amount of leisure time you have
   h. The way you spend your leisure time
4a) Using the same scale how dissatisfied or satisfied are you with your life overall?
4b) Would you say that you are more satisfied with life, less satisfied or feel about the same as you did a year ago?
11. Here are a few questions about your friends. Please choose the three people you consider to be your closest friends starting with the first friend. They should not include people who live with you but they can include relatives.
    − Friend's gender
    − Relative/not relative
    − Age
    − About how long have you known him or her?
    − How often do you see or get in touch with your friend either by visiting, writing or by telephone?
    − About how many miles away does your friend live?
    − Friend's job
    − Friend's ethnic group

The consumer panel survey also collects attitudinal and motivational information. Besides the consumer panel, ACNielsen collects scan data at the retailer level (ScanTrack™).

GfK runs a continuous retail panel in the UK, which audits more than 200 products by monitoring a sample of retail outlets and also runs a quarterly nationally representative consumer panel of 18,000 households (HomeAudit) recording purchases through a home diary. Another weekly survey (Cresta) records food and drink purchasing behaviors.

Information Resources (IRI) developed InfoScan in 1987, a scanner-based supermarket tracking system which provides detailed information on sales, market shares, distribution, pricing and promotion, across retail channels based on probabilistic samples representative by geographic area and retailing sector and channel, covering the grocery, health and beauty, household, impulse, and over-the-counter medicines markets. Data are provided on a weekly basis in the week following the collection.

Taylor-Nelson Sofres (TNS) also runs a continuous household panel in the UK with diaries and home scanners for fast moving consumer goods, while a separate panel is targeted at individuals rather than households for those goods that are likely to be the outcome of an individual rather than a collective decision, like toiletries and cosmetics, textiles, impulse products, tobacco, telecoms, fast food, petrol. TNS also records separately media audience, providing data for assessing an advertising campaigns.

NPD is a specialist in sporting goods and collects data based on electronic point of sales scanners (EPOS) in several European countries including the UK, integrating scan data with those collected through nationally representative samples looking into other uses beside purchases of sports equipment. Another household survey (EuroCrest) monitors through diaries the consumption of snack foods and commercially prepared foods, at home and away from home including the workplace.

Ipsos is a specialist in *Omnibus Surveys,* which are regular surveys sponsored by multiple clients on the same representative sample where one section is stable over time, while other sections are based on the information required by the clients. In the UK, Ipsos runs regularly a survey on 2,000 households (Capibus™) with CAPI interviews. Omnibus surveys are relatively cheaper than other marketing research options, as the various clients share the survey cost, but the sample is usually independent from the research objectives which may reduce representativeness in some circumstances. A CATI alternative (Express™) and a web-based version (i:omnibus™) are also provided. IPSOS also offers the opportunity of administering the same questionnaire in various countries.

Summing up

The action of measurement provides the key ingredient for statistical analysis,- the data. Data can be collected directly for the purposes of research (primary data) or consist in already available data not explicitly built for the research scopes (secondary data). This chapter provides a set of rules for choosing between primary and secondary data, based on the quality-cost trade-off. There are a multitude of sources for secondary data useful to marketing researchers. National statistical offices provide time series expenditure data as the UK consumer trends and run regular household surveys collecting data on household incomes and expenditures, like the UK Expenditure and Food Survey or the US Consumer Expenditure Survey. While these surveys are repeated every year, they do not maintain the same sample. Instead panel surveys, like the British Household Panel Survey, give the opportunity of monitoring dynamics in behaviors and attitudes. Official sources provide valuable and statistically reliable information, but they do not allow one to explore brand dynamics and often information is not released in a timely way. Thus, private companies offer a wide range of continuous surveys, especially those based on bar-code scanning technologies at the retail outlet or with home scanning facilities for individual and household data collection.

EXERCISES

1. Suppose you need to estimate the average weekly petrol consumption of London citizens. Consider the following three options
    a. A magazine called *Londoners on the road* publishes an article stating that they have interviewed 2,000 of their subscribed readers by telephone last Saturday and recorded an average expenditure on petrol of £15. Considering that the price of petrol is about £0.90 per liter, one can estimate that the weekly consumption of petrol is: $15/0.90 \cdot 7$ days $= 116.67$ liters per citizen per week. The magazine providing secondary data costs you £2.
    b. Using secondary data from three retail chains' petrol stations one can build a data-set recording liters purchased in 20 randomly chosen London outlets each day of the last month. This produces a daily average of 78.3 liters per day per drive. Obtaining the set of secondary data from the three retailers costs you £1,000.
    c. You can select a random sample of 500 London citizens and record their petrol consumption every day over one month. You get a final estimate of 71.5 liters per day. Primary data collection costs you £40,000.

    1   Which one is the most reliable measure? Why?
    2   Without prior knowledge of the size of difference in the final estimate and considering the rules of box 2.1, which secondary or primary data source would you choose? Why?

2. Retrieve the following data from the web-site indicated
    a. www.statistics.gov.uk (Her Majesty Statistical Office, UK): Monthly consumer price index for tobacco from 1996 to 2007 (D7CB)
    b. www.bls.gov (Bureau of Labor Statistics, US) Consumer units with reference person under age 25 by income before taxes: Average annual expenditures and characteristics, Consumer Expenditure Survey, 2004–2005
    c. ec.europa.eu/eurostat (Eurostat, European Union) Overall structure of consumption expenditure by detailed COICOP level (per thousand), 1999: coicop cp0321 Shoes and other footwear – all countries
3. Open the EFS data-set in SPSS
    a. Write down all of the variables measuring income. Can you indicate why they differ?
    b. By using the information available on www.data-archive.ac.uk (see e.g. User Guide Vol. 1 of EFS 2004–5), comment on the quality of the data (% of refusals, no contacts, etc.)

Further readings and web-links

❖ **Use of secondary data** – An interesting exemplification of the issues surrounding the choice between primary and secondary data is provided by Rabianski (2000) in relation to the property market. Houston (2004) argues that secondary proxies can be used for marketing constructs and discusses a method for validating this usage.

❖ **Understanding consumers through scanner data** – Some marketing researchers argue that the use of scanner data could help validate consumer behavior theories; for an interesting exemplification see Winer (1999).

❖ **Panel surveys** – Those with a good understanding of statistics may have a look at the work of Baumgarten and Steenkamp (2006), which explores the issue of taking repeated attitude or personality measurements through panel surveys. If measurements change over time, what is the contribution of survey error?

❖ **Commercial data providers** – Web-links for some of the main providers of commercial data for marketing research: ACNielsen (www.acnielsen.com), GfK (www.gfk.com), IRI International (www.infores.com), Taylor-Nelson Sofres (www.tnsofres.com), NPD (www.npd.com) and Ipsos (www.ipsos.com).

Hints for more advanced studies

☞ What are the secondary data for TV audience in your country? How is TV audience measured? How reliable are available data?

☞ Scan data only records purchases, thus it does not fully reflect consumer choice. Why? What sort of bias does this imply?

☞ Most of household budget surveys are based on weekly or daily diaries. What are the implications for recording durable goods like refrigerators? Is there a risk of memory biases with diaries?

Notes

1. Information on UK surveys are provided by the Economic and Social Data Service (ESDC), www.esdc.ac.uk. Authorized users can access data through the same web-site.

2. www.statistics.gov.uk/statbase/tsdtimezone.asp

3. www.data-archive.ac.uk/

4. www.bls.gov

5. Data can be accessed at www.esds.ac.uk/longitudinal/access/bhps/

6. See epp.eurostat.ec.europa.eu for more details.

CHAPTER 3

Primary Data Collection

T HIS CHAPTER illustrates the primary data collection steps and provides a set of rules for designing a survey properly. The chapter also covers the main pros and cons of different types of administration methods and lists with examples the large number of issues to be faced when preparing a questionnaire. Finally, a classification of primary surveys in consumer and marketing research is proposed, according to four broad data-types:

1. socio-demographics;
2. economics and behaviors;
3. psychographics, lifestyles and personality;
4. response to marketing mix variables.

While ideal surveys are a mixture of the above data types, each of these data-types has specific characteristics that require adjustments to the research design, from sampling to questionnaire issues.

Section 3.1 lists error sources for primary surveys and illustrates the research design
Section 3.2 looks at the pros and cons of the main types of administration methods
Section 3.3 explores issues in designing a questionnaire
Section 3.4 provides a classification and exemplification of primary surveys

THREE LEARNING OUTCOMES

This chapter enables the reader to:

➡ Become aware of the multiple sources of errors that may affect the survey results
➡ Appreciate the range of issues that arise when collecting primary data
➡ Gain knowledge of survey administration methods and questionnaire issues

PRELIMINARY KNOWLEDGE: A review of the basic measurement issues (sections 1.1 and 1.2) is advisable

3.1 Primary data collection: surveys errors and the research design

When the conditions for using secondary data (as listed in box 2.1) are not met, or in cases where budget availability suggests running an ad-hoc survey, researchers need alternative approaches to collect data. The one discussed here is primary data collection through a survey, since most statistical methods require data collection based on strictly quantitative grounds and surveys best serve the purpose in most (but not all) situations. However, in psychological research there are many other available routes like those of box 3.1.

As mentioned, the primary data collection process not only implies greater flexibility in targeting the relevant research objectives, but it also gives a great responsibility to the researcher, since the quality of collected data has a key impact on any statistical processing. In chapter 1 it was shown how measuring things is all but straightforward and researchers need to live with measurement errors (with a clear preference for their random component over systematic sources of error). There the emphasis was on errors associated with the use of measurement scales, but when running primary surveys there are many other potential sources of systematic and random error.

To name a well-known source of error that is thoroughly discussed in chapter 5, a sometime relevant portion of survey error can be imputed to sampling, that is, observing a sub-set of the target population instead of collecting information on the whole set. Sampling error can be dealt with through probabilistic sampling which

---

BOX 3.1   *Research methods and alternatives to survey data collection*

For an appropriate discussion of the range of research methods in social research see, for example, Creswell (2003) or Bernard (2000). Very briefly and with no pretence to being exhaustive, here we list some examples of research designs alternative to the use of secondary data (and meta-analysis) discussed in chapter 2 and primary data collection through a survey as illustrated in this chapter. *Experimental methods* are used to test a specific hypothesis by running experiments under controlled conditions. For example, one might recruit 20 students and close them in a room for one month. Their knowledge of statistics is first measured, then they are compelled to read this book and their knowledge is measured again. The hypothesis to be tested could be that their knowledge of statistics improves after reading the book, although they might not be that happy afterwards. *Naturalistic Observation* occurs where the researcher observes consumer behavior in its 'natural' setting, rather than trying to control the conditions. For example, one might place a CCTV in a supermarket to observe behaviors in front of a new product. *Ethnographic methods* are targeted to the study of 'cultures.' Ethnographers immerse themselves in a cultural community to record behaviors or other interesting variables through observation and other qualitative research methods. For example, an ethnographer could live a term with university students to understand their tastes, habits and behaviors. *Test marketing* consists in actually launching the marketing activity (e.g. advertising, new products, etc.) on a small-scale, see the effects and then decide whether to move to the large-scale step. For example, in 1980 IRI launched 'BehaviorScan,' which consisted in choosing two test markets (Marion, Indiana and Pittsfield, Massachusetts) and following 15 supermarkets and 2,000 households in each town. Besides scanning purchases, in test households IRI could monitor TV viewing and directly control advertising (by substituting the adverts with the test ones), thus creating a small scale market to test the impact of adverts on purchasing behaviors. Furthermore, in some situations other *qualitative research methods* (see box 3.2) can be employed. The frontiers of marketing research are also expanding in new directions. For example, *neuromarketing* exploits technologies such as functional Magnetic Resonance Imaging (fMRI) to measure how different brain areas respond to marketing stimuli (see Plassmann et al., 2007).

Table 3.1 *Sources of survey errors*

	Source of error	Description
A	SAMPLING ERROR	Error associated purely with the fact that we observe a sample rather than the whole population, with probabilistic samples it can be estimated
B	NON-SAMPLING ERROR (B1+B2+B3+B4+B5)	This error includes all other sources of errors that do not depend on the sampling process. Non-sampling errors can be random or non-random (biases), where the latter are more likely to affect the estimation results
B1	Sampling frame errors	Some of the population items are not represented in the sampling frame
B2	Non-response errors (B21+B22)	Some of the sampled units do not participate to the survey
B21	*Not-at-home*	The sampled unit could not be contacted
B22	*Refusals*	The sampled unit refused to co-operate
B3	Researcher errors	All those errors imputable to problems in the research design, such as errors in defining the population, inappropriate administration methods, inconsistencies between the research objectives and the questionnaire, errors in data processing, etc.
B4	Interviewer errors	Errors due to inappropriate actions of the interviewer. These include inappropriate selection of the respondents, errors in asking questions, errors in recording the responses or even fabricating them
B5	Respondent errors	While participating to the survey, the respondent provides (willingly or unwillingly) incorrect answers or doesn't answer to some of the questions
	TOTAL SURVEY ERROR (A+B)	Overall error

allows measurement and control of this error component. However, in many situations the sampling error is quite low compared to non-sampling and potentially systematic errors as those shown in table 3.1, especially when the proportion of non-respondent is high. This poses a serious threat to the successful completion of the primary data collection step (see for example the discussion in Assael and Keon, 1982). Contrary to sampling error, they cannot be quantified prior to the survey and sometimes it is difficult to detect them even after the field work. They represent one of those risks where prevention is definitely better than cure. Prevention is based on an accurate inspection of the issues emerging during *primary* data collection.

The primary research process can be articulated in four main stages:

1. Clearly formulate the research objectives
2. Set the survey research design
3. Design data-collection method and forms
4. Design sample and collect data

3.1.1 Formulating the research objectives

This step is all but trivial, as it lays down the foundations of the whole research process. It is especially delicate when research is commissioned, since communication between the client and the market researcher is not always effective. The two parts may easily convene on the main *research problem*. For instance, a company has developed a new product – a blender built on a novel technology – and wants to identify its target market. However, a target market may be defined by a wide set of characteristics. They may be socio-economic characteristics (income, education) or attitudinal data (inclination to home cooking). Also, the company is interested in the market segment which will allow profit maximization rather than the wider market segment. Also, the client may or may not wish to accompany this stage of marketing research with an evaluation of the potential customer's willingness to pay for the new product. These are only a few examples of the variety of specific issues that need to be clarified before a research design is drawn.

So, a first rule is to agree on a clear definition of the main research problem and an identification of the specific components. This step is achieved with a number of actions.

Discussion with the final user. First, a thorough discussion with the client, decision maker and final user of the research results is a necessary step. The researcher needs to become fully aware of the final use of the expected research results to avoid misunderstanding and incomplete results. It is not infrequent for clients to overestimate the potential of a survey.

Interviews with experts on the topic. While the meetings with the final user may have instructed the researcher, usually more specific knowledge on the market is desirable. The researcher is usually an expert in survey techniques, but not on the specific topic targeted by the survey. To know more about blenders, their use, the typical customer, the point of sale, the usual prices, the competing brands, the seasonality in demand and many more market characteristics, a good practice is to arrange a series of meeting with experts. These can be employees of the commissioning firm or other market analysts.

Analysis of secondary data. As widely discussed in the previous section, a survey of existing data is a must, possibly based on a systematic search for information as the one driven by meta-analysis. While at this stage one might even decide not to proceed with the primary data collection, because of availability of adequate secondary data, even when the direct survey is unavoidable a valuable of information might be gathered through the existing secondary sources. For example, one might find an old survey on the same topic, providing a questionnaire to start from. Secondary sources could also simplify the process of identifying the target survey population or suggest the appropriate sampling frames.

Qualitative research. Qualitative research methods are an extremely useful preliminary to quantitative research. These include not only a number of methods which support the questionnaire design, but also the identification of the specific objective (see box 3.2).

Once the research problem has been defined, it is necessary to set the specific research objectives, or *research questions*, which break down the research problem into components. The specific research questions need to be detailed and precise, as they will

BOX 3.2 *Qualitative research methods*

A systematic review of qualitative research methods is outside the scopes of this book. We advise you to refer to the introductory manual by Flick (2006), to the new edition of the comprehensive work edited by Bryman (2007) and to the synthetic and effective guide by Maxwell (2005). For an up-to-date and comprehensive review of qualitative research issues, see the third edition of the SAGE Handbook of Qualitative Research (Denzin and Lincoln, 2005). Here, it will be useful to point the basis of *exploratory research* as opposed to quantitative methods for *confirmatory research*. Exploratory research is aimed at gathering insights and understanding of the research problem. Contrary to the strict requirements of quantitative research, a loose definition of information needed is all that is required to start, as the research process needs to be very flexible and almost unstructured. Samples do not need to be representative and are usually very small, since inference to target populations is beyond the possibilities (and aims) of qualitative research, which is usually exploited as a basis for further research or is the only viable solution for some complex problem or in situations where quantitative research is impossible (for example in some rural areas of developing countries). Qualitative research is also very useful when the research aims to explore emotional and affective relationship (for example whether a TV advert is funny or not), or when the research topics are so sensitive that people are likely to be unwilling or unable to answer. The objectives of qualitative research are:

1. to obtain an adequate definition of the research problem;
2. to develop specific hypothesis to be tested through quantitative research;
3. to identify key variables which will require a specific quantitative analysis; and
4. to set the priorities for further research.

Without pretending to give a complete list of the extending range of qualitative methods, they are mainly subdivided in two categories, *direct methods* aimed at collecting the required information in a straight and undisguised way and *indirect methods* where the final aim of the research is not made explicit and information is gathered indirectly. Direct methods include *focus groups, in-depth interviews* and *panels* such as the *Delphi method* and *nominal group techniques*. Indirect methods are usually psychology-based projective techniques, like *association* (e.g. word association), *completion* (e.g. sentence completion), *construction* (e.g. ink blot tests) and *expressive techniques* (e.g. role playing).

drive the quantitative analysis. They are derived from the problem definition, but they also depend on the theoretical framework which is adopted and which will eventually determine the questionnaire structure. Considering the blender example, one might want to explore the attitudes toward innovative technologies in different population segments and rely on specific theories such as the Theory of Planned Behavior (see box 15.3), which sets very detailed rules on questionnaire building. This is also the stage when the analytical model and the statistical processing methodologies should be decided. For example one may wish to formulate the research questions in terms of hypotheses to be tested statistically: 'Consumers in the income bracket between £20,000 and £30,000 per year will be available to pay more than £20 for the blender.' This hypothesis, designed to support the decision on the price for the new product, could be rejected or not by statistical testing. A second research questions could be 'What is the effect of age, income and level of education on the probability of purchasing the blender?' To provide a statistical answer to this sort of question one might rely on discrete choice models (see chapter 16) and the choice of the type of model leads to different formulations for some of the questionnaire items.

3.1.2 Set the survey research design

Primary data collection is a process determined by a multiplicity of options. These options are not independent from each other, so that a series of actions need to be taken prior to the fieldwork:

1. Identification of the reference population and sampling frame
2. Choice of sampling criteria
3. Definition of the estimation methodology for making inference on the surveyed parameters
4. Choice of sample size
5. Choice of the data-collection method
6. Questionnaire design
7. Cost evaluation

Note that these decisions are not taken in the order shown in the list, but are rather the result of a series of interlinked considerations. Furthermore, while the most common situation is the one where sampling is necessary or desirable (see chapter 5), this is not necessarily always the case. In some circumstances it is feasible to survey all of the target subjects, for example, when a list of all customers is available and the cost of interviewing all subjects is not prohibitive. In these cases, one can jump from step 1 directly to step 5.

Reference population. This is the complete set of subjects we are interested in. The survey will gather information about characteristics (variables) of the population. A precise definition of the population is required for executing the sampling procedure. As discussed in details in chapter 5, a *sampling frame*, which is a list of all the population units, is generally necessary to guarantee probabilistic extraction. Without a clear definition of the reference population, it is not possible to know the degree of inference that can be achieved from a given sample. The definition of the reference population starts from the identification of the basic unit to be surveyed (the individual consumer, the household, geographic areas, etc.). Once the population unit is defined, the *sampling units* will be those units actually considered for selection at some sampling stage. Note that the population and sampling units do not necessarily coincide with the basic elements of the population. For example, one might consider the household as sampling/population units, then measure a set of variables on the single individual belonging to the sample households. In the blender example, we would be interested in any potential blender purchaser in Britain. However, it is not straightforward to identify the reference population. Is the final decision maker who actually uses the blender (the person in the household who cooks) or the one who takes the purchasing decision, if different? And should we focus on those households without a blender or on all households? An improper definition could lead to misleading results. Furthermore, the ideal reference population might not be a viable solution if a sampling frame does not exist. How do we get a list of people with cooking responsibilities? We will necessarily need to target the households in general and make sure that the questionnaire is filled by the person relevant to the research purpose.

Choice of sampling criteria. The type of sampling (probabilistic versus non probabilistic, stratified versus simple random sampling) has deep implications in terms of costs and precision levels. A thorough discussion is provided in chapter 5, again these criteria depend on other choices. If the sampling frame for our reference population does not contain additional information on the sampling units, we will not be able to adopt a stratified sampling design. The administration method is also closely related to the sampling method; if we decide to interview people through mall-intercepts, the only solution for probabilistic sampling is systematic sampling.

Estimators and methods. The sample estimators derive directly from the sampling criteria. However, it may be necessary to decide whether to rely on proportions or on averages, whether we will run a regression model to explore the influence of explanatory variables on the target variable, and possibly decide to use cluster analysis to define population segments. This is a choice which will drive the questionnaire design on the one hand, but it will also strongly influence decisions on sample size.

Choice of sample size. Using the mathematical rules listed in chapter 5, sample size are determined as a function of the precision levels and sampling design. However, it is necessary to take into account other issues in choosing the target sample size. Non-response rates (which depend on the administration method) should be taken into account to avoid a final sample size which is much smaller than the target one. When non-response rates are very high and especially when non-responses are not random (see chapter 4), the role of sampling error becomes negligible compared to non-sampling errors. Thus, targeting high sample sizes simply to guarantee better statistical performances, but ignoring the role of non-responses, is not a wise strategy (see Groves, 2006). Furthermore, when the research relies on information for sub-groups of the target population, representativeness within those sub-groups will lead to an increase of the overall sample size.

Data collection method. The choice of the administration method (face-to-face interviews, telephone interviews, electronic surveys, postal surveys) is strictly related to the sampling method, the sampling size, the sampling frame and the questionnaire design. Hence, this is one of the first decisions to be taken. Usually, the choice of the administration method is based on the number and type of questions. As explored in section 3.2.1, the duration of telephone interviews should be limited to 10–15 minutes and to easy questions, while for complex and long surveys face-to-face interviews are preferable unless there are sensitive question that are better addressed by a mail survey.

Questionnaire. A bad questionnaire could hinder the whole research and there are plenty of issues to be taken into account in designing the questionnaire and pre-testing it. These are discussed in the next sections, but here it could be useful to underline that besides the research objectives, questions should take into account the administration method, the characteristics of the target population and the methodologies chosen for statistical processing.

Cost evaluation. The ideal research design is the most expensive one. We rarely are in the position of finding it affordable, so that compromise is necessary. After costing the ideal research design, one should prioritize the aforementioned issues and identify cost reductions which could be targeted without thwarting the quality of the research.

It should be now clear that the above seven steps cannot be taken sequentially, but require simultaneous consideration and reciprocal adjustments. Once the research objective has been clearly defined, the sampling technique cannot be chosen independently from the administration method and the administration method strictly depends on the length and contents of the questionnaire (in either direction). On top of that, all of these choices are subject to the same budget constraints, so that the survey planner faces a number of trade-offs and the final survey plan is necessarily a balanced compromise. Generally, the first choice concerns the administration method, as the questionnaire forms vary according to the interview type. More commonly, constrained resources set limits to the questionnaire length and the sample size and influence the choice of the administration method.

3.1.3 Design data-collection forms

Once the research design has been determined in all its steps, the researcher proceeds with drafting the actual questionnaire. There are many aspects to be taken into account at this stage. Consistency with the research objectives and internal coherence, as well as the many potential sources of bias, should be carefully inspected. The questionnaire is subject to a pre-test, usually within a pilot study which also allows testing of the effectiveness of the statistical methodologies. Other adjustments may be required to ensure a proper duration of the interview. Cross-country surveys require translation and back-translation stages to warrant that results are comparable across languages. Details of questionnaire design are discussed in section 3.3, but this step is crucial to minimize non-sampling errors.

3.1.4 Design sample and collect data

The final stage consists in implementing the planned sampling process by executing the random extraction following as outlined within the research design. Then, field work and actual interviews can start.

3.2 Administration methods

Administration methods present extremely different characteristics in terms of their effectiveness of costs (see table 3.2). Not surprisingly, the most effective tend to be the most expensive too. Administration methods are usually classified in four categories:

1. Telephone interviews
2. Face-to-face interviews
3. Mail interviewing
4. Electronic interviewing

Within each of these broad groups, there are quite a number of options, also differing in terms of their costs and drawbacks.

3.2.1 Telephone interviews

It is probably the most exploited interviewing method, especially in its *Computer-Assisted* (CATI) version briefly described in box 3.3, which allows a good control on the quality of the interview and limits a number of potential sources of error and

---

BOX 3.3 *CATI interviews*

The influence of the CATI technique starts from the early stages of the survey, since the questionnaire needs to be defined in a way which is consistent with a computer and telephonic implementation. Hence, the CATI software generally provides guidance in designing the questionnaire without incurring potential problems, for instance lengthy questions and allows a proper exploitation of the CATI features, for instance controlling for inconsistencies. The next step is the sampling frame, which is also defined within the computer package. The call management system follows the sampling procedure adopted. As the survey proceeds, the system is also able to control potential problems due to non-response, for example by recording the characteristics of those who do not respond compared to respondents who do. Furthermore, in not-at-home (or busy line) cases, the call-back can be automatically scheduled. Besides registering the respondent answers and delivering data directly in electronic format, the CATI software also provides indicators of the quality of the survey immediately after its completion, so that countermeasures can be taken in a timely way when problems emerge.

---

bias, including an excellent control of the sampling process and dealing with non-responses. Traditional phone interviewing (where basically the interviewer fills up the questionnaire through a telephone call) is possibly cheaper, but as it is still very sensitive to interviewer biases and all other drawbacks of CATI interviewing (see below), it is less and less utilized.

Setting up a CATI laboratory, with the appropriate IT technologies, a good software and trained interviewers is expensive, but after the initial investment the costs are relatively contained as compared to face-to-face interviews. One of the risks of relying too much on CATI interviews is the standardization of the forms, which may limit the degrees of freedom for the respondent.

Pros of CATI. Most of the desirable features of CATI interviewing lie in the software-controlled mechanism. First, the software controls the sampling procedure. For example it may draw randomly from a list of phone numbers or it could resort to random number dialling, where the digits of the number to be dialled are randomly generated so that no sampling frame is needed and the target population includes all telephone numbers. This is an especially desirable feature as it also allows one to reach those numbers which do not appear in publicly available phone books. CATI also controls better for interviewer bias, as interviewers are guided by their computer screen throughout the interviews and the software records quality parameters such as duration of the interviews or questions more likely to lead to abandoning the interview. Respondent biases and inconsistencies are also mechanically checked by the CATI software. Finally, once the interview is completed, collected data are immediately organized by the software and ready for processing.

Cons of CATI. In some circumstances confining the sample to people with a telephone might already be a limit. This does not necessarily happen in rural or poorer areas only, as many people today gave up the fixed phone and rely on mobile phones, unlisted in phone books. While this might be circumvented by recurring to random dialling, even this has drawbacks, as the numbering rules (often linked to city zones, hence possibly to different social status) might make some numbers more likely to be drawn, violating the assumption of equal extraction probabilities and introducing a bias. Also, random dialled numbers cannot be associated to a name or other demographic characteristics that need to be provided by the respondent if required, making quality checks more difficult.

However, the main deterrent against using CATI interviewing are the relatively high costs, which are comparable to those of face-to-face interviews, and the need for containing the overall interview duration to 12–15 minutes, the upper limit for maintaining adequate quality in telephone interviews. It is also very difficult to record open questions – those where respondents are free to answer in their own words instead of choosing from among multiple options. Finally, when the survey is designed to record response to a stimulus, like commenting on some new packaging or reacting to an advert, telephone interviews are obviously inadequate.

3.2.2 Face-to-face interviews

As it is quite clear from their name, face-to-face or personal interviews are based on a direct encounter between the interviewer and the respondent. Personal interviews are highly dependent on the interviewer's skills and their usefulness depends on the content of the survey. Within the face-to-face approach to interviewing, there are some variants:

a. in-home interviews;
b. mall-intercept interviews;
c. computer-assisted interviews.

While during in-home interviews the interviewer actually visits the houses of the respondents after proper arrangements, in mall-intercepts the respondents are stopped inside or outside shops or in the street and asked whether they are willing to participate to the survey. Computer-Assisted Personal Interviews (CAPI) are similar to CATI and require a personal computer, which can be either used to guide the interviewer or can be used directly by the respondent, with the interviewer available for help.

Pros of face-to-face interviews.   Creating a direct contact with the respondent may greatly improve the quality of the interview in some circumstances. Skilled interviewers will gain the trust of the respondent and are able to assess the quality of the interview; questionnaire issues and misunderstanding can be solved immediately and with personal interviews the duration can be longer, up to 30 minutes. For surveys requiring long questionnaires and not touching sensitive issues this is usually the way to go. Furthermore, the response rate is significantly higher than other types of administration methods, provided the interview is pre-arranged and the interviewing subjects (the interviewer and the organization) are perceived as trustworthy. Mall-intercepts have a slightly lower response rate, but are much cheaper and are sometimes convenient when the object of the survey is related to specific shopping activities, so that the sample will include people who have actually purchased a product or who buy at a specific shop. As compared to traditional personal interviewing, mall-intercepts also allow a better use of stimuli, as the interview can show products and other stimuli directly from the shop. Finally, CAPI has a couple of further advantages. First, it has been shown that in front of a computer time perception varies, so interviews may last longer. Secondly, it becomes possible to use both off-screen stimuli, as for mall-intercepts and on-screen stimuli, like adverts, sounds and visions. Mall-intercept and CAPI may be combined, if the respondent is stopped in or outside the shop, then asked to participate in the

---

BOX 3.4   *Family decisions and interview methods*

Suppose you are an interviewer for a face-to-face survey aimed at eliciting what is the family decision process when deciding the timing and destination of the family holiday. This might be indeed a sensitive situation and results could be seriously affected by the situation of the interview and who is actually responding. For example, if both husband and wife are being interviewed, there might be constraints to stating explicitly how decisions are made, because either component could be reticent in admitting that all decisions are taken by the partner, or the other way round. If only one partner is interviewed, there might be a social desirability issue, especially when respondent is of the same gender, since it might be more difficult to admit a subordinate role in taking decisions. For a review of survey methods targeted at family decisions see Schlesinger (1962) and the application by Kang and Hsu (2004).

---

survey by moving to a separate room adequately equipped. The same holds for in-home interviewing, through the use of a laptop.

Cons of personal interviewing.   The main drawback of personal interviewing – of any type – is that respondents may be wary of talking directly to an interviewer or even letting them inside their place, which can reduce the response rates and may be sometime a source of bias, where busier people (like workers) will avoid this form of contact. Furthermore, given the scarce anonymity guaranteed, this type of administration method is inadequate for questionnaires containing sensitive questions or questions likely to raise social desirability issues (see section 3.3). These issues might be mitigated although with an appropriate use of CAPI, that is, letting the respondent fill sensitive questions directly on the computer. Another problem is generally the high cost, especially when the interviewer is asked to travel to different areas as for in-home interviewing. In such cases, a good sampling plan (like cluster sampling where each interviewer only covers small geographical areas) may greatly reduce costs. As mentioned, these methods rely strongly on the skill of the interviewers and good training is necessary. The quality of the response is highly sensitive to any *interviewer biases*, which can be even more influential when several interviewers are doing the field work. Most respondents will be more available for a nice and attractive interviewer (although probably more subject to social desirability issues) than to a bad-mood interviewer. For some type of surveys, the interviewer's gender might be also an issue (see box 3.4). Although CAPI can reduce some of these biases and it also reduces time perception, the use of a computer generally slows down the interview as well. Finally, personal interviewing may reduce the sampling control and it usually requires specific types of probabilistic sampling such as cluster/area sampling or systematic sampling for mall-intercepts, where the interviews stop every k-th person, k being the size of the sampling step (see section 5.2.1).

3.2.3 Mail interviewing

Mail surveys consist in sending the questionnaire by mail, generally enclosing a stamped envelop for returning the filled questionnaire with no costs for the respondent. For businesses, mail interviewing includes submission of questionnaires through the fax. Apart from those circumstances when mail surveys are regulated by law and conducted directly by public administration, the response rate of this administration method has become so low that it has become less and less popular. The low response rate is also associated with relevant *selection biases*, as those taking the time for filling the surveys are likely in specific segments of the population with a larger availability of time.

A more appealing type of mail survey is based on the *panel* approach, where the same respondent is asked to fill a questionnaire at regular intervals of time to record dynamics and changes. These surveys, usually supported by adequate incentives to the participants, have a higher response rate than their one-off counterpart.

Pros of mail surveys. There are some advantages in administering a questionnaire by snail mail. First, it is possible to guarantee a higher level of anonymity if the questionnaire and the envelope do not contain any sign for recognition, although it is not necessarily easy to convince the respondents about that, given that they have received the questionnaire to their addresses. Another good thing of mail surveys is that they are quite cheap, especially when they are designed for electronic reading, as the most expensive step is the electronic input of the data. While stamps and envelopes are relatively cheap, the low response rate usually requires much larger initial sampling sizes. Since the respondent fills the questionnaire with no need for an interviewer, it is a suitable method for those surveys where the interviewer bias is likely to be an issue.

Cons of mail surveys. Besides the above-cited very low response rates and the selection biases, there are other problems with mail surveys. First, there is no guarantee that the returned questionnaire has actually been completed by the targeted respondents, as other members of the household may have done it and not necessarily correctly. This is a bigger issue for panel surveys. Second, mail surveys are very slow as compared with other methods. Between the mailing and the actual availability of the data-set there is an interval of several weeks.

3.2.4 Electronic surveys

Electronic surveys administered by e-mail or through questionnaires on the World Wide Web have multiplied over the last decade with the increasing diffusion of Internet accesses. E-surveys have in common the use of the Internet as a means to reach respondents, but there are a variety of approaches and many different technologies. The basic e-survey is a questionnaire sent by e-mail, asking the respondent to fill it in and return it. This is not too different from a mail survey in principle, it is obviously much cheaper, but it is quite rare to have an adequate sampling frame of e-mail addresses. More commonly the questionnaires are administered through a web-browser, either with a dedicated web-site that the respondent can access upon invitation or as a pop-up window for visitors of a given web-site. The latter can be regarded as a variant of mall-intercepts, where the location is the web-site, while the former has more in common with traditional mail surveys, as there is no interviewer control when the respondent completes the survey.

Pros of e-interviewing. They are usually a very cheap method, as apart from the fixed costs of programming the questionnaire in a web form and possibly for a software doing consistency checks and organizing the data, the costs for contacting the respondents are close to zero. They are also quick, as replies to e-mails generally come within a

Table 3.2 *Indicative UK costs and response rates for different types of interviews*

Survey type	Fixed costs	Variable costs		Total costs		
		Per questionnaire	Per respondent (assumed response rate)	200 respondents	500 respondents	1,000 respondents
Mail survey	2,000	3	15(20)	5,000	9,500	17,000
Personal	4,000	50	63(80)	16,500	35,250	66,500
CAPI	15,000	60	86(70)	32,143	57,857	100,714
Telephone	4,000	20	40(50)	12,000	24,000	44,000
CATI	15,000	50	83(60)	31,667	56,667	98,333

Note: costs are in British pounds and are purely indicative, based on 2005 rates of private marketing research agencies. Fixed costs basically include sampling frame, hiring of equipment and training of interviewers.

few days or they do not come at all. Web-based questionnaires, for example those built on Java programs, represent computer-assisted interviews as for CATI and CAPI. Thus, Computer-assisted Web Interviews (CAWI) have the usual advantages of allowing for consistency checks, using multimedia stimuli, organizing the data for analysis, controlling the sampling process. If the form is adequately designed, web-based questionnaire might also guarantee anonymity to the respondent in case of sensitive questions.

Cons of e-interviewing.   There's a key selection bias issue to be considered prior to opting for electronic surveys, as reference populations will be limited to people able to use a computer and access the Internet. Depending on the research question, this can be a minor limitation or a major drawback.

For e-mail interviews there are many limitations that have strongly reduced the use of this administration method. First, response rates are now very low and they are quickly decreasing. Secondly, unless sophisticated programming techniques are exploited (which may limit the accessibility for many respondents), the respondents actually answer the questions replying to the e-mail or attaching a filled document, which means that a data inputting and organization step is still necessary prior to statistical processing. E-mail surveys also raise an anonymity issue, and it is difficult to implement an adequate probabilistic sampling plan.

On the other hand web-based questionnaires have relatively higher response rates and benefit of the CAWI features discussed above. However, sampling control is quite low and it is difficult to randomly select respondents and ensure representativeness. In general, the quality of data collected through electronic surveys is low.

3.3 Questionnaire

The crucial question for transforming a good survey design into a successful one is are we actually measuring what we want to measure? While sampling influences representativeness and the quality of the collected data can be affected by the administration method, questionnaire biases are certainly one of the most dangerous sources of non-sampling error. The first threat is the potential discordance between the information which the researcher believes he collects and the information the respondent actually provides. Furthermore, bad questionnaires might prevent respondents from co-operating, hence increasing *non-response errors*. Ill-posed questions also raise *response errors* like inaccurate or mis-recorded answers. Thus, clearly defining the desired information is the initial step in designing a questionnaire. The key passages are eight:

1. Specify the information to be collected
2. Define the information collected by each individual questions
3. Choose structure and measurement scale
4. Determine the wording of each questions
5. Sort the questions (and possibly divide them into sections)
6. Code the questions and simulate statistical processing
7. Write an appropriate introduction/presentation and define the layout
8. Pilot the questionnaire and revise where necessary

Specify the information to be collected. As suggested, good preliminary qualitative research can highlight many of the potential misunderstandings that might affect the use of a questionnaire. For example, a focus group with a small group of potential respondents or in-depth interviews about the survey topic could show potential differences in individual perceptions. A definition of the required information in terms of variables should follow directly from the research objectives and questions specified when laying down the research design. For example, the research objectives are understanding the radio listening habits of some population; this might translate into three research questions:

a. when people listen to radio;
b. where people listen to radio;
c. what radio stations they listen to.

The 'when' question can be articulated in a set of information:

(i) time of the day when they listen to radio;
(ii) days of the week when they listen to radio;
(iii) activities they do while listening to radio;
(iv) seasonal differences in radio listening habits, etc.

The disaggregation of the other research questions is left to the reader.

Together with the research objectives and questions, at this stage the researcher should already take into account the type of statistical techniques to be used in data processing, as some of them may require a more accurate quantification. Furthermore, especially when working with psychographics or economics, it may be extremely helpful to decide on a theoretical framework, which could place requirements, constraints and hints on the type of necessary information. For example, the Theory of Planned Behavior mentioned in chapters 1 and 15 follows a detailed procedure in designing the questionnaires.

Define the information collected by each individual question. The next step follows directly from the definition of the broad information content. It is not necessarily true that one piece of information corresponds to a single question. It may be very useful to start thinking in terms of a spreadsheet which will contain the variables for subsequent analysis. Considering the radio listening example, let's consider the information content on 'time of the day when they listen to radio' and 'days of the week when they listen to radio.' It is not straightforward to define the content and structure of the questions; no two questions could satisfy the information needs. One could first check the frequency of radio listening (days per week), then – provided the answer is different from 'never' – the following question could ask about prevalence of listening in specific days of the week (for example, weekdays versus week-ends). A third question could enquire about the preferred time of the day (morning, afternoons, evenings, night...). Clearly, information might not be very precise if collected through these questions, nor very clear to the respondent. If one needs more accurate details, it is probably better to have a single question recording for each day of the past week, whether the respondent listened to the radio, and at what time. Hence, questions that are too aggregate to be properly understood ('when do you listen to the radio' might refer to the frequency or to the time of the day) need to be separated to specific items. On the other hand, it might be convenient to aggregate to the 'when' question also the information on the radio station they listened to, so that it might be associated with the day of the week. This would allow to capture those that listen to a specific radio on a given weekday or time of the day because they like a specific transmission.

Choose structure and measurement scale.   The key issue of measurement scales is treated more extensively in chapter 1. There are a number of aspects to be considered before choosing the measurement scale. First, one should choose between unstructured open-ended questions (where respondents are free to give any answer they wish) or structured questions (dichotomous, multiple choice, scales), where the answer is bounded by a closed set of possibilities. Dichotomous questions just allow for two outcomes (plus possibilities for not knowing or no answers); the typical is yes/no. The issue with these type of questions is that the positive or negative wording may influence the answer by leading the answer. If this is the case, a solution is to split the sample in two and differentiate the wording. For example 'Do you think that listening to radio is preferable to watching television' may be reworded in the opposite way (television preferable to radio) for half of the sample. Multiple choice questions extend the choice among three or more possible answers. The bias here might be related to the order of question and the solution could be to sort randomly the answers in each questionnaire. Finally, for those types of structures requiring a quantification, the definition of a measurement scale is not always straightforward. On the one hand, one should target a measurement scale which is easily understood by the respondent, but this needs to be balanced with the desired level of precision and the statistical methodologies that the researcher plans to use to achieve the research objective. Sticking to the radio example, it is probably easier for someone to say whether they listen to radio more on week-days or week-ends than specifying how many hours they listened to radio for each single day of the previous week. If the answer to this question is expected to be an explanatory variable in a regression model, then measuring the hours per week might provide more variability than simply asking about days per week, albeit leaving room for a larger memory bias and quantification effort. For some sensitive questions, like income levels, it is advisable to avoid direct quantification and provide relatively broad categories which wouldn't be interesting to the tax authorities. There are other ingenious techniques to elicit sensitive answers, like the one illustrated in box 3.5.

Determine the wording of each question.   Each question should be readily understood by the respondent and calibrated to the administration method. For example, long and elaborate questions are not appropriate for telephone interviews, while they may fit

BOX 3.5 *Sensitive questions and randomized techniques*

A relatively frequent issue in consumer surveys is the possibility of incurring in biases due to the sensitiveness of some questions. This problem is particularly relevant, because a respondent facing a question perceived to be too personal or subject to social judgment could quit the interview or provide a false response. An interesting statistical trick for avoiding embarrassment or refusals while getting aggregate answers to the sensitive question is based on *randomized techniques* (see Greenberg et al., 1971; Kuk, 1990 and references therein). The principle is quite easily explained by considering a yes-no question. Suppose we are running a survey with our readers asking whether they like this book. They might be cautious in openly declaring they did not, so we will probably get an overestimate of those who appreciate this book. If we really want to circumvent this problem, we need an alternative yes/no question, whose percentage of yes (no) answer is known. For example, if we know the portion of respondents that are female (say it is 60%), the alternative question could be: 'Are you a female?'

Suppose we are running the survey by mail interviewing. The questionnaire asks the respondents to flip a coin, but not to state whether they got a head or a tail. Instead, the respondent must answer to either question A (the sensitive one) if they got a head, or question B (the gender question) if they got a tail. The answer is recorded, but not the question. Once the survey is completed, there will be a percentage of 'Yes' answer, without knowing how many respondents answered question A. However, we may reasonably expect – with a large sample – that half of the respondents got a head and the other half a tail. This piece of information enables us to dissect the overall percentage of yes answers into the two sub-groups A and B. As a matter of fact we know that:

Total % of yes answers = 50% × (% of yes answers to A) + 50% × (% of yes answers to B)

The only unknown quantity in the above equation is the % of yes answers to A. Suppose we got a total percentage of yes answers of 40%. Thus, we may compute:

% of yes answers to A = [Total % of yes answers − 50% × (yes answers to B)]/50%, which means
% of yes answers to A = (40% − 50% × 60%)/50% = 20%

Only one reader out of five likes this book. Of course, this is simply a hypothetical example.

mail surveys or face-to-face surveys. The exact wording of the question also depends on the measurement scale. It is important that the question is phrased in a way which ensures consistency between the answer and the targeted variable. Questions should be designed to reduce non-sampling errors. Common rules are the use of ordinary words, avoiding words with ambiguous meanings ('generally,' 'frequently,' etc.), phrasing which suggests the answer ('Do you think that people should listen more to radio?'). Any relevant aspect should be made explicit and the respondent should be not asked to make difficult computations or generalization of estimates ('On average, how many hours a month did you listen to radio last year?'). For example, one should avoid asking questions that might be difficult to answer because of lack of information ('What is the frequency of your favorite radio station?') or because of memory problems ('What radio stations did you listen to when you were 15?'). These sorts of problems frequently lead to inaccurate answers rather than non-responses. Very broad questions that require a certain elaboration effort before answering are also to be avoided: 'Why do you like listening to radio?' might generate a high variety of responses and a multiple choice set of answers is preferable to open-questions. Good phrasing could also reduce the risk of missing responses to sensitive questions. For example, the 'third person' technique helps reducing the social desirability bias ('Some people haven't read a book in the last 6 months. In your opinion, what are the reasons?').

Sort the questions (and possibly divide them into sections).   The order of questions in a questionnaire can heavily influence the response rates and minimize the non-response errors. For example, when confronted with ill-posed questions which are difficult to be answered or when the content of the question is perceived as sensitive or inappropriate, the respondent may decide to abandon the whole questionnaire. When sensitive or complex questions cannot be avoided, these should be placed at the end of the questionnaire or among a group of easy and non-sensitive questions. Instead, ice-breaking questions ('What is your current favorite radio hit?') usually go in the initial section of the questionnaire. On the other hand, if some information is central to the objective of the research, it should be asked at the beginning of the questionnaire to avoid missing data due to those who abandon the questionnaire before they reach the end. Furthermore, there must be logic in the flow of the questionnaire, general questions should precede specific questions and questions on the same topic should be grouped together. Finally, the sequence of the questions might make transparent the objectives of the research and sometimes this might introduce biases for social desirability or sensitiveness. It is not infrequent to add questions with the sole purpose of disguising the purpose of the survey.

Code the questions and simulate statistical processing.   Most of the statistical techniques explored in this book are successful when there is enough variability in the measured variables. Sometime, a multiple choice question or a measurement scale might turn out to be inappropriate for the desired processing method. A good strategy is to anticipate the problem by coding the questions and translating them in a spreadsheet structure. By simply simulating (making up) a sufficient number of questionnaires prior to the survey, one may test the statistical methodologies on the generated data-set and pick up measurement issues or other inadequacies of the questionnaire.

Write an appropriate introduction/presentation and define the layout.   Quoting the promoter of the research and the objectives might be either an advantage or a limitation. If an explicit declaration of the sponsor and/or objectives is likely to improve the respondent willingness to co-operate because of enhanced trust or perception of its relevance, then this could be stated in the introduction. However, when the sponsor is a private brand or a vested interest is perceived by the respondent, one should try to disguise this sort of information where possible. Another issue requiring careful consideration is the provision of incentives (money or gifts) to encourage participation. While this clearly increases the response rates, it may cause biases by inducing a selection bias (for example, low income respondents will be more sensitive to monetary incentives) and possibly also influence responses. The actual layout and physical appearance of the survey may impact non-sampling errors in self-administered questionnaires (mail, e-surveys). For example, questionnaires should have a professional look and the placement of each question in the page is relevant (for example, questions should not be split across two pages and questions at the bottom of pages are more likely to be ignored).

Pilot the questionnaire and revise where necessary.   An accurate piloting and pretesting can greatly improve the survey effectiveness. Besides controlling for specific quality parameters (length of the questionnaire), all issues listed here can be checked by piloting the questionnaire on a small number of respondents. If possible, different versions of the questionnaire should be tested to check which one works better. Regardless of the chosen interview method, personal piloting interviews are more likely to highlight issues so part of the pilot survey should be conducted through face-to-face interviews. When the chosen survey method is actually face-to-face interview, piloting with a variety of interviews could point to specific interviewer biases to be considered and the abilities of each of them checked. Another technique for testing self-administered surveys is to observe the respondents and ask them to think aloud when filling the forms. After compilation it is also very informative to go through the questionnaire with the respondent to check whether the interpretation is actually consistent with the objectives. Piloting may lead to minor adjustments as well as radical changes in the choices made in previous step, like switching to a different administration method or revising the sampling design.

3.4 Four types of surveys

Consumer behavior is a complex discipline, the main reason being that it is not a discipline. It is rather a cross-section of disciplines that provide different perspectives on the set of factors expected to explain and predict how people behave and why. A shared definition of consumer behavior (see for example Solomon, 2007) involves several actions (purchase, use and disposal), several targets (products, services, ideas, experiences) and a couple of non-trivial outcomes (satisfying needs and desires). Table 3.3 extends the range of consumer actions to 10 behaviors. This would be enough to fill one book on psychology, one on economics, one on sociology and one on marketing...

Here, however, the objective is 'measuring consumers' and translating such a complex set of behaviors and determinants into a spreadsheet of numbers and – hopefully – a couple of new hints on how consumers behave, after an adequate bit of statistical processing. The more primary data collection is able to exploit this multi-disciplinary perspective, the more effective will be the information processing stage. With the sole purpose of simplifying this tortuous path among data needs and theoretical frameworks, here is a tentative classification of the types of data that a consumer researcher is interested in when planning primary data collection:

1. Socio-demographics
2. Behaviors and economics
3. Psychographics, lifestyle and attitudes
4. Response to marketing actions

Table 3.3 *Ten consumer behaviors*

1. Searching for information
2. Learning
3. Recognizing
4. Remembering
5. Choosing among alternatives
6. Purchasing
7. Consuming/Using
8. Disposing
9. Being satisfied/dissatisfied
10. Complaining/word of mouth

The ideal survey would cover all of the above and would be contained in a 15-minutes questionnaire. The ideal survey obviously does not exist, which means that the best possible survey is mainly based on one of those four categories, drawing on the three remaining ones where necessary and possible. Box 3.6 illustrates each of the four primary survey types with the aid of real examples and following the scheme for primary data collection introduced in section 3.1.2.

Socio-demographic information. It is rarely the case that a consumer survey is only focused on socio-demographic information, but it is even more rare that such information is ignored. Almost all of the surveys contain a socio-demographic section, which allow testing of whether different behaviors, attitudes or needs can be directly related to specific segments of the population. As is made clearer in chapter 5, socio-demographics are often at the core of the sampling process and are used as a benchmark to test the representativeness of a sample. Among the few surveys mainly targeted at collecting socio-demographic information, the most known one is certainly the population census, which all countries run every 10 years to get a picture of their changing populations. By looking at the UK census, one may get a good idea of the sort of socio-demographic information that could be used to produce a demographic segmentation of consumers. While a census is usually outside the options for a marketing researchers (see chapter 5 for a detailed discussion on censuses and sampling), it still represents the most ancient survey for primary data collection and the ancient Greeks had started the habit of running censuses well before the well-known census which preceded the birth of Jesus. Censuses are carried out every decade, but between two censuses the national statistical offices run surveys to maintain information on household socio-demographic characteristics. Since 2004 in the United Kingdom, census data are updated yearly through the Annual Population Survey (APS).

Behavioral and economic surveys. These surveys are useful to monitor consumer purchasing decision and to relate behaviors to its determinants. The typical economic survey looks into recorded expenditures for different products or services, by brand or category (*expenditure surveys*). Other purchasing decision information refers to the frequency of purchase, the point of purchase and brand switching. But behavioral surveys are not limited to purchasing, as purchasing does not necessarily translate into consuming or use and consumption surveys may return very different result from expenditure surveys. When the research question focuses on 'consumer use or disposal,' the questionnaire and administration methods will be different. Let's take a simple example – we purchase food, we bring it home, we store it in the fridge. Some of this food ends up in our stomach, other in the bin. If the objective of a study is to see how much we spend, then an expenditure survey will do. But if we are investigating the relationship between food consumption and health, some other form of survey is necessary, like dietary surveys which require the respondent to keep a diary of consumed food and weigh it before consumption. Expenditure data can be collected in a relatively straightforward manner and are generally accompanied by information on socio-demographic variables and possibly questions on lifestyle determinants allowing one to explore their relationship with purchasing choices.

Attitudinal, psychographic and lifestyle studies. Prices and incomes are losing power in explaining consumer decisions. Lifestyle, social pressure, individual attitudes and habits are becoming the main factors in explaining different purchasing and consumption behaviors. This is why market research surveys are increasingly integrated

BOX 3.6 *Four types of surveys*

	Socio-demographic surveys	Expenditure surveys	Attitudinal, psychographic and lifestyle studies	Response to marketing actions
Research objectives	Population studies like the UK Annual Population Surveys respond to multiple objectives: (1) to count population by area and socio-demographic characteristics; (2) to gather information on housing; (3) to update information on ageing and relate it to socio-economic characteristics; (4) to monitor employment; (5) to collect information on transport usage and pressures on transport systems; (6) to monitor migration and update information on minority ethnic groups.	Expenditure surveys can be used for a variety of purposes, like, just to give some examples (a) monitoring category and brand expenditures; (b) evaluate responsiveness to prices through the estimation of elasticities; (c) quantify income effects and classify goods as luxuries or necessities; (d) evaluate distribution issues looking at expenditure across different sub-groups of the population.	Broadly speaking, attitudinal studies cover consumer behavior determinants not classified under socio-economic characteristics. They aim to identify personality traits that influence actual behavior, including: (a) lifestyle and psychographic determinants; (b) personal interests and opinions; (c) individual preferences toward objects and actions. These traits are seen as forerunners of behaviors and their measurement opens the way to explain behavioral intentions and possibly behaviors.	In measuring response to advertising, the specific objectives of marketing research depend on the aims of the advertising campaign, to list a few, the evaluation of: (a) the direct effect of advertising (sales, market shares and/or price support); (b) increases in consumption from stock and future purchase; (c) increases in retailer demand and availability on shelf; (d) the impact on repeated purchases and brand loyalty; (e) the impact on brand image; (f) the impact on brand awareness; (g) the impact on brand attitude.
Reference population and sampling frame	The APS is expected to represent all of the UK resident population, so the reference population is defined by all persons resident in the UK in private households, and young people living away from the parental home in student halls of residence or similar institutions during term time. The sampling frame is mainly taken from the 'small users' sub-file of the Royal Mail Postcode Address File (PAF), that is delivery points receiving less than 50 mail items a day. This includes 97% of private households.	It is usually quite difficult to target the individual consumer with these surveys and it might not be the right way to proceed. Most purchases are shared among household members, so that recorded expenditures are the final outcome of a group decision process. Hence, the typical target population for expenditure surveys is the household, although collecting further information on individual allocation of purchased goods may explain the process driving the collective decision.	Here, contrarily to the case of expenditure survey, the population unit is necessarily the individual, since the final aim is to measure personality traits.	This depends on the type of advertising being evaluated. For a campaign to support loyalty, the reference population is given by existing customers. If the aim is the launch of a new product, the RP is extended to all potential customers. A definition of the latter population and sampling are less straightforward. A clear definition of the RP and access to an adequate sampling frame are the main difficulties in running surveys to elicit response to advertising.

(Continued)

BOX 3.6 Cont'd

	Socio-demographic surveys	Expenditure surveys	Attitudinal, psychographic and lifestyle studies	Response to marketing actions
Sampling criteria	A multi-stage stratified random sample (see chapter 5) is adopted for the APS survey. The survey also has a panel element, as some of the households are interviewed annually over four waves.	This depends on the specific survey objectives and size. A major issue in expenditure surveys is to cover all relevant population sub-groups, including low-income groups or remote and rural areas. Thus, official expenditure surveys are usually based on multi-stage stratified sampling processes, where geographical locations are the first stage sampling units and households are extracted at a later stage. Since personal interviews are usually needed for accurate data collection, a good strategy to minimize costs is to adopt an area sampling design.	Again, there is no general rule. A further complication is that it may be difficult to warrant representativeness of personality segments, as personality is not necessarily related to measurable variables. Qualitative research is a key element in developing a proper research and sampling design in attitudinal studies. Simple random sampling might be a safer solution (albeit more expensive) than using improper stratification variables.	These strictly depend on the availability of a sampling frame. The typical strategy is based on phone book listing, as computer-assisted telephone interviews are a suitable method for a quick survey on advertising response (timing is a key element in the aftermath of a campaign). An issue is the exclusion of contacts that are not relevant to the target population. It may be useful to start with simple random sampling, then exploit post-stratification for inference. Furthermore, response to advertising needs to be monitored over time. Panel surveys are the preferred way.
Estimation methodologies	Apart from the obvious computations of totals (sizes of population groups by socio-demographic characteristics), the usual research question is whether a variable is statistically different across population groups. There are a number of multivariate statistical techniques to answer to such question, all discussed in this book. A popular analysis is segmentation, which means the subdivision of the population in homogeneous segments according to socio-demographic or other characteristics. Cluster analysis is one of the techniques allowing for socio-demographic segmentation.	Statistical processing of expenditure data is usually based on multivariate statistics, such as systems of equations to explore budget allocation across expenditure categories or cluster analysis to identify population clusters. To this purpose, it is advisable that where possible answers are recorded through interval or ratio scales, that is with continuous measurement units, which minimizes the issues raised by modeling discrete data.	Attitudinal surveys cannot be treated as material for data mining. At the core of any attitudinal survey it is necessary to put one or more theoretical paradigms, which will drive the questionnaire design and the data processing method. For example, if the theoretical background is given by the Theory of Planned Behavior (box 3.7), then the most appropriate statistical methods will be discrete choice models (probit and logit) or structural equation models (confirmatory factor analysis), as shown in chapters 16 and 15, respectively.	There is a vast modeling toolbox, from indicators of ads perception and memorization to models linking advertising and purchases. Methods can be classified into three categories: (a) cognitive response (perception and memorization of ads); (b) affective response (perception of product attributes, attitudes toward product, product\brand preference); (c) behavioral response (purchase and re-purchase behaviors). With panel surveys, dynamic models look at the duration of the advertising effect (short-term vs. persisting impact). Data fusion, combines data from two sources using common elements, e.g. purchases and TV viewing through consumer viewing habits. The large body of research on consumer processing of information can support questionnaire design and data analysis.

Sample sizes	The 2005 APS sample included 495,097 individuals. It is a quite large size as compared to other surveys, but this is a common drawback for population studies aiming at collecting statistically significant information for all potential sub-population groups. Furthermore, the APS aims to be representative at the Local Authority District level.	Expenditure surveys may require relatively long interviews and a certain effort by the respondent, like filling diaries or retaining receipts. In absence of incentives, this generates high non-response rates and sample sizes should take this into account. Furthermore, since the interest is in exploring disparities across sub-groups of the population, sub-sample sizes (stratum sizes in stratified samples) assume great importance.	Questions of attitudinal studies, as compared to expenditure surveys, may be more sensitive or difficult to understand and respond, thus rates of non-response and abandonment can be high. With sensitive issues, face-to-face and telephone interviews are not a viable solution, while anonymous mail surveys have low response rates. The average size of an attitudinal study is generally larger than expenditure surveys.	In some respects, this issue is not as relevant as in expenditure surveys or lifestyle studies. If the target population has been clearly defined and the sampling strategy works, relatively small sampling sizes could provide a good approximation of consumer response. This is especially true when the aim is to monitor dynamics (that is when a panel survey can be used), so that the focus is on variations rather than absolute measures.
Data-collection methods	The APS questionnaire is administered through face-to-face and telephone interviews. All interviews in wave 1 at an address are carried out face-to-face, while recall interviews are by telephone if the respondent agrees to it. Overall, including wave 1, around 62% of interviews are by telephone, and 38% are face-to-face.	Usually expenditures are recorded through a weekly diary, distributed to the households for self completion. At the end of the recording period, an interviewer visits the household to administer a broader questionnaire and check the accuracy of diary keeping. Recently diaries have been substituted by home scanning facilities, bar-code scanners are distributed to easily record expenditures, prices and quantities. This – however – rules out expenditures for fresh foods with no bar code or durables, so that an interview integration is still desirable.	The administration method strongly depends on the study object. If the target of measurement is the use of drugs or tax evasion, the use of face-to-face or telephone interviews is ruled out, while it should not constitute a problem if the survey is about inclinations toward ethnic foods. Sensitive issues do not always prevent from using telephone or face-to-face interview, but in these situation a great care must be taken in designing the data collection forms.	Generally, response to advertising is not a sensitive issue. This opens the way to the use of the most cost-effective methods. And if the focus is on a single product\category, the questionnaire can be easily restricted to a few items, so that the interview will fit with the constraints of computer-assisted telephone interviewing (CATI). This is the most widely used administration method, although mall-intercepts are a viable option when the questionnaire is more complex.

(Continued)

BOX 3.6 Cont'd

	Socio-demographic surveys	Expenditure surveys	Attitudinal, psychographic and lifestyle studies	Response to marketing actions
Questionnaire design	The APS questionnaire collects information on socio-demographic characteristics of the household components. These questions are grouped in main categories: (a) demographic information (genders, ages, marital status, length of residence); (b) household composition; (c) ethnic group (ethnic group, national identity, religion, country of birth); (d) tenure (ownership, mortgages, rented, etc.); (e) socio-economics (economic status, employment status, occupation); (f) education; (g) household motor vehicles.	The questionnaire size of expenditure surveys is usually quite large (the UK-EFS reference document is 341 pages long), but the single questions are generally easy to be understood. What could be confusing is the classification of single goods into categories, which is not always intuitive. This is why the generalization effort is usually demanded to the interviewer, in this case assisted by a computer software.	Psychology-based and attitudinal questionnaires are not a friendly environment for practitioners. It is not straightforward to ensure correspondence between the meaning given by the researcher to a question and the respondent understanding. This can lead to the formulation of extremely articulated and complex questions, with non-response problems and incompatibility with telephone interviews. Qualitative research, pre-testing and piloting of the questionnaire are the starting basis. Reliability and consistency checks can be also based on quantitative measures as the Cronbach's alpha (chapter 1). Attitudinal studies are subject to a series of issues, from sensitivity to social desirability, discussed in this chapter 3.	Given that the specific research objective is one from the aforementioned list, a limited number of simple questions could suffice for monitoring advertising impact. Still, since information processing and memory are complex psychological processes, caution is necessary to ensure that the question actually responds to the object of our study. For example, one must be very careful in distinguishing between ad-recall and ad-recognition, where the former refers to unassisted recall of the advert and the second implies some assistance from the interviewer. There are many reference textbooks on the theories explaining the impact of advertising and its measurement (see for example Tellis, 2004 or Jones, 1998).
Cost evaluation	When the aim of a survey is explicitly to collect information on the socio-demographic composition of the target population, costs may rise quickly depending on the size of the population and the desired level of detail. If one wants to guarantee representativeness for minorities, the overall sample size must be large enough to guarantee an adequate sub-sample size for that group. All population studies are based on stratified sampling or its combination with other sampling techniques.	Expenditure surveys of this type are highly expensive. First, respondent are asked to spend a considerable amount of time, both in keeping the diary and in answering to a long questionnaire. Furthermore, household expenditure survey usually require the same household to participate more than once. To ensure an acceptable quality of the collected information, face-to-face interview are generally necessary at the end of the recording period. This adds further costs.	Apart from sample size, an additional source of costs for attitudinal studies is interviewer training, often a key component of successful attitudinal studies.	Costs will depend on the required level of precision. While in general the possibility to exploit CATI and the relatively small sampling sizes reduce costs, surveys are unlikely to be helpful if run at a single point in time. The most effective way to explore response to advertising is to monitor the same target population before and after the campaign and possibly to follow it over time. This can be a pricey process.

BOX 3.7 *Attitudes and behavior in Ajzen and Fishbein theories*

The Theory of Planned Behavior (TPB) framework, devised from the Theory of Reasoned Action (TRA), (Ajzen & Fishbein, 1980), defines human action as a combination of three dimensions, behavioral beliefs (attitudes to behavior), normative beliefs (subjective norm) and control beliefs (perceived behavioral control) (Ajzen, 2002). Attitude to behavior is noted as being principally different from the broader concept of attitude toward an object. For example, one may like chocolate (an attitude to chocolate), yet chose not to purchase chocolate because they are currently on a diet (attitude to buying chocolate). However, an attitude to the behavior, such as buying chocolate, depends on a set of behavioral beliefs that may include its relevance to a diet. Subjective norm is a concept based on how one should 'act' in response to the views or thoughts of others, like friends, family members, colleagues, doctors, religious organizations, etc. The passage from TRA to TPB is given by the additional component of perceived behavioral control (PCB) (Ajzen, 1991), which can be described as 'the measure of confidence that one can act' (East, 1997). The combination of attitudes to behavior, subjective norm and perceived behavioral control drives the intentions to behave. The TPB framework, which has been successfully applied to a huge range of behaviors (for a long list of examples see Ajzen web-site, www.people.umass.edu/aizen/tpb.html) consists in detailed instruction on how to build a questionnaire and process the data prior to modeling. Further details are provided in chapter 15. Since their introduction, TRA and TPB have been the subject of many extension and adaptations, such as the inclusion of self-identity, self-efficacy and moral norms.

by psychology-based questionnaire sections. While academics are still pursuing a successful theoretical basis merging psychological with economic theories,[1] practice is anticipating theory and many empirical studies have focused on the role of attitudes in explaining purchasing behaviors. The most popular tool for collecting attitudinal data and linking it to behaviors (or rather behavioral intention) is probably the Theory of Planned Behavior (an evolution of the Theory of Reasoned Action), developed by Ajzen and Fishbein (see box 3.7). As an example of lifestyle measurement, see box 3.8 showing food lifestyles.

BOX 3.8 *Brunsø and Grunert approach to measuring lifestyles*

An area where researchers are still struggling to find consensus is the measurement of lifestyles. Brunsø, Grunert and other authors have been testing the lifestyle construct in several theoretical and applied studies, suggesting data collection forms and statistical analysis tools. In Brunsø and Grunert (1998), food-related lifestyles are grouped in five major life domains: *Ways of shopping, Cooking methods, Quality aspects, Consumption situations* and *Purchasing motives*. Starting from these five domains, 69 items have been identified for measurement, which provide the basis for designing questionnaires (by asking the level of agreement with the listed sentences):

Ways of shopping
Importance of product information

- To me product information is of major importance. I need to know what the product contains.
- I compare labels to select the most nutritious food.
- I compare product information labels to decide which brand to try.

(Continued)

BOX 3.8 *Cont'd*

Attitudes to advertising

- I have more confidence in food products that I have seen advertised than in unadvertised products.
- I am influenced by what people say about a food product.
- Information from advertising helps me to make better buying decisions.

Enjoyment from shopping

- Shopping for food does not interest me at all.
- I just love shopping for food.
- Shopping for food is like a game to me.

Speciality shops

- I do not see any reason to shop in speciality food shops.
- I like buying food products in speciality food shops where I can get expert advice.
- I like to know what I am buying, so I often ask questions in shops where I shop for food.

Price criteria

- I always check prices, even on small items.
- I notice when products I buy regularly change in price.
- I watch for ads in the newspaper for shop specials and plan to take advantage of them when I go shopping.

Shopping list

- Before I go shopping for food, I make a list of everything I need.
- I make a shopping list to guide my food purchases.
- I have a tendency to buy a few more things than I had planned.

Quality aspects
Health

- I prefer to buy natural products, i.e. products without preservatives.
- To me the naturalness of the food that I buy is an important quality.
- I try to avoid food products with additives.

Price/quality relation

- I always try to get the best quality for the best price.
- I compare prices between product variants in order to get the best value for money.
- It is important for me to know that I get quality for all my money.

Novelty

- I love to try recipes from foreign countries.
- I like to try new foods that I have never tasted before.
- Well-known recipes are indeed the best.

BOX 3.8 *Cont'd*

Organic products

- I always buy organically grown food products if I have the opportunity.
- I make a point of using natural or ecological products.
- I don't mind paying a premium for ecological products.

Taste

- I find the taste of food products important.
- When cooking, I first and foremost consider the taste.
- It is more important to choose food products for their taste rather than for their nutritional value.

Freshness

- I prefer fresh products to canned or frozen products.
- It is important to me that food products are fresh.
- I prefer to buy meat and vegetables fresh rather than pre-packed.

Cooking methods

Interest in cooking

- I like to have ample time in the kitchen.
- Cooking is a task that is best over and done with.
- I don't like spending too much time on cooking.

Looking for new ways

- I like to try out new recipes.
- I look for ways to prepare unusual meals.
- Recipes and articles on food from other culinary traditions make me experiment in the kitchen.

Convenience

- I use a lot of frozen foods in my cooking.
- We use a lot of ready-to-eat foods in our household.
- I use a lot of mixes, for instance baking mixes and powder soups.

Whole family

- The kids always help in the kitchen; for example they peel the potatoes and cut the vegetables.
- My family helps with other mealtime chores, such as setting the table and washing up.
- When I do not feel like cooking, I can get one of the kids or my husband to do it.

Planning

- What we are going to have for supper is very often a spontaneous decision.
- Cooking needs to be planned in advance.
- I always plan what we are going to eat a couple of days in advance.

(Continued)

BOX 3.8 *Cont'd*

Woman's task

- I consider the kitchen to be the woman's domain.
- It is the woman's responsibility to keep the family healthy by serving a nutritious diet.
- Nowadays the responsibility for shopping and cooking lies with the husband as much as the wife.

Consumption situations
Snacks versus meals

- I eat before I get hungry, which means that I am never hungry at meal times.
- I eat whenever I feel the slightest bit hungry.
- In our house, nibbling has taken over and replaced set eating hours.

Social event

- Going out for dinner is a regular part of our eating habits.
- We often get together with friends to enjoy an easy-to-cook, casual dinner.
- I do not consider it a luxury to go out with my family to have dinner in a restaurant.

Purchasing motives
Self-fulfillment in food

- Being praised for my cooking adds a lot to my self-esteem.
- Eating is to me a matter of touching, smelling, tasting and seeing, all the senses are involved. It is a very exciting sensation.
- I am an excellent cook.

Security

- I dislike everything that might change my eating habits.
- I only buy and eat foods which are familiar to me.
- A familiar dish gives me a sense of security.

Social relationships

- I find that dining with friends is an important part of my social life.
- When I serve a dinner to friends, the most important thing is that we are together.
- Over a meal one may have a lovely chat with friends.

Response to marketing actions. Evaluating consumer response to a marketing and communication actions is probably the area with the highest private and public demand for marketing intelligence. While consumer response is certainly measured in terms of behaviors, a proper evaluation of the impact of marketing actions such as advertising requires a more complex and focused design. Table 3.4 shows how the effects of some of the most common marketing activities and the number of consumer variables that may change in response to these actions.

Considering the case of advertising, even in the frequent situation where advertising has simply a defensive purpose as mature markets do not allow for consumer expansion and market shares are pretty settled, it is crucial to evaluate the effectiveness of

Table 3.4 *Extended marketing mix and consumer response variables*

Marketing mix	Example action	Examples of response variables
Price	Cut prices	Sales Profits New customers gained Market shares Perceived quality Stock reduction
Product	Launch a new product	Consumer acceptance Brand image effect Target customer profile Price positioning
Promotion	Advertising	Brand/product awareness Brand loyalty Brand image Sales Market shares New customers gained Customer retention Product positioning Willingness to pay
Place	Launch e-commerce	New customers gained Sales Profits Brand image Market shares
Participants	Improve customer contact management	Customer satisfaction Brand loyalty Customer retention Perceived quality
Process	Introduce a customer complaining procedure	Customer satisfaction Customer retention Brand loyalty
Physical evidence	Change selling environment	Consumer acceptance Sales New customers gained Perceived quality Willingness to pay Time spent on the shop Customer satisfaction Brand loyalty

advertising strategies. These represent a major part of a company's budget. Nowadays, advertising is not simply promoting the purchase of a product, it can be directed at a variety of objectives. This makes market research on the response to advertising more difficult and more specific. It is not simply demand response to advertising levels; demand may remain unchanged, but loyalty of existing customers increase. The other major issue in investigating advertising effect is that the impact of one company's advertising strictly depends on advertising behaviors of competing companies. Thus it may be very difficult to isolate the impact of a specific campaign.

Summing up

While primary data collection allows more control and consistency with the research objectives, there are many potential sources of error. Besides the statistical error due to the fact that a sub-set of units is sampled from the target population, another major source of inaccuracy for data collected through surveys is caused by non-sampling errors which cannot be quantified through statistical procedures and may induce severe biases in the data. Errors can be minimized with a proper research design, which can be articulated in seven essential steps for conducting primary research:

a. defining a reference population;
b. setting the sampling criteria;
c. considering the estimation methodologies;
d. determining sample sizes;
e. choosing the data-collection methods;
f. preparing the questionnaire;
g. calculating costs.

These seven steps assume varying relevance and raise different issues according to the main focus of the survey. Socio-demographic surveys (as e.g. the Annual Population Survey) require larger samples and are more expensive, behavioral and economic surveys (as e.g. the Expenditure and Food Survey) are often based on time-consuming interviews and raise a number of questionnaire issues, psychographic and lifestyle surveys (as e.g. those based on the Theory of Planned Behavior) require a clear theoretical framework. Finally, surveys investigating consumer response to marketing need a careful thinking of the actual objectives of the marketing action and the recording of variables that sometimes are difficult to be measured.

Many non-sampling errors are generated by the interview process and the administration method, including problems in the administration, form the questionnaire. Administration methods are characterized by relevant trade-offs and the optimal technique depends on the situation and the objective of the survey. Face-to-face interviews are preferred for longer interviews, but are subject to biases due to the interaction with the interviewer. Telephone and mail interviews are better for situations where anonymity is expected to be guaranteed, but the former suffers from limitations in terms of interview length and the latter is characterized by an extremely low response rate. The use of a computerized system can help both personal (CAPI) and telephone (CATI) interviews. The increasing diffusion of personal computers has also favored the growth of e-interviews, through e-mail and web-sites, but these raise issues in terms of accuracy of responses and self-selection biases, since they tend to rule out those who do not have access to Internet. Another key element prior to fieldwork is the design of an appropriate questionnaire. This requires many aspects to be taken into account, which we summarized into eight steps:

1. a clear definition of the information to be collected;
2. the design of individual questions in terms of broad content;
3. the choice of the measurement scale;

4. the wording of each question;
5. the sorting of the questions;
6. the coding of the questions for data inputting;
7. an appropriate presentation and layout;
8. piloting and pre-testing of the questionnaire.

EXERCISES

1. Consider the following surveys and questionnaire items. Which administration method would you consider more appropriate in terms of (a) reduction of non-sampling error and (b) cost?

Survey objective/questionnaire item	Administration method
Binge drinking during weekdays; over last month, how often did you have more than five drinks in one evening?	
Lunch habits; where have you had lunch yesterday (canteen, restaurant, etc.)?	
Taxes: how much did you earn last year?	
Attitudes toward football; describe your feelings when you watch a football match	
Customer satisfaction for train users; how satisfactory do you rate your last train trip?	
Use of Internet at work; how many times did you check your personal e-mail this morning?	

With a budget of $20,000, you are asked to run a survey. Consider the following fixed costs, variable costs and response rates:

Survey type	Fixed cost	Cost per questionnaire	Response rate
E-mail	1,000	0.50	10%
CATI	10,000	10.00	50%
Mail	100	1.00	1%
Mall-intercept	2,000	15.00	40%

  a. Compute the expected number of valid responses for each of the administration types.
  b. Your client pretends that you use mall-intercept on at least part of the sample and that the sample size should not be less than 700. What is the most effective way to proceed with mixed administration methods?
2. Open the Trust data-set in SPSS.
  a. Check the content of individual questions by looking at the LABEL and VALUES columns in the variable view
  b. Compare responses on attitudes toward chicken as measured by q9, q10, q12a, q12d. Do responses look similar? How relevant are differences in the wording of questions?

Further readings and web-links

❖ **Measurement errors and surveys** – A comprehensive discussion of the issue of measurement errors for a range of survey types (including business surveys) can be found in the book edited by Biemer et al. (2004).

❖ **Measuring lifestyles** – A short but advanced discussion on measurement of lifestyles is provided in Brunso et al. (2004). An interesting approach in marketing is the association of brands with a 'personality' construct (see Aaker, 1997).

❖ **Designing questionnaires for households in developing countries** – The World Bank promotes multi-purpose household surveys in many developing countries (www.worldbank.org/LSMS/) and the measurement issues are quite relevant to data quality. The book by Grosh and Glewwe (2000) provides useful suggestions based on the World Bank experience.

❖ **Web-surveys** – There many attractive features in web-surveys, but at least as many dangers. See Couper (2000) for a good review and Braunsberger et al. (2007) for a more recent discussion.

Hints for more advanced studies

☞ Recall the discussion on measurement scales of chapter 1 and the concept of reliability. What are the main routes to measure attitudes in literature? How can attitude measurement be validated?

☞ In many consumer research studies, visual or tactile stimuli might be important. Discuss the potential use of stimuli according to the administration method.

☞ There are many different measures to evaluate the impact of an advertising campaign. What is the difference between recall and recognition? Which one is more appropriate and when?

Notes

1. Arrow launched a clear challenge to integrate utility with attitudes to explain choice in 1956. Many attempts to achieve a comprehensive view of consumer behavior linking economics and psychology exist, but none of them has gained popularity.

CHAPTER 4

Data Preparation and Descriptive Statistics

T HIS CHAPTER introduces the actions required to move a further step toward statistical analysis. After primary data have been collected or secondary data obtained, they need to be checked and prepared for subsequent statistical analysis. First, the quality of the collected data needs to be carefully assessed, then the data-set is built with careful consideration for data organization and variable coding. When data inconsistencies or missing data exist, the post-editing statistical processing may improve the accuracy of the subsequent analysis. Prior to more advanced analysis, it is also a good idea to get to know the data through charts and descriptive statistics. The tools described in this chapter allow the researcher to get acquainted with the data through plots and tables and to identify data pitfalls that might affect further data processing, like missing and anomalous values (outliers).

Section 4.1   discusses the editing and post-editing processes in relation to data quality
Section 4.2   provides a review of the main graphical and tabular representations
Section 4.3   summarizes the main issues in detecting and treating missing values and outliers

THREE LEARNING OUTCOMES

This chapter enables the reader to:

➡ Perform the preliminary data processing steps to improve the quality of the data-set
➡ Effectively summarize, plot and describe the variables in the data-set
➡ Acknowledge the relevance of missing data and outliers and learn how to detect them

PRELIMINARY KNOWLEDGE: A review of the basic descriptive statistics (frequencies, central tendency measures, variability measures, see appendix) is essential. This chapter exploits concepts on measurement, administration methods and questionnaire issues introduced in chapters 1 and 3. Data-sets for SPSS examples are those presented in section 1.3.

4.1 Data preparation

There is a relevant step between the completion of the field work and the statistical processing of the collected data, which consists in transferring the questionnaire information into an electronic format which allows and facilitates subsequent data processing. Not only is this passage rich in complex issues, but it has a large impact on the quality and costs of the following research phases. In the *editing* process, data preparation must take into account the following questions:

1. Are non-response errors within acceptable limits?
2. Does the questionnaire meet the basic respondent requirements?
3. Are the responses in the questionnaire complete?
4. Are they consistent and clear?
5. Should low quality questionnaires be replaced or discarded?
6. How should the database be organized?
7. How are the questions coded?
8. How is the transcribing process organized?

Non-response errors. As seen in chapter 3, response rates may vary from 15% to 90% according to the administration method and other survey characteristics. This large differential emphasizes the need for countermeasures when refusals (or not-at-home) affect the field work. A portion of the non-response error can be avoided with appropriate measures decided before the field work. *Refusals* are reduced with contacts prior to the interview, providing incentives and follow-ups, while *not-at-homes* are reduced with a good plan for call-backs. Once the questionnaires have been collected, some further adjustment can occur in the editing phase. One strategy is to extract a sub-sample of non-respondents and proceed to a new interview, usually changing the administration method (preferably exploiting personal interview). Alternatively, the non-respondents can be substituted by other respondents, generally from the same sampling frame that are similar to the non-respondents with respect to some key characteristics. Other solutions concern the *post-editing* phase, where weighting and statistical imputation can reduce the impact of non-response and are discussed in the section 4.1.1.

Basic requirements. A questionnaire may not meet the key conditions to be valid for the survey. If the targeted respondent is an adult and a child fills the questionnaire, considering this observation may bias the subsequent analysis. Questionnaires often contain some filter question, to ensure that the respondent is qualified to answer. If these questions are not answered or some of the answers are clearly deceptive, then the whole questionnaire is likely to be invalid.

Completeness of questionnaires. Questionnaires often generate missing responses to some items. Consider a phone-interview questionnaire with very sensitive questions in the final section; it is likely that many respondents quit the interview at that stage. In another situation, some of the returned mail questionnaires may lack one or more pages.

Consistency. Questionnaires can be structured in order to check the quality of key answers, in order to detect inconsistencies and ambiguous answers. Take the case of a questionnaire on potential customers for a car industry, where a respondent declares to drive 6 hours a week and in another section states to have no driving license. While these situations are theoretically possible, one would expect that either the first answer or the second one is wrong and re-contacting the respondent should allow validation of the questionnaire. Another outcome could be the ambiguity in answers to some questions, for example in a multiple choice question where only one tick is allowed, two or more have been chosen. Finally, some answers might simply be illegible.

What to do with unsatisfactory responses and questionnaires. When possible, the best solution is to go back to the respondent and clarify the issues, generally through a different interview method. For example, missing or otherwise unsatisfactory responses from an e-mail questionnaire could be treated through a telephone interview. This obviously increases the costs of the survey, but could lead to significant improvements in the quality of the data. A second option is to assign *missing values* to the unsatisfactory responses, especially when those variables are not essential and the frequency of unsatisfactory responses is low, both within the same questionnaire and across the whole survey. However, if there is a large proportion of unsatisfactory responses for the same respondent, it is advisable to discard the whole questionnaire. One should also consider the percentage of unsatisfactory questionnaires; if they are a small proportion, then it might be preferable to discard those questionnaires with unsatisfactory answers. The discarding procedure is extremely delicate, as it may reintroduce biases that the sampling procedure was expected to avoid. For example, if all low quality questionnaires come from the same socio-economic group, discarding them would reduce the representativeness of the sample. When unsatisfactory responses are non-random, they are the symptom of some problem in the survey procedure; thus, it is advisable to explore the issue to a further extent rather than using some automatic procedure and move to the statistical analysis. On the other hand, if unsatisfactory responses are signalled through missing data, there are statistical procedures (see sections 4.1.1 and 4.5) to check for their relevance and possibly adjust the sample.

How should the database be organized? Before entering the data in electronic format, the structure of the records and variables needs to be clearly defined. This procedure may require a compromise among different factors of the research process. On the one hand, some structures may be easier to enter as they simply reflect the questionnaires as they are; on the other hand, a different organization could be more adequate for the statistical analysis and the software used by the researchers in the following stage. For example, consider a multiple choice question with more than one potential answer, as a question asking 'what transport means do you use to reach your workplace?,' where several options are available, car, scooter, bike, walking ... it is not trivial to decide whether each of the options should become a 0-1 variable or a single qualitative variable with multiple response categories should be preferred. Most of the time, statistical packages are flexible and allow to pass from one option to the other. However, when multiple possibilities are available, a single variable needs to take into account all potential combination of transport means, making the entering procedure more complex. Another aspect is the treatment of open-ended questions. They may be transcribed as they are or coded into categories in the data entering process.

The codebook. Assignment of variable names, or preferably decision on some systematic rule for assigning variable names, is the first step of the coding process. Other choices which precede translation of the questionnaires into coded questions concern:

a. the type of variable corresponding to each question (qualitative/non-metric or quantitative/metric as discussed in chapter 1);

    b.   if qualitative, (b1) whether it is on a nominal or ordinal scale; and (b2) coding for each of the response categories of the qualitative variable, a number or a letter is necessary;

    c.   if quantitative, the number of meaningful digits and decimals;

    d.   the width of the variable in the spreadsheet depending on the outcomes of (b) or (c); and

    e.   a rule for identifying missing values (like the number -999 may refer to non-response, -888 to unreadable, etc.).

A typical electronic data-set will contain a row for each questionnaire (corresponding to a *case* or *observation*) and a variable number of single-digit columns for each of the variables. Opened in a word-processor it only looks as an unintelligible sequence of numbers and/or letters, while the codebook provides the key to interpret the electronic file, the list of response categories and the link to the questionnaire items (see figure 4.1).

Transcribing process. For computer-assisted administration methods such as CAPI, CATI or CAWI, the interviewer keys in the answers directly and the software automatically produces the data-set, so that the previous step needs to be defined prior to the field work. The software also performs consistency checks to validate the data-set and anomalies are signalled to the interviewer. For other questionnaires, there is an option for electronic reading using appropriate optical scanners and software. In some circumstances, however, the questionnaires are keyed into the computer. This is a potential source of error and random checks on the entered data are advisable.

4.1.1 Post-editing statistical treatments

Once the data-set is available on electronic format (the *editing* step), a series of operations can improve the quality of the data prior to the actual statistical analysis (the *post-editing* process). The statistical treatments performed by most statistical packages (see box 4.1 for post-editing in SPSS), consist of:

    1.   Further consistency checks
    2.   Missing data treatment
    3.   Weighting cases
    4.   Transforming variables
    5.   Creating new variables

Consistency checks. The first consistency checks were carried out in the editing step. Once the data-set is complete, however, it becomes possible to run more complex automatic checks. For example, it is possible to test whether some of the recorded quantitative values are too high or too low by looking for *outliers*. Multiple choice variables might be out of the established range as well. Some packages also allow to set rules for consistency checks. For example, an automatic search for those who declare to drive but have no driving license. When possible, a useful procedure consists in crossing the collected data with some other data from an external source (mainly census and other demographic data), which allows to detect anomalies.

Treating missing data. Missing data are a source of error for two different reasons. First, they reduce the sample size – hence the precision level of statistical estimates. Second, they can introduce systematic biases if there is a specific link between the missing

Data-set

```
00000000011111111112222222222333333333
12345678901234567890123456789012345678
1James      37abae4231831London      42
2Lucy       24bcbd30212330xford      32
3Daniel     28ddad177-999Brighton    10
```

Codebook

Position	Variable	Description (Question)	Coding
Col 1	CODE	Respondent code	Quantitative
Col 2-10	NAME	Respondent name (Q1)	String
Col 11-12	AGE	Respondent age (Q2)	Quantitative -8=non response -9=refusal
Col 13	JOB	Employment(Q8)	Qual. Nominal a=employed b=unemployed c=retired d=other 8=non response 9=refusal
Col 14	EDUC	Education level (Q9)	Qual. Ordinal a=no formal ed. b=primary sc. c=secondary sc. d=University dg. 8=non response 9=refusal
Col 15	GEND	Gender (Q3)	Qual. Nominal a=male b=female 8=non response 9=refusal
Col 16	MS	Marital status (Q4)	Qual. Nominal a=single b=married c=other 8=non response 9=refusal
Col 17-19	FOOD	Monthly food exp.(Q10)	Quantitative -88=non response -99=refusal
Col 20-23	INC	Monthly income (Q11)	Quantitative -88=non response -99=refusal
Col 24-35	CITY	City (Q5)	String
Col 36	SIZE	Household size (Q6)	Quantitative -8=non response -9=refusal
Col 37	SIZE	Number of children(Q6)	Quantitative -8=non response -9=refusal

Figure 4.1 *Data file and codebook*

data and other characteristics of the cases with missing data. For example, if those who refuse to indicate their income level are all in the highest income range, simply discarding the missing data would lead to a downward bias in the final estimate of the average sample income. There are situations where these two sources of error are negligible. If the sample size is reasonably large and non-responses are random, which means not driven by a particular reason, missing data can be simply omitted from the analysis. These data are called MAR (missing at random). In this case, the researcher must decide how to discard the data, whether through *case-wise* (or *list-wise*) *deletion*, where all other data referring to the same case are omitted from the analysis, or *pair-wise deletion*, where for each estimation all valid cases are considered, so that one case might enter in one estimation and not in another because of missing data.

When missing data are non-random, hence linked to some specific pattern, ignoring them is potentially dangerous. In these situations, the better remedy is prevention, since appropriate research designs and questionnaires (see chapter 3) reduce the incidence of systematic biases. Some solutions can be also adopted in the editing phase (as discussed in the previous section). Other countermeasures can be taken during the data processing steps which are preliminary to the actual statistical analysis. Thus, the first step is testing whether the missing data can be regarded as MAR. For each variable with missing data, one can split the sample into two groups, one where all cases are with missing data, the other with complete data. Then mean comparison tests (see chapter 6) indicate whether statistically significant difference exist between the two groups in terms of other relevant (control) variables. For example, if respondents in the highest income bracket refuse to state their income, this could be detected by testing differences between the sample of non-respondents and the sample of those who responded. If the latter shows significant higher means for other variables, like the number of cars owned, the expenditure in luxury goods, the level of education, then the difference between non-respondents and respondents is clearly non-random. Alternative approaches to detecting non-randomness exist in literature (see Little and Rubin, 1989).

Once non-random missing data are detected, these cannot be ignored. There are various procedures to deal with this sort of missing data, some of them are based on weighting (see page 83), other on *statistical imputation*, where missing data are replaced using other information available in the sample (Schafer and Graham, 2002). The more elemental approach consists in assigning to the missing values of a variable the mean of all valid observations for the same variable (*mean substitution*), although this leads to an undesirable reduction in the data variability. A more sophisticated method generates the missing data using *regression analysis* (see chapter 9) or through *multiple imputation* (Rubin, 2003). In regression analysis, it is assumed that the variable with missing data is related to other variables. Considering the above example, it can be assumed that household income is related to the number of cars, to the level of education of its members, to the expenditure in luxury goods, etc. More details on regression analysis are provided in chapter 8; here it may suffice to know that regression quantifies the relationship between a variable (the dependent variable) and other variables (the explanatory variables) by estimating the parameters of a linear function (see *parameters* in the appendix). Hence, once the regression model is estimated, one can exploit the values of the explanatory variables (the non-missing data) to estimate the dependent variable (the missing data). This method also reduces variability in the variables with missing data and reinforces the assumed relationship. Thus, it is important to ensure that the relationship is statistically valid to avoid biases. This method can be improved by adding an artificial error term generated randomly from the distribution of the stochastic error of the regression model, so that two missing values with identical values for their explanatory covariates are replaced by different values. To these individual replacement methods, *multiple imputation* is often preferred. With multiple imputation, the missing value is estimated using two or more different methods, then it is replaced by an average of these estimates. In addition to these strategies, statistical packages generally provide more complex methods, like the iterative EM (expectation + maximization) algorithm. The algorithm starts from the assumption that each variable follows a specific distribution (usually normal), which allows an estimate for the missing value. Then some parameters of the distribution (for example, mean and variance) are estimated through the *maximum likelihood* method. Once these maximum likelihood estimates are available, the missing values are replaced by other values extracted from the distribution. The process continues until the estimated value and the distribution parameters become stable. Finally, missing values can be dealt with during the statistical analysis. For example, in some statistical analyses, one can assume that all the observations with missing values belong to a single homogeneous sub-group of the population, so that they are processed like the other observations without missing values. For example, if one wishes to test the effect of income on some other characteristics like food expenditure, one may consider non-respondents as a stand-alone income group. This allows one to maintain larger sample sizes leading to more efficient statistics, and makes it possible to elicit the separate effect of missing values on the income-food expenditure relationship, whatever their actual value.

Weighting cases. Another issue that needs to be considered before computing any statistic on the data-set is whether *weighting* is needed. The question is whether some particular sub-groups of the population might be under- or over-represented in the sample, so that applying a weight to each case restores the representativeness of the sample. In theory, the decision whether to apply a weighting system depends on the sampling strategy. With stratified sampling or cluster sampling, weights are necessary and depend on the sampling design, while in simple random sampling all units are expected to be selected proportionally to their probability of inclusion (see chapter 5). In probabilistic cases, samples are expected to be self-weighted. However, the *self-weighting* property of probabilistic samples could be undermined by non-responses and weights allow an adjustment the sample. For example, if the non-response error is 50% in one stratum (a sub-group of the sample) and 25% in another stratum, by assigning a double weight to the cases in the first stratum the original proportions are restored. This is clearly a risky procedure. If some selection bias exists within a stratum, it will be made worse by the weights. Another situation where weighting helps is the case where the results can be improved with respect to the final objective of the analysis. As an example, consider an opinion poll administered to the spectators of a movie theater run in a single night. The manager could be more interested in the opinions of those who declared to go to the movies every night than those that go once a month, so the researcher could decide to assign a larger weight to the former, provided this decision is made explicit in the research report.

Transforming variables. Other treatments before the actual statistical analysis involve a transformation of the original variable. The main transformation involves:

   a. mathematical transformations;
   b. banding;
   c. recoding;
   d. ranking.

Mathematical transformations are useful in some situation to improve the statistical analysis and the interpretation of results. For example, in demand analysis, if the quantity, price and income variables are transformed into logarithms, the coefficients of a regression equation with quantity as a dependent variable will be the elasticities which have a direct economic interpretation (the percentage change of consumption in response to a 1% change in prices). The logarithmic transformation is also useful to reduce the variability of a given variable. Other potential mathematical transformations include transforming a variable measured in absolute level into a percentage, the

BOX 4.1 *Post-editing statistical treatment in SPSS*

Consistency. For a variety of variable screening techniques, SPSS provides the function EXPLORE (menu ANALYZE/DESCRIPTIVE STATISTICS/EXPLORE), which returns the preliminary descriptive statistics and graphs and other useful indicators for the detection of outliers.

Missing data. Imputation of missing data through mean replacement, median replacement or other interpolation techniques which only exploit data from the target variable is achieved through TRANSFORM/REPLACE MISSING VALUES, then selecting the chosen replacement option in the apposite box. It is also possible to use only a small number of neighboring values in case the cases are sorted to some specific rule (as in systematic sampling). For more complex imputation methods using more variables, there are several possibilities using the menu ANALYZE/MISSING VALUE ANALYSIS, which allows to detect pattern in missing values (and use *t*-tests to check for randomness) and to apply more complex methods like the EM algorithm or regression to replace the missing values.

Weighting cases. Using DATA/WEIGHT CASES it is possible to assign a specific weight to each of the cases through an auxiliary variable available in the data-set.

Transforming and creating new variables. There are several possibilities in the TRANSFORM menu to modify variables or create new variables from the original ones. The COMPUTE function allows to perform mathematical computations and transformations, with the possibilities of making these computations conditional to an 'If' condition. The RECODE function allows to re-classify a nominal or ordinal variable into a smaller or larger set of categories. In order to categorize a quantitative (metric) variable, there is a VISUAL BANDER option which contains a number of useful features to obtain the most appropriate split into categories. There are also function to count values and create a count variable (COUNT), to rank cases and create ranking variables (RANK) and for automatic recoding, like the transformation of string variables into nominal variables (AUTOMATIC RECODE).

change of the measurement unit, currency changes, etc. Instead, *banding* refers to the categorization of numeric variables. For tabulation purposes it may be more useful to have a variable measuring food expenditure split into four expenditure categories, rather than a series of individual values. A third transformation consists in *recoding*, which is especially useful for changing the response categories of a qualitative variable. For example, one might ask the education level distinguishing between primary school, lower secondary school, upper secondary school, undergraduate university degree and postgraduate university degree. In order to simplify the analysis (and possibly improve the results) it might be useful to recode the variable into three categories instead of five, primary school, secondary school and university degree. A special type of recoding involves *ranking* the variable, that is substituting the absolute value with some other value which reflects how the original value is ranked across the sample. Thus, if one is more interested in the income percentile rather than the actual income, cases need to be sorted and divided into percentiles of equal size; then the transformed variable will simply indicate to which percentile the original value belongs.

Creating new variables. Similarly to transformations, mathematical operations can be exploited to create new variables. For example, the sum of time spent reading, going to movies, playing sports, etc. could create a variable for leisure time.

4.2 Data exploration

Once the data are available in electronic form and quality checks and cleaning procedures have been performed, the researcher must resist the temptation of proceeding directly to the advanced statistical analysis. Getting familiar with the data-set through graphical and tabular exploration can greatly improve the effectiveness of the statistical analysis. Sometimes a good descriptive analysis may suffice to the objectives of the research.

While one should avoid providing a large amount of graphs and tables, an effective research report should always include a selection of the most informative descriptive statistics and graphical representation of the original data-set. Data exploration is especially useful when the data-set is large and subsequent statistical processing will involve multivariate analysis of the relationships among a set of variables. Data exploration and descriptive statistics are standard procedures in commercial statistical packages and difficulties lie in selecting the most informative and effective graphs and tables. The main actions in data exploration can be summarized in the following points:

a. *Graphical plots of the data*, to get a first overview of the main characteristics of the data-set, especially the distribution of the original variables across the whole sample and for sub-samples;

b. *Univariate descriptive statistics and one-way tabulation*, which synthesize the main characteristics of each of the variables in the data-set;

c. *Multivariate descriptive statistics and cross-tabulation*, to get a first understanding of the relationship existing between different variables and enabling the joint examination of two or more variables; and

d. *Missing data and outliers detection*, to allow an early detection of potential issues in subsequent analysis.

While the problem of missing data has been already discussed in the previous sections, it is common to maintain some proportion of cases with missing values in the data-set. When the distribution of missing values differs across variables, there may be problems in multivariate analysis, which range from the loss of observation to statistical biases. Early detection is essential. Preliminary data exploration is also crucial to detect *outliers*, observations which show a large difference from other observations in the data-set. Outliers are often the result of recording or transcription errors and when ignored may induce biases in the statistical analysis.

The graphical or tabular representation of the data obviously requires some synthesis. To this purpose, it is essential to recall the basic statistical concepts, as frequency, central tendency and variability measures. For a quick reference, readers are encouraged to refer to the appendix.

4.2.1 Graphical representations

Statistical packages and common spreadsheets offer a wide range of options for plotting the data. This section reviews the basic alternatives according to the following classification scheme:[1]

1. Univariate plots of qualitative or discrete data
2. Univariate plots of quantitative data

3. Bivariate and multivariate plots of quantitative data
4. Bivariate and multivariate plots of quantitative versus qualitative data

Univariate plots of qualitative or discrete data. When a variable is categorical, that is measured through one of the qualitative scales discussed in section 1.2, the choice is generally among bar charts, pie charts, line charts. *Bar charts* are diagrams where each response category of the variable is represented by a rectangle whose length is proportional to its frequency. *Pie charts* are useful to emphasize proportion or shares and consist in circles divided into slices. The area (or the angle) corresponding to each response category is proportional to the frequency for such response category. *Line charts* have a similar use to bar charts, but instead of rectangles, the frequency of each category of a qualitative variable is proportional to the height of a single point. A line may connect the points corresponding to the frequencies of each response category. These types of charts can be applied to metric variables as well. For example, line charts are the typical representations for metric variables observed over time.

Univariate plots of continuous data. If the data are not categorical, but measured through quantitative (metric) scales, a wider range of graphs becomes available besides those listed in the previous section. The most frequently used diagrams are histograms, box plots and error bars, but more complex plots as Pareto charts or the Q-Q plot may be very informative and become essential preliminary plots for subsequent analyses. *Histograms* are a generalization of bar charts, where the rectangles may have different width depending on the size of the class for the target variable. Thus, the frequency is proportional to the area rather than the length of the rectangles. *Box plots*, also known as *box-whisker diagrams*. The variable (usually within a sub-sample) can be represented by a box with two lines (the 'whiskers'), which allows to summarize several information on the frequency distribution. The box is a rectangle drawn from the value corresponding to the lower quartile up to the highest quartile, while the line corresponds to the median values and splits the box in two parts. Whiskers are lines drawn from the box margins to the minimum and maximum values. *Error bar charts* plot the standard errors (a measure of precision introduced in chapter 4), standard deviations or confidence intervals (see chapter 6) of individual variables. Two error bars are drawn from the point representing the average value on a bi-dimensional axis for each of the sub-samples. *Pareto* charts are similar to bar charts, but the bars are ordered in decreasing order of the frequencies they represent. The same graph also shows a line indicating the cumulative proportion. Pareto charts are especially useful in quality control, (and the latest versions of SPSS includes them under the ANALYZE/QUALITY CONTROL menu), as they are used to look at the frequencies of different sources of problem. For example, one might exploit a Pareto chart to explore the relevance of different causes of missing data (refusal, not-at-home, etc.).

An example of more complex univariate chart is the *Q-Q plot*, which allows comparison of the empirical (observed) data distribution and some theoretical distribution. When the observed distribution is close to the theoretical one, the plotted values tend to lie on a straight line.

Bivariate and multivariate plots of continuous data. There are several graphs involving two or more variables. Frequently employed plots include several types of scatter diagrams and Pareto plots. Graphs listed for univariate plots (histograms, pie charts, line charts) can be exploited to show more variables at the same time, but separately. Pie charts can also combine two or more variables in the same chart (provided they are measured with the same measurement unit). For example, using the EFS survey, one might create a pie chart where each slice is determined by a different expenditure variable for each macro-category (food, clothing, restaurants, etc.). Instead, *scatter diagrams (scatterplots)* are bi-dimensional (sometime tri-dimensional) charts, where data are plotted as a series of points using cartesian coordinates. If one uses a bi-dimensional representation, a third variable can be represented by varying the size of the points. Or, if one wants to represent multiple pairs on the same cartesian diagram, different colors could be adopted. For example, using the EFS data-set one might want to plot clothing expenditure against household income. If one want to consider total expenditure as an alternative to income in relation to clothing expenditure, these two pairs can appear in the same graph with different colors. *Pareto plots* can also be extended to account for multiple variables with the same measurement units, comparing their sum as it has been suggested for multiple-variable pie charts. In SPSS it is also possible to obtain stacked Pareto plots, where stacks depend on a given variable (for example income range in categories).

Bivariate and multivariate plots of quantitative versus qualitative data.   The above example of stacked Pareto charts introduces the last combination of charts. As a matter of fact *stacked charts*, that is those charts where bars, histograms or Pareto charts are generalized to account for different categories of an additional qualitative or discrete variable, follow directly from univariate and multivariate plots of quantitative variables. Besides stacked charts, SPSS also offers a choice of *clustered charts*, where bar charts, box plots or error bar charts show 'clusters' of one or more quantitative variables within clusters defined by a qualitative (categorical) variable. For example, one could draw a *clustered box plot* where several expenditure variables are represented and clusters represent different income quartiles.

Figure 4.2 shows examples of charts that can be produced with SPSS. Charts *a*, *b* and *c* are alternative views of the frequency distribution for the household size variable from the EFS data-set. While the bar chart and the line chart emphasize when frequency decreases for larger household size, pie charts are more effective in giving an idea of the proportion and chart *c* shows that households with one and two members represent almost two thirds of the sample. The histogram in *d* shows income distribution within the sample, which allows to note asymmetries in the income distribution and the presence of some outliers. The box-whiskers diagram in *e* also provides some information on the distribution of income, since the box represents values between the first and third quartile and the line within the box is the median value and is closer to the lowest quartile, which indicates a left-skewed distribution. The median is also closer to the minimum value of the variable than the maximum. Instead the error bar chart in *f* shows the mean household income in each quartile and the error lines show the size of the standard deviation, in excess and defect. Household incomes are much more concentrated in the first two quartiles and variability is highest in the upper quartile.

The simple Pareto chart in *g* compares the distribution of incomes across the four quartiles and allows an immediate evaluation of the concentration of income. On the one hand, the sum of revenues from medium-low and low incomes together account for almost 80% of total wealth (which is shown by the line), while boxes provide information on each class, showing total revenues in monetary terms and percentages as well. The Q-Q plot in *h* compares the distribution of food and drink expenditure with a normal distribution. The closer the points to the straight line, the closer the distributions. In general, the distribution seems to be pretty close to the normal one and some outliers emerge from the plot.

When more than one variable is taken into consideration, one may refer to a clustered bar chart, such as the one in *i*, which emphasizes differences in expenditure levels for various items across income quartiles. While expenditures on all items increase with income, the graph highlights small differences in recreation expenditure for the first three quartiles, then a major step upward for the highest quartile. If one wants to compare the distribution of total expenditures for the various items of the EFS

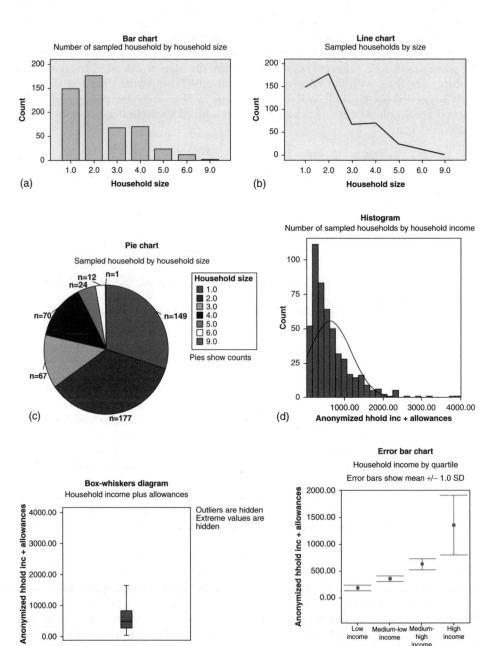

Figure 4.2   *Charts with SPSS*

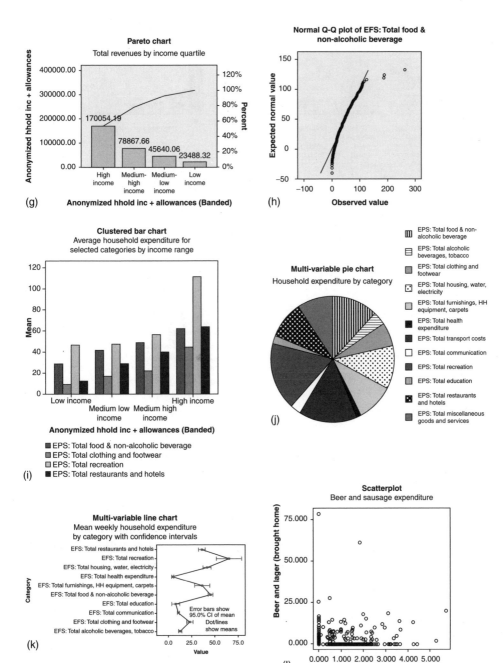

Figure 4.2 *Continued*

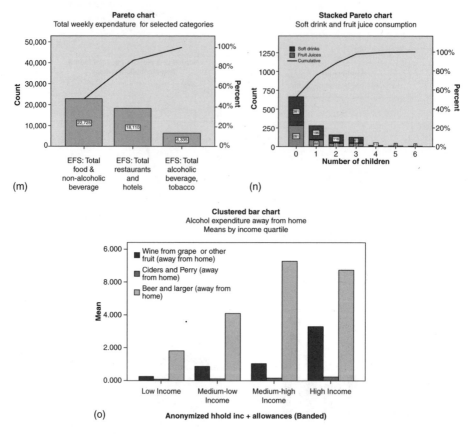

Figure 4.2 *Continued*

survey, a multiple variable pie chart, such as the one in *j*, does the job. As one would always like to find, recreation seems to be the largest slice and health expenditure the thinnest one.

A good way to look into average values and variability is the multi-variable line chart in *k* which adds some information. Recreation is indeed the category with the highest average expenditure, but it also shows the largest variability as indicated by the confidence intervals (see chapter 6). If one wants to look for a relationship between beer and sausage expenditures, a good way is the scatter-plot shown in *l*, where each household has a co-ordinate on the *x*-axis depending on expenditure on sausages and another on the *y*-axis based on expenditure on beer. Ignoring all those households lying on the two axes (which means that they bought beer only or sausages only), it becomes possible to check whether the two variables are positively related by checking whether the points concentrate on an imaginary upward leaning line. If points are randomly scattered, there is no relationship and the graph shows weak links between the two expenditure items.

The Pareto chart in *m* is simply an extension to the case of more than one variable and allows comparison with the relative weight of various expenditure items. Considering food and non-alcoholic drinks, eating out and alcoholic drinks, it shows that the first category represents more than 40% of total expenditure, while the line presents the cumulated proportion after sorting the expenditure items. Pareto charts may be stacked as in n. The graph shows household expenditure in soft drinks and fruit juices according to the number of children. Household with no children represent more than half of the total expenditure and the stacked bar shows that soft drinks and fruit juices are almost equally important. As the number of children increases, the weight of total expenditure decreases (mainly because there are fewer families), but the chart emphasizes that the reduction is smaller for soft drinks, suggesting that children are a more influential factor for soft drink than fruit juice expenditure. The last chart in o is a clustered bar chart, which plots different expenditure items against income quartiles and shows how the richest portion of the sample decreases beer consumption in favor of a much higher expenditure on wine.

4.2.2 Tabulation

Getting to know the data by simply looking at them is a step that is too often neglected by researchers. Instead, misleading and unexpected results from complex analyses can be anticipated; and if necessary, avoided by a careful data exploration. The first step is obviously a closer look to single variables. After plotting their frequency distributions (bar chart, histograms) as shown in the previous section, simple tables allow to take a closer look, especially to assess the risk of data problems. For example, if one runs a multivariate model using ten variables, where nine of them have no missing values while one only covers half of the sample, the consequence could be that the analysis is only based on half of the available observations unless countermeasures are taken.

First, one should make the distinction between qualitative (non-metric) and quantitative (metric) variables. For qualitative variables (or metric variables with a small number of discrete values), *frequency tables* are the most appropriate summary tool. For each variable, one may derive a table where the absolute and/or relative number of occurrences for each response category is shown, as in the example from the Trust data-set (see table 4.1).

For quantitative data and continuous variables, frequency tables are only meaningful if the variable is divided into classes. For example, the Trust data-set records weekly chicken consumption in kilograms, with a range from zero kgs to 25 kgs, although 98% of the recorded values are below 3 kgs. Hence, one could band chicken consumption into three categories, less or equal to 1 kg, between 1 and 2 kgs and above 2 kgs and obtain the frequency table 4.2.

Table 4.1 *Frequency table for variable q1 in the Trust data-set*

Response category	How many people do you regularly buy food for home consumption (including yourself)?	
	Count	%
1 – Extremely unlikely	91	18.3
2	176	35.4
3	100	20.1
4 – Neither	94	18.9
5	21	4.2
6	13	2.6
7 – Extremely likely	2	0.4
Total	497	100.0
Missing values	3	

Table 4.2 *Trust data-set: In a typical week how much fresh or frozen chicken do you buy for your household consumption (kg)? (Banded)*

		Frequency	Percent	Valid Percent	Cumulative Percent
Valid	Less or equal to 1 kg	319	63.8	71.5	71.5
	Above 1 kg and less or equal to 2 kgs	90	18.0	20.2	91.7
	Above 2 kgs	37	7.4	8.3	100.0
	Total	446	89.2	100.0	
Missing		54	10.8		
Total		500	100.0		

Table 4.2 also shows a separate relative and cumulative frequency for valid cases.

With quantitative variables, either continuous or discrete, it becomes possible to compute a set of additional central tendency (mean, median, etc.) and variability (variance, standard deviation, etc.) measures as those described in the appendix. Table 4.3 shows the descriptive statistics for the chicken consumption variable, for the chicken expenditure variable and for the respondent's age variable.

The 'average' respondent is aged between 45 and 46, buys 1.05 kgs of chicken in a typical week for an expenditure of about €5.67. The standard error of the mean provides an indication of the precision of the estimator (this concept is explained in full detail in chapter 5), while the standard deviation is a measure of variability. The median values are those splitting the sample into two halves and identify a respondent aged 45, buying 0.91 kgs of chicken and spending €5.00.

Table 4.3 also shows the value corresponding to each quartile, for example 25% of respondents are aged below 32 and 25% are aged above 57.

The next step is the construction of multivariate tables, where the association between two or more variables is explored. When the variables are categorical, discrete or arranged into classes (as chicken consumption in table 4.2), *cross-tabulation* displays

Table 4.3 *Descriptive statistics*

		In a typical week how much fresh or frozen chicken do you buy for your household consumption (kg)?	In a typical week how much do you spend on fresh or frozen chicken (Euro)?	Age
N	Valid	446	443	500
	Missing	54	57	0
Mean		1.0582	5.6677	45.582
Std. error of mean		0.06843	0.19640	0.7100
Median		0.9100	5.0000	45.000
Mode		1.00	3.00	45.0
Std. deviation		1.44514	4.13383	15.8763
Variance		2.088	17.089	252.055
Minimum		0.00	0.00	18.0
Maximum		25.03	30.00	87.0
Percentiles	25	0.5000	3.0000	32.000
	50	0.9100	5.0000	45.000
	75	1.3600	7.5000	57.000

the frequencies for each combination. These combined frequencies represent the *joint distribution* of two or more variables. Using the EFS data-set as an example, one may wish to check the relation between the number of children (*childband*, a discrete variable, with an aggregated category) and the income quartiles. Table 4.4 shows the absolute and relative frequencies.

More variables may be cross-tabulated through a *nesting* approach. (For example, one may want to consider the age group of the reference person (an ordinal variable) besides income quartiles in connection to the number of children. The relative frequencies are shown in table 4.5.

While cross-tabulation refers to joint frequencies for categorical, discrete or grouped variables, a number of tables can be created to explore quantitative (continuous) variables in relation to other non-continuous variables. To put it in plain terms, bivariate or multivariate tables allow one to explore how a continuous variable varies across sub-samples. To accomplish this task, one must refer to some descriptive statistics. For example, still referring to EFS data, if the researcher assumes that there is a relationship between expenditure for specific food categories and the age of the household reference person, it could be useful to look at the average values and standard deviations of expenditure by age group as in table 4.6. SPSS offers many options for detailed table construction, as summarized in box 4.2.

4.3 Missing values and outliers detection

Missing value treatment is discussed in the initial section of this chapter in terms of post-editing statistical treatment. However, even once the data-set has been finalized, missing values might still pose a serious problem for some statistical techniques. As has been emphasized, the key issue is whether missing values occur randomly or are related to some respondent characteristic, as the latter circumstance may affect the quality of results.

Consider question about income in the Trust data-set. There is a 'direct' question (q60), where the respondent is asked to indicate the gross household income choosing between six categories and with an explicit 'refusal to respond' option. There is also an 'indirect' question, where respondents are asked to answer to a more innocuous classification of their household financial situation in qualitative terms and no explicit option for non-response is provided (q61). The non-response rate is 31.6% (almost one third) for the direct question and only 2.8% for the indirect one. A cross-tabulation provides a first exploration of the characteristics of missing values according to the indirect question.

Table 4.7 shows that the percentage of non-respondents to question q61 (direct) clearly increases as the financial situation is perceived to improve. The non-response rate for the direct income question ranges from 11.5% for those that declared a 'not very well off financial situation' to percentages well above 30% in the 'reasonable' and 'well-off' categories. The non-response bias is likely to affect results when using the direct variable.

If the variable with missing values is a quantitative one, it becomes possible to run more specific statistical tests. If the indicator variable is quantitative, it becomes possible to perform mean comparison tests (see chapter 6) and ANOVA tests (see chapter 7) to check for statistical significance of the difference across groups with missing values. For example, if the target variable is the weekly amount spent in chicken (q5), one may test whether non-respondents to q60 show a significantly different chicken expenditure through the hypothesis testing techniques discussed in the following chapters.

Table 4.4 Cross-tabulation

Food & non-alcoholic beverage (Binned)* Anonymized hhold inc + allowances (Banded) Cross-tabulation

			Anonymized hhold inc + allowances(Banded)				Total
			Low income	Medium-low income	Medium-high income	High income	
Food & non-alcoholic beverage (Binned)	£20 or less	Count	47	19	18	4	88
		% of Total	9.4%	3.8%	3.6%	0.8%	17.6%
	From £20 to £40	Count	57	48	24	22	151
		% of Total	11.4%	9.6%	4.8%	4.4%	30.2%
	From £40 to £60	Count	17	31	45	40	133
		% of Total	3.4%	6.2%	9.0%	8.0%	26.6%
	More than £60	Count	4	27	38	59	128
		% of Total	0.8%	5.4%	7.6%	11.8%	25.6%
Total		Count	125	125	125	125	500
		% of Total	25.0%	25.0%	25.0%	25.0%	100.0%

Table 4.5 Three-variables frequency table

Children, income and age of HRP

Anonymized hhold inc + allowances (Banded)	Age of HRP – Anonymized (Binned)	Number of children (Banded)				Total
		No children	One children	Two children	More than two children	
		Table %	Table %	Table %	Table %	Table %
Low income	Less than 30 years	1.4	0.8	0.2		2.4
	From 30 to 55 years	2.6	1.0	0.4		4.2
	More than 55 years	18.4			0.2	18.4
Medium-low income	Less than 30 years	1.0		0.2	0.2	1.4
	From 30 to 55 years	4.8	2.2	2.6	1.6	11.2
	More than 55 years	12.0	0.4			12.4
Medium-high income	Less than 30 years	0.6	0.4	0.8	0.2	2.0
	From 30 to 55 years	7.6	2.6	3.8	2.4	16.4
	More than 55 years	6.4		0.2		6.6
High income	Less than 30 years	1.6	0.4			2.0
	From 30 to 55 years	9.2	2.6	4.6	1.6	18.0
	More than 55 years	4.8		0.2		5.0

Table 4.6 *Expenditure by age group in the EFS survey*

	Age of HRP – Anonymized (Binned)					
	Less than 30 years		From 30 to 55 years		More than 55 years	
	Mean	Standard deviation	Mean	Standard deviation	Mean	Standard deviation
Books	0.589	(1.59)	2.701	(8.26)	1.112	(3.78)
Ice cream	0.000	(0.00)	0.067	(0.39)	0.010	(0.11)
Internet subscription fees	0.263	(0.96)	0.396	(1.80)	0.139	(0.84)
Cinemas	1.240	(5.78)	0.644	(3.37)	0.107	(0.84)

Table 4.7 *Missing values in q61 of the Trust data-set*

				q61					
			Total	Not very well off	Difficult	Modest	Reasonable	Well off	Missing SysMis
Income	Present	Count	342	23	33	114	117	53	2
		Percent	68.4	88.5	76.7	74.5	63.9	65.4	14.3
	Missing	% SysMis	31.6	11.5	23.3	25.5	36.1	34.6	85.7

While missing data have a clear meaning for researchers, dealing with *outliers* is less univocal. Outliers are observations appearing to be anomalous because they have values that are quite different to other observations in the same data-set. The most frequent cause for outliers is a recording error, which should be corrected for in the post-editing phase. However, outliers might be correct observations determined by exceptional circumstances. In this case, it is responsibility of the researcher to decide whether they should be included or not in the analysis. On the one hand, an exceptional occurrence might be important to explore the phenomenon. For example, a supermarket surveying its customers may value information about a small number of bulk customers purchasing very large amounts. By looking to these outliers, one might find a potential and profitable new target market, for example restaurants. As a matter of fact, in this example the outlier is relevant to consumer research because in economic terms a very large customer might be worth 100 small customers. As a counter example, consider a study run by a small local shop on the weekly purchases of soft drinks in relation to prices. If a price increase occurs in an extremely hot week when the competing local supermarket is on strike, this observation could strongly bias the evaluation of the impact of price changes.

These two examples lead to two main considerations – first, one should not ignore outliers as they might hide relevant information or pose a serious risk to the reliability of statistical analysis and second, once outliers have been identified, there is no univocal rule on how to treat them.

Detecting outliers in a data-set requires some preliminary choice about how to define them. Considering a single variable, an outlier is typically defined[2] as an observation with a value that is more than 2.5 standard deviations from the mean. An alternative definition is a value that lies more than 1.5 times outside the interquantile range[3] beyond the upper or the lower quartile. It should be noted that as the sample size increase, the probability of having anomalous values increases and their impact on statistical analysis decreases. Hence, one may wish to raise the above coefficient to higher values (for example 4 standard deviations instead of 2.5). Scatter-plots and box-plots as those described in section 4.2 allow one to graphically detect outliers.

A second outlier detection approach relies on multivariate analysis. In some cases, observations show ordinary values for variables taken individually, but the combination of these variables might show exceptional values. If the cause of these outliers is related to the data collection and recording process, consistency checks in the editing phase should allow one to detect and correct them. Otherwise bivariate scatter-plots help to detect isolated points representing unusual combination of values for two variables.

Once an outlier has been identified, the researcher must choose what to do with it – delete, correct or retain it. This clearly depends on knowledge about the cause of the outlier. If the cause is a recording error and this cannot be corrected, the observation is deleted or treated as missing data. If the cause is external and can be readily known, one can decide whether it is convenient or not to retain the data according to the objectives of the analysis, as discussed above. The situation when the cause is unknown is the most delicate and decision usually depends on whether these value might reflect some 'real' although infrequent situation (which should lead to retention) or it is more likely that they are a consequence of data quality problems (which should lead to deletion).

4.3.1 Detecting outliers using SPSS

Consider the Trust data-set and the objective of identifying outliers in the chicken consumption variable (q4kilos). Using the EXPLORE menu, we obtain the following descriptive statistics:

Mean:	1.06 kg
Standard Deviation:	1.45 kg
Upper quartile:	0.50 kg
Lower quartile:	1.36 kg
Interquartile range:	0.86 kg

Outlier definition 1 (mean plus or minus 2.5 times the standard deviation) defines outliers only those values that are above 4.69 kgs, as the lower extreme would be negative, and zero consumption is quite common. The alternative definition of outlier (lower quartile minus 1.5 times the interquartile range or upper quartile plus 1.5 times the interquartile range) sets the minimum value for detecting outliers at a much more conservative 2.65 kgs. Using the first definition, we use the DATA/SELECT CASES menu and filter out cases below 4.69 kgs Then we can summarize (REPORT/CASE SUMMARIES) the cases that are left in the data-set, accompanied by two other variables that could explain the outliers. We have five outliers, as shown in table 4.8.

The first two observations are likely to be the outcome of some recording error or misleading response, since no expenditure is recorded and the values are somehow too precise to be reliable with two decimals, so one could feel inclined to discard them.

Table 4.8 *Outlier detection*

Case Summaries

		In a typical week how much fresh or frozen chicken do you buy for your household consumption (kg)?	In a typical week how much do you spend on fresh frozen chicken (Euro)?	How many people do you regularly buy food for home consumption (including yourself)?
1		25.03	0.00	5
2		6.81	0.00	6
3		5.00	17.00	3
4		7.00	30.00	5
5		8.00	30.00	5
Total	N	5	5	5
	Mean	10.3680	15.400	4.80

The remaining three values are relatively consistent with the stated expenditure value and the number of household components, so they could be retained for subsequent analysis.

A scatter-plot showing q4kilos (consumption) and q5 (expenditure) is shown in the left pane of figure 4.3. The plot highlights the same anomalous cases. Case 78 is confirmed to be quite far from the usual values, while some doubt remains about cases 163, 236 and 434. A meticulous researcher may choose to compute (TRANSFORM/COMPUTE) a price variable as the ratio between q5 and q4kilos to check for consistency between these two variables. The plot showing these new variables against q4 is the right pane of figure 4.3.

Although some anomalous prices clearly emerge on the price axis, if one maintains the focus on anomalous values for the q4 variable, we still find observation 78 to be quite

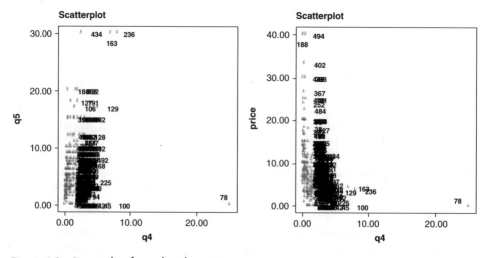

Figure 4.3 *Scatterplots for outlier detection*

BOX 4.2 *Tables and statistics in SPSS*

SPSS offers a wide range of tools for tabulation, which can be used alternatively to obtain descriptive statistics. The number of equivalent routes to obtain similar data description may be a source of confusion for those unfamiliar with SPSS. Data exploration can be pursued within the ANALYZE menu, which allows to produce tables through the REPORT item (for groups of continuous variables), the DESCRIPTIVE STATISTICS item (for single variables) or the TABLES item (any type of table). The same table can be obtained by exploiting alternative routes.

Reports

OLAP cubes where OLAP stands for on-line analytical processing, represent a data exploration procedure where statistics for one or more scale variables are computed for different combinations of one or more categorical variables. The output of OLAP cubes in SPSS shows a single output table, with a drop-menu for each of the categorical variables defining the sub-samples, so that one can easily explore statistics for different sub-samples by simply selecting different values of the grouping variables.
Case summaries allow to compute the same descriptive statistics as for OLAP cubes, but all of the combinations are shown in a single and comprehensive output table. It is also possible to display each of the single cases, although this is rarely a useful operation.
Report summaries in rows (columns) produce text reports (no table formatting) where the descriptive statistics for are structured in rows (columns) and the break (grouping) variables are the columns (rows).

Descriptive statistics tables

Frequencies allow to display relative and absolute frequency tables, cumulative frequencies and bar charts/histogram for individual categorical variables. For continuous variables, it is possible to compute the main descriptive statistics, including percentiles and shape indicators.
Descriptives provide the basic univariate descriptive statistics for quantitative variables. Variables can be ordered according to their means. Within this command, it is also possible to compute and save 'standardized' variables, variables rescaled to have a 0 mean and a unity variance (see appendix).
Explore is a command targeted to tabulation and graphs for a preliminary description of variables. It is possible to compute most of the descriptive statistics for separate sub-samples according to one or more categorical variable, to look at the distribution through frequency tables and percentiles and to display a number of charts, including those aimed at testing whether data follow a normal distribution.
Crosstabs display joint frequency tables for two or more categorical variables. This command also provides statistics to test for significant associations between the variables (see section 9.1). Clustered bar charts (see section 4.2.1) can be also obtained.
Ratio allows to compute more complex statistics for the ratio of two quantitative variables.
P-P Plot is a graphical tool to check whether a variable follows a given theoretical distribution. When the empirical data fit the theoretical distribution (and many options are available), the points follow a straight line.
Q-Q Plot is similar to P-P plots, but instead of comparing each point, this option plots the empirical quartiles against those of the theoretical distribution. Again, points around a straight line indicate conformity to the theoretical distribution.

Tables

Custom tables is probably the most flexible interface in SPSS to design bi-dimensional tables, choose the categorical variable(s) for rows and columns and a range of descriptive statistics defining the content of the cells. It is similar to pivot tables in Excel. It is also possible to define further grouping variables (layers), similarly to OLAP cubes.

(Continued)

BOX 4.2 *Cont'd*

Multiple response sets can be defined when a single attribute is measured through a set of binary or categorical variables. This is typical when respondents face multiple choices where they can tick more than an answer box. For example, in the Trust data-set, the question on the typical point of purchase for chicken is measured through a set of 8 binary dummies for each of the potential retailer (discount outlet, supermarket, etc.). SPSS allows to combine these separate variable into a combined set, for use in **Multiple response tables**, where sets can be cross-tabulated with other categorical variables. In the Trust example, one could create a table where relative or absolute frequencies of each point of purchase in different countries are displayed. Multiple response sets can also be used in General Tables.

---

anomalous, having a zero price. Instead, observations 163, 236 and 434 do not show apparent inconsistencies in terms of price, so they should be retained for subsequent analysis.

Summing up

After data collection, preparation of the data-set requires some *editing* steps and *post-editing* statistical processing. Editing involves checking for the data quality, by looking into non-response errors, questionnaire issues, completeness and consistency of responses. Decision on the organization of the database and variable coding are taken during this step, before actual transcription. Once the data are available in electronic form it is necessary to perform post-editing steps, such as missing data treatment and other preliminary variable processing steps. Further data quality checks are desirable. Once the data-set has gone through the editing and post-editing phases, it is time to start with the statistical analysis. However, a sensible choice is to start with the elemental data description, through frequencies, univariate descriptive statistics, plots and tables. By looking at the frequencies for one or more variable, it is possible to learn more about the distribution of the data across the sample, for nominal, ordinal or quantitative variables classified into groups. Descriptive statistics for quantitative variables add information about the central tendency (mean, median, mode, quartiles) and about data variability (variance, standard deviation, coefficient of variation). A visual grasp of the data is provided by a range of graphical tools, both for univariate exploration (for example histograms, pie charts) and multivariate plots (for example scatter-plots). By looking at charts and tables it becomes possible to hypothesize relationships between variables. These descriptive steps should be accompanied by a careful consideration of the potential bias linked to missing data and outliers. Missing data are dangerous when their occurrence is related to non-random causes, outliers could be the consequence of recording error and they should be deleted from the data-set when they affect the reliability of statistical analysis.

EXERCISES

1. Open the Trust SPSS data-set and perform the following operations:
   a. The variable q64 (country) is a nominal variable with five response categories, each corresponding to a country. Transform it into five dummy variables, one for each country, by using the COMPUTE command (you may find it easier by exploiting SPSS syntax
   b. Create a MULTIPLE RESPONSE SET (see SPSS Help and the DATA menu) for the dummy variables recording the point of purchase (q8a to q8d) and name it as POP
   c. Using the CUSTOM TABLE command under the ANALYZE/TABLE menu, create a frequency table to show the proportion of points of purchase across countries (POP by q64)
   d. Create a pie chart showing the proportion of points of purchase for the whole data-set.

Case ID	Math ability	Attitude to driving (1 to 9)	Car value ($)
John	Poor	3	12,000
Amanda	Medium	7	13,225
Ben	Poor	5	8,000
Frank	Excellent	8	25,000
Johanna	Excellent	6	16,450
Steve	Medium	2	5,000
Matilda	Poor	9	29,000
Patrick	Excellent	6	14,000
Claire	Excellent	8	19,000
Andrea	Poor	4	7,000

2. Create a new data-set in SPSS and name it Chap4.sav
   a. Insert the four variables (including CASE ID) and ten observations, shown in the table assigning properly TYPE, LABEL, VALUES, MISSING VALUES. WIDTH, DECIMALS and MEASURE
   b. Create a binned variable with three categories for car value
   c. Create a frequency table for the binned variable
   d. Create a table showing the average attitude to driving for each category of the car value, using two decimals
   e. Create a histogram showing the attitude to driving and car value for each level of math ability.
3. Open the EFS data-set in SPSS:
   a. Consider variable a094 (NS-SEC 8 Class of Household Reference Person)
   b. Create a frequency table also showing the percentage of missing values by type
   c. Create a Pareto chart showing the distribution of different types of missing values
   d. Create a table with the mean of p600 (EFS-Total Consumption Expenditure) for the various categories of missing values plus a single category of non-missing values. Do you presume that the missing values may be biasing the sample?
   e. Perform the outliers checks described in chapter 4

 f. Compute the average of p600 for the whole sample

 g. Weight the observation using weight a (Annual Weight). Compute the sample mean of p600 again. Does it change?

Further readings and web-links

❖ **Missing data and statistical imputation** – The art of imputing missing values is full of risks, but the consequences of ignoring the presence of missing data might be as bad as mishandling them. A classic (but advanced) book discussing statistics for missing data is Little and Rubin (2002). Statistical imputation is generally controlled by the software, for a review see Horton and Lispitz (2001).

❖ **Outliers** – The detection and impact of outliers generally depends on the statistical technique being used. Outliers may have different outcomes with different techniques and specific methods have their own way to deal with them. A recent review is provided by Hodge and Austin (2004). Clark (1989) reviews the issue with reference to marketing research.

❖ **Descriptive statistics, tables and charts in SPSS** – The Department of Statistics of the Texas A&M University has an excellent web-page with tutorials on using a range of SPSS tools, illustrated through short and effective movies and based on SPSS sample data-sets. The link is www.stat.tamu.edu/spss.php

❖ **SPSS syntax** – Syntax written by others may help sorting out very quickly SPSS problems … for example see Raynald Levesque's web-page (www.spsstools.net) which contains more than 400 programs.

Hints for more advanced studies

☞ When is deletion of missing data preferable to imputation? When should one choose list-wise deletion and when is pair-wise deletion preferable?

☞ Statistical quality control to support quality management is a major area where simple tabular or graphical tools can provide major support to the detection and correction of problems. Explore the main routes to statistical quality control, especially focusing on charts.

☞ When a questionnaire contains control questions to ensure the consistency of responses, it becomes necessary to develop measures for consistency and decide what to do when inconsistencies arise. How is internal consistency tested?

Notes

1. For definitions of specific charts, see 'A Dictionary of Statistics' by Upton and Cook (2006).
2. 'Outlier' as in Upton and Cook (2006).
3. The difference between the upper and lower quartile.

PART II

Sampling, Probability and Inference

The second part of the book looks into the probabilistic foundation of statistical analysis, which originates in probabilistic sampling, and introduces the reader to the arena of hypothesis testing.

Chapter 5 explores the main random and controllable source of error, sampling, as opposed to non-sampling errors, potentially very dangerous and unknown. It shows the statistical advantages of extracting samples using probabilistic rules, illustrating the main sampling techniques with an introduction to the concept of estimation associated with precision and accuracy. Non-probability techniques, which do not allow quantification of the sampling error, are also briefly reviewed. **Chapter 6** explains the principles of hypothesis testing based on probability theory and the sampling principles of previous chapter. It also explains how to compute confidence intervals and how statistics allow one to test hypotheses on one or two samples. **Chapter 7** extends the discussion to the case of more than two samples, through a class of techniques which goes under the name of analysis of variance. The principles are explained and with the aid of SPSS examples the chapter provides a quick introduction to advanced and complex designs under the broader general linear modeling approach.

CHAPTER 5

Sampling

T HIS CHAPTER provides an introduction to sampling theory and the sampling process. When research is conducted through a sample survey instead of analyzing the whole target population, it is unavoidable to commit an error. The overall survey error can be split into two components:

(a) the sampling error, due to the fact that only a sub-set of the reference population is interviewed; and
(b) the non-sampling error, due to other measurement errors and survey biases not associated with the sampling process, discussed in chapters 3 and 4.

With probability samples as those described in this chapter, it becomes possible to estimate the population characteristics and the sampling error at the same time (inference of the sample characteristics to the population). This chapter explores the main sampling techniques, the estimation methods and their precision and accuracy levels depending on the sample size. Non-probability techniques, which do not allow quantification of the sampling error, are also briefly reviewed.

Section 5.1   introduces the key concepts and principles of sampling
Section 5.2   discusses technical details and lists the main types of probability
              sampling
Section 5.3   lists the main types of non-probability samples

THREE LEARNING OUTCOMES

This chapter enables the reader to:

➡ Appreciate the potential of probability sampling in consumer data collection
➡ Get familiar with the main principles and types of probability samples
➡ Become aware of the key principles of statistical inference and probability

PRELIMINARY KNOWLEDGE: For a proper understanding of this chapter, familiarity with the key probability concepts reviewed in the appendix at the end of this book is essential.

This chapter also exploits some mathematical notation. Again, a good reading of the same appendix facilitates understanding.

5.1 To sample or not to sample

It is usually unfeasible, for economic or practical reasons, to measure the characteristics of a population by collecting data on all of its members, as *censuses* aim to do. As a matter of fact, even censuses are unlikely to be a complete survey of the target population, either because it is impossible to have a complete and up-to-date list of all of the population elements or due to *non-response errors,* because of the failure to reach some of the respondents or the actual refusal to co-operate to the survey (see chapter 3).

In most situations, researchers try to obtain the desired data by surveying a sub-set, or *sample,* of the population. Hopefully, this should allow one to generalize the characteristics observed in the sample to the entire target population, inevitably accepting some margin of error which depends on a wide range of factors. However, generalization to the whole population is not always possible – or worse – it may be misleading.

The key characteristic of a sample allowing generalization is its probabilistic versus non-probabilistic nature. To appreciate the relevance of this distinction, consider the following example. A multiple retailer has the objective of estimating the average age of customers shopping through their on-line web-site, using a sample of 100 shopping visits. Three alternative sampling strategies are proposed by competing marketing research consultants:

1. (convenience sampling) The first 100 visitors are requested to state their age and the average age is computed. If a visitor returns to the web site more than once, subsequent visits are ignored.
2. (quota sampling) For ten consecutive days, ten visitors are requested to state their age. It is known that 70% of the retailer's customers spend more than £50. In order to include both light and heavy shoppers in the sample, the researchers ensures that expenditure for seven visits are below £50 and the remaining are above. The mean age will be a weighted average.
3. (simple random sampling) A sample of 100 customers is randomly extracted from the database of all registered users. The sampled customers are contacted by phone and asked to state their age.

The first method is the quickest and cheapest and the researcher promises to give the results in three days. The second method is slightly more expensive and time consuming, as it requires 10 days of work and allows a distinction between light and heavy shoppers. The third method is the most expensive, as it requires telephone calls.

However, only the latter method is probabilistic and allows inference on the population age, as the selection of the sampling units is based on random extraction.

Surveys 1 and 2 might be seriously biased. Consider the case in which daytime and weekday shoppers are younger (for example University students using on-line access in their academic premises), while older people with home on-line access just shop in the evenings or at the week-ends. Furthermore, heavy shoppers could be older than light shoppers.

In case one, let one suppose that the survey starts on Monday morning and by Tuesday lunchtime 100 visits are recorded, so that the sampling process is completed in less than 48 hours. However, the survey will exclude – for instance – all those that shop on-line over the week-end. Also, the sample will include two mornings and only one afternoon and one evening. If the week-end customers and the morning customers have different characteristics related to age, then the sample will be biased and the estimated age is likely to be lower than the actual one.

In case two, the alleged 'representativeness' of the sample is not guaranteed for similar reasons, unless the rule for extracting the visitors is stated as random. Let one suppose that the person in charge of recording the visits starts at 9 a.m. every day and (usually) by 1 p.m. has collected the age of the three heavy shoppers, and just three light shoppers. After 1 p.m. only light shoppers will be interviewed. Hence, all heavy shoppers will be interviewed in the mornings. While the proportion of heavy shoppers is respected, they're likely to be the younger ones (as they shop in the morning). Again, the estimated age will be lower than the actual one.

Of course, random selection does not exclude bad luck. Samples including the 100 youngest consumers or the 100 oldest ones are possible. However, given that the extraction is random (probabilistic), we know the likelihood of extracting those samples and we know – thanks to the normal distribution – that balanced samples are much more likely than extreme ones. In a nutshell, sampling error can be quantified in case three, but not in cases one and two.

The example introduces the first key classification of samples into two main categories – probability and non-probability samples. *Probability sampling* requires that each unit in the sampling frame is associated to a given probability of being included in the sample, which means that the probability of each potential sample is known. Prior knowledge on such probability values allows *statistical inference*, that is the generalization of sample statistics (*parameters*) to the target population, subject to a margin of uncertainty, or *sampling error*. In other words, through the probability laws it becomes possible to ascertain the extent to which the estimated characteristics of the sample reflect the true characteristics of the target population. The sampling error can be estimated and used to assess the precision and accuracy of sample estimates. While the sampling error does not cover the overall survey error as discussed in chapter 3, it still allows some control over it. A good survey plan allows one to minimize the non-sampling error without quantifying it and relying on probabilities and sampling theory opens the way to a quantitative assessment of the accuracy of sample estimates. When the sample is *non-probabilistic*, the selection of the sampling units might fall into the huge realm of subjectivity. While one may argue that expertise might lead to a better sample selection than chance, it is impossible to assess scientifically the ability to avoid the potential *biases* of a subjective (non-probability) choice, as shown in the above example.

However, it can not be ignored that the use of non-probability samples is quite common in marketing research, especially quota sampling (see section 5.3). This is a controversial point. Some authors correctly argue that in most cases the sampling error is much smaller than error from non-sampling sources (see the study by Assael and Keon, 1982) and that efforts (including the budget ones) should be rather concentrated on eliminating all potential sources of biases and containing non-responses (Lavrakas, 1996).

The only way to actually assess potential biases due to non-probability approaches consists of comparing their results from those obtained on the whole population, which is not a viable strategy. It is also true that quota sampling strategies such as those implemented by software for computer-assisted surveys (see chapter 3) usually guarantee minor violations of the purest probability assumptions. The example which opened this chapter could look exaggerated (more details are provided in section 5.3). However, this chapter aims to emphasize a need for coherence when using statistics. In the rest of this book, many more or less advanced statistical techniques are discussed. Almost invariably, these statistical methods are developed on the basis of probability assumptions, in most cases the normality of the data distribution. For example, the probability basis is central to the hypothesis testing, confidence intervals and ANOVA techniques described in chapters 6 and 7. There are techniques which relax the need for data obtained through probability methods, but this is actually the point. It is necessary to know why and how sample extraction is based on probability before deciding whether to accept alternative routes and shortcuts.[1]

5.1.1 *Variability, precision and accuracy: standard deviation versus standard error*

To achieve a clear understanding of the inference process, it is essential to highlight the difference between various sources of *sampling error*. When relying on a sub-set of the population – the sample – it is clear that measurements are affected by an error. The size of this error depends on three factors:

1. The *sampling fraction*, which is the ratio between the sample size and the population size. Clearly, estimates become more precise as the sample size grows closer to the population size. However, a key result of sampling theory is that the gain in precision marginally decreases as the sampling fraction increases. Thus it is not economically convenient to pursue increased precision by simply increasing the sample size as shown in detail in section 5.3.
2. The *data variability* in the population. If the target variable has a large dispersion around the mean, it is more likely that the computed sample statistics are distant from the true population mean, whereas if the variability is small even a small sample could return very precise statistics. Note that the concept of *precision* refers to the degree of variability and not to the distance from the true population value (which is *accuracy*). The population variability is measured by the population variance and standard deviation (see appendix to this chapter). Obviously these population parameters are usually unknown as their knowledge would require that the mean itself is known, which would make the sampling process irrelevant. Estimates of the population variability are obtained by computing variability statistics on the sample data, such as the *sample standard deviation* and *sample variance*.
3. Finally, the success of the sampling process depends on the *precision of the sample estimators*. This is appraised by variability measures for the sample statistics and should not be confused with the sample variance and standard deviation. In fact the objective of measurement is not the data variability any more, but rather the variability of the *estimator* (the sample statistic) intended as a random variable distributed around the true population parameter across the sample space. For example, if the researcher is interested in the population mean value and a mean value is computed on sample data, the precision of such estimate can be evaluated through the *variance of the mean* or its square root – the *standard error of the mean*.

The distinction between standard deviation and standard error should be apparent if we think that a researcher could estimate the population *standard deviation* on a sample and the measure of the accuracy of the *sample standard deviation* will be provided by a statistic called *standard error of the standard deviation*.

Note that a *precise* sampling estimator is not necessarily an *accurate* one, although the two concepts are related. Accuracy measures closeness to the true population value, while precision refers to the variability of the estimator. For example, a sample mean estimator is more accurate than another when its estimated mean is closer to the true population mean, while it is more precise if its standard error is smaller. Accuracy is discussed in section 5.2.2.

5.1.2   The key terms of sampling theory

Let one refer to a sub-set or *sample* of n observations (*sample size*), extracted from a *target population* of N elements. If we knew the true population value and the probability of extraction associated to each of the N elements of the target population (the *sampling method*), given the sample size n, we could define all potential samples, that is, all potential outcomes of the sampling process. Hence we could derive the exact distribution of any sample statistic around the true population value; the *sampling distribution* would be known exactly. To make a trivial example, consider the case where $N = 3$ and $n = 2$, where A, B and C are population units. There are three potential samples (A,B), (B,C) and (C,A). Together they constitute the *sampling space*.

Clearly enough, extracting all potential samples would be a useless exercise, given that the researcher could directly compute the true value of the population. So, rather than solving the above *direct problem*, the statistician is interested in methods to solve the *indirect problem*, which means that:

(a)   only one sample is extracted;
(b)   only the *sample statistics* are known;
(c)   a *sampling distribution* can be ascertained on the basis of the *sampling method*, this is the so-called specification problem; and
(d)   estimates of the true values of the desired statistics within the target population are obtained from the sample statistics through *statistical inference*.

5.2   Probability sampling

Before exploring specific probabilistic sampling methods and inference, it is useful to introduce some basic notation (beyond the basic mathematical notation of the appendix) to simplify discussion. If we define X_i as the value assumed by the variable X for the i-th member of the population, a *population* of N elements can be defined as:

$$P = (X_1, X_2, \ldots, X_i, \ldots, X_N)$$

whereas if one refers to a specific sample i of n elements extracted from P, this can be identified by referring to the population subscripts

$$S = (x_{i_1}, x_{i_2}, \ldots, x_{i_j}, \ldots, x_{i_n})$$

where i_j indicates the population unit which is included as the j-th element of the sample i, for example $i_1 = 7$ means that the first observation in the sample corresponds to the 7th element of the population (or that $x_{i_1} = X_7$).

Sampling is needed to infer knowledge of some *parameters* of the population P, usually the population mean, the variability and possibly other distribution features (shape of the distribution, symmetry, etc.). As these are unknown, estimates are obtained through the corresponding *sample statistics*.[2] In the rest of the discussion it

---

BOX 5.1 *Sample statistics and inference in simple random sampling*

Population parameters	Sample statistics

$$\mu = \frac{1}{N}\sum_{i=1}^{N} X_i \text{ (mean)}$$

$$\bar{x} = \frac{1}{n}\sum_{j=1}^{n} x_j \text{ (sample mean)}$$

$$\sigma = \sqrt{\frac{\sum_{i=1}^{N}(X_i - \mu)^2}{N}} \text{ (standard deviation)}$$

$$s = \sqrt{\frac{\sum_{i=1}^{n}(x_i - \bar{x})^2}{n-1}} \text{ (sample standard deviation)}$$

$$\sigma^2 = \frac{\sum_{i=1}^{N}(X_i - \mu)^2}{N} \text{ (variance)}$$

$$s^2 = \frac{\sum_{i=1}^{n}(x_i - \bar{x})^2}{n-1} \text{ (sample variance)}$$

---

will be assumed that variables in capital letters (Greek letter for statistics) refer to the population, while small letters are used for sample observations and statistics. Box 5.1 summarizes the equations for computing the basic population parameters and sample statistics, mean, standard deviation and variance. The population statistics are the usual central tendency and variability measures detailed in the appendix and discussed in chapter 4. The sample statistics are obviously quite similar (apart from the notation).

There are obvious similarities in the equations computing the population parameters (which are the usual descriptive statistics illustrated in the appendix) and the sample statistics. There is also a major difference for the variability measures, as the population standard deviation and variance have the number of population elements N in the denominator N, while for the sample statistics the denominator is $n − 1$, which ensures that the estimator is *unbiased*. This means that the expected value of the estimator[3] is equal to the actual population parameter. This is a desirable property, together with *efficiency* and *consistency*. Efficiency refers to precision and means that the estimator is the most precise (has the lowest variance) among all unbiased estimators for the same population parameters. Consistency refers to accuracy and means that as the sample size becomes very large (tends to infinity), the probability that there is a difference between the estimator value and the actual population parameter tends to zero.

5.2.1 Basic probabilistic sampling methods

Since the only requirement for a probabilistic sampling method is that the probability of extraction of a population unit is known, it is clear that the potential range of sampling methods is extremely large. However, for many surveys it will be sufficient to employ one of the elemental methods, or a combination of two or more of them. Box 5.2 summarizes the key characteristics of the sampling methods discussed in this chapter. Those interested in the equations for estimating population statistics and their precisions are referred to the appendix to this chapter.

The methods listed in box 5.2 guarantee that extraction is probabilistic, hence inference from the sample statistics to the population. However, while inference from a simple random sampling is quite straightforward, more complex sampling method like stratified and cluster sampling require that probabilities of extraction and sample allocation rules are taken into account to allow correct generalization of the sample statistics to the population parameters.

1. **Simple Random Sampling** has the desirable characteristic of being based on a straightforward random extraction of units from the population. It does not require any information on the population units but on their list or *sampling frame*. When this is available, the researcher's only issue is to guarantee

---

BOX 5.2 *Probabilistic sampling methods*

	Description	Notation
Simple Random Sampling	Each population unit has the same probability of being in the sample. It follows that all potential samples of a given size n have the same probability of being extracted.	N population size; n sample size; $\pi_i = n/N$ probability of extraction for each population unit i
Systematic sampling	The population (sampling frame) units are sorted from 1 to N according to some characteristic, only the first sampling unit is extracted randomly, while other units are extracted systematically following a step k. The *sampling step k* is given by the ratio N/n and must be an integer. The starting unit r is selected randomly among the first k unit of the sample. Then, this method requires the systematic extraction of every k-th unit thereafter, i.e. units $r+k$, $r+2k$, $r+3k$, ..., $r+(n-1)k$. Note that the sorting order should be related to the target characteristic, while cyclical sorting should be avoided.	$k = N/n$ sampling step r starting unit $\pi_i = n/N = 1/k$ probability of extraction for each population unit i
Stratified Sampling	The population is subdivided into S complementary and exhaustive groups (*strata*) with respect to some relevant population characteristic, so that units are expected to be relatively similar (homogeneous) within each stratum and different (heterogeneous) across different strata, with respect to the variable being studied. Random extraction (as in simple random sampling) occurs within each stratum, so that adequate representativeness is ensured for all the identified population sub-groups. The sample size allocation across strata can be either proportional to the stratum size or follow some alternative allocation criterion aimed to increase precision, such as *Neyman allocation*	S number of strata N_j number of population units in stratum j n_j sample size within stratum j $\pi_{ij} = n_j/N_j$ probability of extraction for each population unit i within stratum j $N = \sum_{j=1}^{S} N_j$ $n = \sum_{j=1}^{S} n_j$

BOX 5.2 *Cont'd*

	Description	Notation
Cluster Sampling (Area Sampling)	When a population is naturally or artificially subdivided into G complementary and exhaustive groups *(clusters)* and reasonable homogeneity across these clusters and heterogeneity within each of them can be assumed, this method is based on the random selection of clusters. In one-stage cluster sampling, all the population units belonging to the sampled clusters are included in the sample. In two-stages cluster sampling, some form of probability sampling is applied within each of the sampled clusters. This method is especially useful (and economically convenient) when the population is subdivided in geographic areas *(Area sampling)*, so that a sample of areas (for example regions) is extracted in the first stage, and simple random sampling is performed within each area.	G number of clusters g number of sampled clusters N_k number of population units in cluster k n_k sample size within cluster k **One-stage cluster sampling** $(n_k = N_k)$ $\pi_{ik} = g/G$ probability of extraction for each population unit i within cluster k **Two-stage cluster sampling** $(n_k < N_k)$ $\pi_{ik} = (g/G)(n_k/N_k)$ probability of extraction for each population unit i within cluster k $$N = \sum_{j=1}^{G} N_j \quad n = \sum_{j=1}^{G} n_j$$

perfect randomness. Sample statistics such as mean values, standard deviations and variances, can be computed by applying the equations listed in box 5.1. However, some limitations of this basic probability method should be pointed out. First, ignoring any potentially available additional information on the population units reduces the precision of estimates. Among the class of probability methods, simple random sampling is the least *efficient* method, as for a given sample size its sample statistics are those with the highest standard errors (that is, the lowest precision).

2. **Systematic Sampling** has very similar characteristics to simple random sampling and insofar as the units are listed randomly in the sampling frame, it only represents an extraction method for simple random sampling. This method becomes more interesting when the sorting order of the population units in the sampling frame is expected to be related to the target characteristic. The stricter is this relation, the higher is the gain in efficiency. Ideally, the sampling frame contains at least one known variable which is related to the target one and can be used to sort the population units. Let one refer to the example of section 5.1. Suppose that the database of registered users includes information on the time of registration. As explained, one may expect that the age of the customers is related to the time of the day and weekday they shop, so it is also likely that it is related to the registration information. If customers are sorted according to the time of registration, systematic sampling will guarantee that customers registering throughout the day are included in the sampling, hence increasing the probability that all potential ages will be represented in the sample with a gain in representativeness and precision. However, sorting the population units also involves some risks, as if the sorting order is cyclical the consequence could be a serious bias. Let's say that the customers are sorted out by weekday

and time of registration, so that the sampling frame starts with customers registering on a Monday morning, then goes on with other Monday's registrants until the evening, then switches to Tuesday morning customers and so on. An inappropriate sampling step could lead to the inclusion in the sample of all morning registrants skipping all those who registered in afternoons or evenings. Clearly if the assumption that age is correlated to the time of registration is true, the application of systematic sampling would lead to the extraction of a seriously biased sample. In synthesis, systematic sampling should be preferred to simple random sampling when a sorting variable related to the target variable is available, taking care to avoid any circuity in the sorting strategy.

A desirable feature of systematic sampling is that it can also be used without requiring a sampling frame, or more precisely, the sampling frame is built while doing systematic sampling. For example, this is the case when only one in every ten customers at a supermarket till is stopped for interview. The sample is built without knowing the list of customers on that day and the actual population size N is known only when the sampling process is terminated.

3. **Stratified sampling** is potentially the most efficient among elemental sampling strategies. When the population can be easily divided into sub-groups (*strata*), random selection of sampling units within each of the sub-groups can lead to major gains in precision. The rationale behind this sampling process is that the target characteristic shows less variability within each stratum, as it is related to the stratification variable(s) and varies with it. Thus, by extracting the sample units from different strata, representativeness is increased. It is essential to distinguish stratified sampling from the non-probability quota sampling discussed in the example of section 5.1 and in section 5.4. Stratified sampling requires the actual subdivision of the sampling frame into subpopulations, so that random extraction is ensured independently within each of the strata. While this leads to an increase in costs it safeguards inference. Instead, quota sampling does not require any sampling frame and only presumes knowledge of the proportional allocation of the sub-populations. Even with a random quota sampling, the sampling units are extracted from the overall population subject to the rule to maintain the prescribed percentages of units coming from the identified sub-populations. Referring to the example, in order to implement a stratified sampling design, the researcher should:

   (1) associate the (average) amount spent to each of the customers in the database;
   (2) sub-divide the sampling frame into two sub-groups, one with those customers who spend more than £50, the other with those than spend less;
   (3) decide on the size of each of the strata (see box 5.3; for example 70 light shoppers and 30 heavy shoppers if the population proportions are maintained);
   (4) extract randomly and independently from the two sub-populations; and
   (5) compute the sample mean through a weighted average.

   Besides increasing representativeness, stratified random sampling is particularly useful if separate estimates are needed for each sub-domain, for instance estimating the average on-line shopper age separately for heavy and light shoppers.

4. **Cluster sampling** is one of the most employed elemental methods and is often a component of more complex methods. Its most desirable feature is that it does not necessarily require a list of population units, and it is hence applicable with

BOX 5.3 *Stratum sample size in stratified sampling*

An issue that emerges from the choice of stratifying a sample is whether the allocation of the sample size into the strata should be proportional to the allocation of the target population. This has major consequences on the precision of the estimates and the representativeness of sub-domain estimates. While sometimes it may be convenient to assume the same sample size for each of the sub-domains (for example, when the objective is comparing means across sub-domains), in most situations the choice is between *proportional* and *optimal* allocation. Proportional allocation simply requires that the sample size allocation corresponds to the distribution of population units across the sub-domains:

$$n_j = N_j \cdot \frac{n}{N}$$

In this case the sample becomes self-weighting, as the selected units can be regarded as a unique sample independently from the existence of data and population parameters can be estimated without the need for applying weights. Proportional allocation greatly simplifies the estimation process, provided that the response rate is the same across the sub-domains. However, non-response rates may greatly differ for some stratification variables (for example income), so that the computational gain disappears. Maintaining proportionality does not necessarily lead to the highest degree of precision, especially when the variability of the target variable is very different across strata. In such case, the allocation process should be based on information on the variability of the sub-domains, so that larger sample sizes are allocated to the strata with higher variability and smaller sample sizes to those with lower variation:

$$n_j = n \cdot \frac{N_j \sigma_{X_j}}{\sum\limits_{j=1}^{s} N_j \sigma_{X_j}}$$

where σ_{X_j} is the standard deviation of the target variable within the j stratum. However, it is unlikely that such information is available to the researcher prior to the survey and it is usually derived from previous or pilot studies. The optimal allocation described above (also known as Neyman's allocation) ignores any additional information on sampling cost, which may differ across strata. Nevertheless, it is valid whenever the unit sampling cost is the same for all observations from any stratum. In some circumstances, sampling costs are different across strata. For instance, it may be more expensive to reach individuals in rural areas. More general equations can be obtained to maximize precision assuming that total cost C is a constraint (the most frequent case) and C_j is the unit sampling cost within stratum j (see Levy and Lemeshow, 1999, p. 161):

$$n_j = C \cdot \frac{N_j \sigma_{X_j} / \sqrt{C_j}}{\sum\limits_{j=1}^{s} \left(N_j \sigma_{X_j} / \sqrt{C_j} \right)}$$

In a rare and ideal situation where the budget is not a problem and the objective is simply to maximize precision, for a given sample size n, the allocation is obtained as follows:

$$n_j = n \cdot \frac{N_j \sigma_{X_j} / \sqrt{C_j}}{\sum\limits_{j=1}^{s} \left(N_j \sigma_{X_j} \sqrt{C_j} \right)}$$

More complex optimizations take into account multivariate sampling designs, which means optimizing allocation on a set variables rather than only one. These have been the subject of research (see Solomon and Zacks, 1970 or Bethel, 1989). When there is some flexibility on the budget and precision targets, complex optimization algorithms help choosing the best balance after weighting the relevance of costs versus precision through a system of weights. See for example Gentle (2006, ch. 10 and references therein) or Narula and Wellington (2002).

more convenient survey methods (as mall-intercepts) and to geographical (area) sampling. In many situations, while it is impossible or economically unfeasible to obtain a list of each individual sampling unit, it is relatively easy to define groups (*clusters*) of sampling units.

To avoid confusion with stratified sampling, it is useful to establish the main difference between the two strategies. While stratified sampling aims to maximize homogeneity within each stratum, cluster sampling is most effective when heterogeneity within each cluster is high and the clusters are relatively similar.

As an example, consider a survey targeted to measure some characteristics of cinema audience for a specific movie shown in London. *A priori*, it is not possible to distinguish between those that will go and see that specific film. However, the researcher has information on the list of cinemas that will show the movie. A simple random sampling extraction can be carried out from the list to select a small number of cinemas, then an exhaustive survey of the audience on a given week can be carried out. Cluster sampling can be articulated in more stages. For instance, if the above study had to be targeted on the whole UK population, in a first stage a number of cities could be extracted and in a second stage a number of cinemas could be extracted within each city.

While the feasibility and convenience of this method is obvious, the results are not necessarily satisfactory in terms of precision. Cluster sampling produces relatively large standard errors especially when the sampling units within each cluster are homogeneous with respect to some characteristic. For example, given differences in ticket prices and cinema locations around the city, it is likely that the audience is relatively homogeneous in terms of socio-economic status. This has two consequences – first, it is necessary that the number of selected clusters is high enough to avoid a selection bias and second, if there is low variability within a cluster, an exhaustive sampling (interviewing the whole audience) is likely to be excessive and unnecessarily expensive. Hence, the trade-off to be considered is between the benefits from limiting the survey to a small number of clusters and the additional costs due to the excessive sampling rate. In cases where a sampling frame can actually be obtained for the selected clusters, a convenient strategy is to apply simple random sampling or systematic sampling within the clusters.

5.2.2 Precision and accuracy of estimators

Given the many sampling options discussed in the previous sections, researchers need to face many alternative choices, not to mention the frequent use of mixed and complex methods, as discussed later. In most case, the decision is made easy by budget constraints. There are many trade-offs to be considered. For instance, random sampling is the least efficient method among the probability ones, as it requires a larger sample size to achieve the same precision levels of, say, stratified sampling. However, stratification is a costly process in itself, as the population units need to be classified according to the stratification variable.

As mentioned, efficiency (precision) is measured through the *standard error*, as shown in the appendix. In general, the standard error increases with higher population variances and decreases with higher sample sizes. However, as sample size increases, the relative gain in efficiency becomes smaller. This has some interesting and not intuitive implications. As shown below, it is not convenient to go above a certain sample size, because while the precision gain decreases the additional cost does not.

Standard errors also increases with population size. However, the required sampling size to attain a given level of precision does not increase proportionally with population size. Hence, as shown in the example of box 5.4, the target sample size does not vary much between a relatively small population and a very large one.

A concept that might be useful when assessing the performance of estimators is *relative sampling accuracy* (or *relative sampling error*), that is a measure of accuracy for the estimated mean (or proportion) at a given level of confidence. If one considers simple random sampling, relative accuracy can be computed by assuming a normal distribution for the mean estimate. The *significance level* α (further discussed in chapter 6) is the probability that the relative difference between the estimated sample mean and the population mean is larger than the relative accuracy level r:

$$\Pr\left(\left|\frac{\bar{x} - \bar{X}}{\bar{X}}\right| < r\right) = 1 - \alpha$$

where $\bar{X}$ is the population mean and $1-\alpha$ is the level of confidence.[4] The above equation means that the probability that the difference between the sample mean and the population mean (in percentage terms) is smaller than the fixed threshold r is equal to the level of confidence. Higher levels of confidence (lower levels of α) imply less accuracy (a higher r), which means that there is a trade-off between the confidence level and the target accuracy level. Thus, one first fixes α at a relatively small level, usually at 0.05 or 0.01, which means a confidence level of 95% or 99%, respectively. Then it becomes possible to determine the relative accuracy level r as a function of the population and sample size of the standard error and to a coefficient which depends on the value chosen for α.

The exact equation to compute the relative accuracy level is (Cochran, 1977):

$$r = \pm\frac{t_{\alpha/2}S_x}{\sqrt{n}\bar{X}}\sqrt{1 - \frac{n}{N}}$$

where $t_{\alpha/2}$ is a fixed coefficient which depends from the chosen confidence level α and from the size of the sample. The meaning of $t_{\alpha/2}$ will be made clearer with the discussion of confidence intervals and hypothesis testing in chapter 6; here it may suffice to notice that the value of $t_{\alpha/2}$ (which is fixed and tabulated) becomes larger as the level of confidence increases (that is α becomes smaller). Thus, if one wants higher confidence on the estimate, then a larger r must be accepted, which means accepting a lower accuracy level. The above equation also shows that when the sampling fraction n/N becomes larger (that is the sample becomes larger compared to the population size), then the level of accuracy is higher. Relative accuracy is directly related to the standard deviation.

When a proportion of cases p, rather than a mean is concerned, the equation becomes.

$$r = \pm t_{\alpha/2}\sqrt{\frac{1-p}{p \cdot n}\left(1 - \frac{n}{N}\right)}$$

which can be interpreted exactly as the equation for the mean.

5.2.3 Deciding the sample size

When it is not a straightforward consequence of budget constraints, sample size is determined by several factors. It may depend on the precision level, especially when the research findings are expected to influence major decisions and there is little error tolerance. While box 5.4 shows that a size of 500 could do for almost any population size, another driving factor is the need for statistics on sub-groups of the population, which leads to an increase in the overall sample size. Also, if the survey topic is liable to be affected by non-response issues, it might be advisable to have a larger initial sample size, although non-response should be treated with care as discussed in chapters 3 and 4. Sometimes, the real constraint is time. Some sampling strategies can be implemented quickly, others are less immediate, especially when it becomes necessary to add information to the sampling frame.

Since the precision of estimators depends on the sampling design, the determination of sample size also requires different equations according to the chosen sampling method. Furthermore, since the sampling accuracy is based on a probability design and distributional assumption, it is necessary to set *confidence levels*, (introduced in

---

BOX 5.4   *SRS: Relative accuracy of a mean estimator, population and sample sizes*

The following table provides an useful comparison of relative accuracies with varying sample and population sizes, for a fixed population mean and standard deviation. Consider simple random sampling to measure average monthly household expenditure on pizza. Let's fix the 'true' population mean at $20 and the population standard deviation at 9 and the confidence level at $\alpha = 0.05$.

Sample size	Population size						
	100	1000	2000	5000	100,000	1,000,000	100,000,000
30	11.75%	16.28%	16.53%	16.78%	16.77%	16.78%	16.78%
50	6.39%	12.14%	12.46%	12.65%	12.78%	12.78%	12.78%
100	0.00%	8.04%	8.48%	8.75%	8.92%	8.93%	8.93%
200		5.02%	5.65%	6.02%	6.26%	6.27%	6.27%
500		1.98%	2.97%	3.56%	3.93%	3.95%	3.95%
1000		0.00%	1.40%	2.23%	2.76%	2.79%	2.79%
2000			0.00%	1.18%	1.93%	1.97%	1.97%

With a sample of 30 units on a population of 100 units, we commit an 11.75% relative error (in excess or in deficiency), i.e. about $2.35. With the same sample and a population of 1000 units, the error only rises to 16.28% ($3.26). And even with a population of 100 millions of units, the error is basically unchanged at 16.78% (£3.36). These are, however, quite big relative errors. Let's say that we need to stay under 5%. With a sample size of 200, we achieve a 5.02% error on a population of 1000, while for all other population sizes, the error is slightly higher.
With a sample size of 500, we commit a 1.98% error on a population of 1000 (but we have sampled half of the population units!), an error 2.97% with 2000 units, then we manage to stay under 4% with any population size between 5000 and 100,000,000. For any sample size, there is virtually no difference whether the population is made by 100 thousands units or 100 millions units. In marketing research, for estimating sample means, a sample of 500 is usually more than acceptable, whatever the population size.

previous section and further explained in section 6.1), which should not be confused with precision measures. The latter can be controlled scientifically on the basis of the method and the assumptions. But a high level of precision does not rule out very bad luck, as extracting samples randomly allow (with a lower probability) for extreme samples which provide inaccurate statistics. The confidence level specifies the risk level that the researcher is willing to take, as it is the probability value associated with a *confidence interval*, which is the range of values which will include the population value with a probability equal to the confidence level (see chapter 6). Clearly, the width of the confidence interval and the confidence level are positively related. If one wants a very precise estimate (a small confidence interval), the lower is the probability to find the true population value within that bracket (a lower confidence). In order to get precise estimates with a high confidence level, it is necessary to increase sample size. As discussed for accuracy, the confidence level is a probability measure which ranges between 0 and 1, usually denoted with $(1-\alpha)$, where α is the *significance level*, commonly fixed to 0.05, which means confidence level of 0.95 (or 95%), or to 0.01 (99%) when risk aversion toward errors is higher.

Another non-trivial issue faced when determining the sample size is the knowledge of one or more population parameters, generally the population variance. It is very unlikely that such value is known beforehand, unless measured in previous studies. The common practice is to get some preliminary estimates through a pilot study, through a mathematical model (see for example Deming, 1990, chapter 14) or even make some arbitrary (and conservative) assumption. Note that even in the situation where the population variance is underestimated (and hence the sample size is lower than required), the sample statistics will still be unbiased although less efficient than expected.

In synthesis, the target sample size increases with the desired accuracy level and is higher for larger population variances and for higher confidence levels. In simple random sampling, the equation to determine the sample size for a given level of accuracy is the following:

$$n = \left(\frac{t_{\alpha/2}\sigma}{r\mu} \right)^2$$

As before, $t_{\alpha/2}$ is a fixed value (the t statistic) which depends on the chosen confidence level $1-\alpha$. If the size is larger than 50, the fixed value $t_{\alpha/2}$ can be replaced with the one taken from normal approximation ($z_{\alpha/2}$), σ is the known *standard deviation* of the population (usually substituted by an estimate s), r is the *relative accuracy* level defined above and μ is the *population mean*. Since the latter is also usually unknown (very likely, since most of the time it is the objective of the survey), it is also estimated through a pilot study or a conservative guess can be used.

Equations for determining sample size as a function of population parameters and target accuracy vary with the sampling method. For example, with stratified sampling the sample size for a pre-determined accuracy r is given by:

$$n = \left[\left(\sqrt{N} \frac{r\mu}{z_{\alpha/2} \sum_{j=1}^{S} \sqrt{N_j}s_j} \right)^2 + \frac{1}{N} \right]^{-1}$$

5.2.4 Post-stratification

A typical obstacle to the implementation of stratified sampling is the unavailability of a sampling frame for each of the identified strata, which implies the knowledge of the stratification variable(s) for all the population units. In such a circumstance it may be useful to proceed through simple random sampling and exploit the stratified estimator once the sample has been extracted, which increases efficiency. All that is required is the knowledge of the stratum sizes in the population and that such post-stratum sizes are sufficiently large. The advantage of post-stratifications is two-fold:

- It allows to correct the potential bias due to insufficient coverage of the survey (incomplete sampling frame); and
- It allows to correct the bias due to missing responses, provided that the post-stratification variable is related both to the target variable and to the cause of non-response

Post-stratification is carried out by extracting a simple random sample of size n, and then units are classified into strata. Instead of the usual SRS mean, a post-stratified estimator is computed by weighting the means of the sub-groups by the size of each sub-group. The procedure is identical to the one of stratified sampling and the only difference is that the allocation into strata is made ex-post. The gain in precision is related to the sample size in each stratum and (inversely) to the difference between the sample weights and the population ones (the complex equation for the unbiased estimator for the standard error of the mean is provided in the appendix to this chapter). The standard error for the post-stratified mean estimator is larger than the stratified sampling one, because additional variability is given by the fact that the sample stratum sizes are themselves the outcome of a random process.

5.2.5 Sampling solutions to non-sampling errors

A review of the non-sampling sources of errors is provided in chapter 3, but it may be useful to discuss here some methods to reduce the biases of an actual sample which is smaller than the planned one or less representative than expected. This is generally due to two sources of errors – *non-response errors*, when some of the sampling units cannot be reached, are unable or unwilling to answer to the survey questions and *incomplete sampling frames*, when the sampling frame does not fully cover the target population (*coverage error*). In estimating a sample mean, the size of these errors is

(a) directly proportional to the discrepancy between the mean value for the actually sampled units and the (usually unknown) mean value for those units that could not be sampled and
(b) inversely proportional to the proportion of non-respondents on the total sample (or to the proportion of those that are not included the sampling frame on the total population size).

The post-stratification method is one solution to non-response errors, especially useful when a specific stratum is under-represented in the sample. For example, a telephone survey might be biased by an under-representation of those people who spend less time at home, like commuters and those who often travel for work as compared to pensioners and housewives. The relevance of this bias can be high if this characteristic is related to the target variable. Considering the example of box 5.5, if the target variable is the level of satisfaction with railways, this group is likely to be very important. If the proportion of commuters on the target population is known (or estimated in a separate survey), post-stratification constitutes a *non-response* correction as it gives the appropriate weight to the under-represented subgroup.

Similarly, post-stratification may help in situations where the problem lies in the sampling frame. If the phone book is used as a sampling frame, household in some rural areas where not all households have a phone line will be under-represented. Another solution is the use of dual frame surveys, specifically two parallel sampling frames (for example white pages and electoral register) and consequently two different survey methods (personal interviews for those sampled through the electoral register). Other methods (see Thompson, 1997) consist in extracting a further random sample out of the list of non-respondents (to get representative estimates for the group of non-respondents), in using a *regression estimator* or a *ratio estimator* or in weighting sample units through same demographic or socio-economic variables measured in the population. This latter method is similar to post-stratification, but not identical as unlike post-stratification it does not require homogeneity within each stratum and heterogeneity across strata.

5.2.6 Complex probabilistic sampling methods

The probability methods discussed in this chapter can be combined and developed into more complex sampling strategies, aimed at increasing efficiency or reducing cost through more practical targeting.

A commonly used approach in sampling household is *two-stage sampling*. Most of the household budget surveys in Europe are based on this method. Two-stage methods imply the use of two different sampling units, where the second-stage sampling units are a sub-set of the first-stage ones. Typically, a sample of cities or municipalities is extracted in the first stage, while in the second stage the actual sample of households is extracted out of the first-stage units. Any probability design can be applied within each stage. For example, municipalities can be stratified according to their populations in the first stage, to ensure that the sample will include small and rural towns as well as large cities, while in the second stage one could apply *area sampling*, a particular type of cluster sampling where:

(1)   each sampled municipality is subdivided into blocks on a map through geographical co-ordinates (or post codes);
(2)   blocks are extracted through simple random sampling; and
(3)   all households in a block are interviewed.

Clearly, if the survey method is personal interview, this sampling strategy minimizes costs as the interviewers will be able to cover many sampling units in a small area.

As complex methods are subject to a number of adjustments and options, the sample statistics can become very complex themselves. Box 5.6 below brings real examples of complex sampling design in consumer research, while box 5.7 illustrates how samples can be extracted from sampling frames in SPSS and SAS.

BOX 5.5   *Sample size determination: an example*

Consider a survey whose target is to estimate the level of satisfaction of travelers on the train service between London and Reading, on a population of 13,500 season ticket holders. The target variable is measured on a 1–7 Likert scale, the required precision level is ±10% ($r = 0.1$) and the required confidence level is 95% ($\alpha = 0.05$). First, we need an estimate (or a conservative guess) of the population mean and standard deviation. If we want a conservative estimate, we may choose the population average and standard deviation that maximize the ratio (σ/μ). The average μ must be included between 1 and 7 and the *maximum* standard deviation depends on the distribution we expect from the variable (Deming, 1990, ch. 14). If we assume a **normal distribution** around the mean (where the mean is 4), we can estimate that the standard deviation is about 1.15, in fact:

- within the normal distribution that 99% of values are between $4 - 2.6\sigma$ and $4 + 2.6\sigma$
- 100% of values are between 1 and 7
- With a good degree of approximation we can assume that $4 - 2.6\sigma = 1$ and $4 + 2.6\sigma = 7$, which means that $\sigma = $ **3/2.6 = 1.15**
- An alternative rule of thumb from statistical literature tells us that when the distribution is normal, then $\sigma = $ **(max–min)/6 = 6/6 = 1**

Other distributions are possible, though. The **binomial distribution** assumes that the data are concentrated in the two extreme values and the maximum variability is given in the case where the respondents are exactly split in two halves, that is half of the respondents say 1 and the other half say 7. The average is still 4 and the standard deviation is given by $\sigma = $ **(max–min)/2 = 6/2 = 3**
The **rectangular distribution** assumes that the values are equally distributed across the various options (1/6 of the respondents say 1, 1/6 say 2, 1/6 say 3, etc.). Again the average is 4, but the standard error is given by $\sigma = $ **(max–min)** × **0.29 = 6 × 0.29 = 1.74**
   All other potential distributions have a standard error smaller than in the case of the **binomial distribution**. To summarize: if we assume a normal distribution, then $\sigma = $ **1.15**. If we assume a binomial distribution, then $\sigma = $ **3**. The last assumption we need is related to the level of confidence α. This is different from the precision level and depends on the fact that we could still get very peculiar samples. The lower is α, the less likely is that our results are biased by the extraction of very unlikely samples. Usually one sets $\alpha = 0.05$.
   Now we have all elements to estimate our sample size. The most conservative estimate will be based on the binomial distribution. One remaining issue regards the choice between the *t* (student) distribution and the *z* one. Initially, one can be very conservative and assume that the sample is going to include around 20 units. Hence, with $\alpha = 0.05$, we use the statistic $t_{0.025}(20) = 2.09$
The sample size for simple random sampling is then computed as follows:

$$n = \left(\frac{t_{\alpha/2}\sigma}{r\mu}\right)^2 = \left(\frac{2.09 \cdot 3}{0.1 \cdot 4}\right)^2 = 246$$

Since the sample size is much higher than 50, we might have used the normal approximation. In such case $z_{0.025} = 1.96$:

$$n = \left(\frac{z_{\alpha/2}\sigma}{r\mu}\right)^2 = \left(\frac{1.96 \cdot 3}{0.1 \cdot 4}\right)^2 = 216$$

Hence, in the **most conservative** case, a sample size of 216 guarantees the desired precision.
(Continued)

BOX 5.5 *Cont'd*

This sample size varies if we use alternative assumption on the data distribution of the target variable (not to be confused with the distribution used in the equation, which depends on the sample size). Examples are:

Rectangular (uniform) distribution of the target variable

$$n = \left(\frac{z_{\alpha/2}\sigma}{r\mu} \right)^2 = \left(\frac{1.96 \cdot 1.74}{0.1 \cdot 4} \right)^2 = 73$$

Normal distribution of the target variable

$$n = \left(\frac{z_{\alpha/2}\sigma}{r\mu} \right)^2 = \left(\frac{1.96 \cdot 1.15}{0.1 \cdot 4} \right)^2 = 32$$

In the above case the sample size is low enough to suggest the use of the *t* distribution in the sample size equation, even if this has a minor impact:

$$n = \left(\frac{t_{\alpha/2}(30)\sigma}{r\mu} \right)^2 = \left(\frac{2.04 \cdot 1.15}{0.1 \cdot 4} \right)^2 = 34$$

We could also get different sample sizes by setting a higher confidence level. For a 99% confidence level ($\alpha = 0.01$) these are the results

Binomial distribution $\qquad n = \left(\dfrac{z_{\alpha/2}\sigma}{r\mu} \right)^2 = \left(\dfrac{2.58 \cdot 3}{0.1 \cdot 4} \right)^2 = 374$

Rectangular distribution $\qquad n = \left(\dfrac{z_{\alpha/2}\sigma}{r\mu} \right)^2 = \left(\dfrac{2.58 \cdot 1.74}{0.1 \cdot 4} \right)^2 = 126$

Normal distribution $\qquad n = \left(\dfrac{z_{\alpha/2}\sigma}{r\mu} \right)^2 = \left(\dfrac{2.58 \cdot 1.15}{0.1 \cdot 4} \right)^2 = 55$

According to different assumptions, we derived desired sampling sizes ranging from 34 to 374. So, what is the choice?

The first question would be: is it safe to assume a normal distribution of the target variable in the population? While this is not an extremely strong assumption for some variable observed in nature (height, weight, etc.), for attitudinal studies it might be risky. It could be that the population follows a uniform distribution or even that it is exactly split into the two extremes (think about a perfectly bipolar political system). So, if we want to stay on 'safe grounds,' we assume a binomial distribution. For the remaining choice, the level of confidence, the researcher has the freedom (arbitrariness) to choose among the alternatives, usually based on the available budget. Usually a level of confidence of 95% is considered to be acceptable. In our case, a reasonable proposal could be to build a sample of 216 season ticket holders.

BOX 5.6   *Examples of complex sampling designs*

Expenditure and Food Survey (United Kingdom)	*Method*: Two-stage stratified random sample with clustering

Expenditure and Food Survey
(United Kingdom)

Method: Two-stage stratified random sample with clustering
Interview methods: Face-to-face and diaries
Sample size: 7,048
Basic survey unit: household
Description: The EFS household sample is drawn from the Small Users file of the Postcode Address File - the Post Office's list of addresses. Postal sectors (ward size) are the primary sample unit, 672 postal sectors are randomly selected during the year after being arranged in strata defined by Government Office Regions (sub-divided into metropolitan and non-metropolitan areas) and two 1991 Census variables – socio-economic group and ownership of cars.

Consumer Expenditure Survey
(United States)

Method: Two-stage stratified random sampling
Interview methods: Face-to-face and diaries
Sample size: approximately 7,600 per quarter (interviews) and about 7,700 per year (diaries)
Basic survey unit: household
Description: The selection of households begins with the definition and selection of primary sampling units (PSUs), which consist of counties (or parts thereof), groups of counties, or independent cities. The sample of PSUs used for the survey consists of 105 areas, classified into four categories: 31 'A' PSUs, which are Metropolitan Statistical Areas (MSAs) with a population of 1.5 million or greater, 46 'B' PSUs, which are MSAs with a population less than 1.5 million, 10 'C' PSUs, which are non-metropolitan areas used in the BLS Consumer Price Index program (CPI), 18 'D' PSUs, which are non-metropolitan areas not used in the CPI. Within these PSUs, the sampling frame is generated from the census file.

Chinese Rural Household Survey

Method: Three-stage sampling method with systematic PPS (systematic Probability Proportional to size) in the first two stages
Interview method: diaries filled with the interviewer help
Sample size: about 67,000
Basic survey unit: household
Description: At the first stage, systematic PPS drawing of 18 to 30 counties (PSUs) from each province (about 850 counties). At the second stage, administrative villages (about 20,000) are drawn from counties as SSUs with a systematic PPS method. At the third stage, households are drawn from each selected village with equal probability (simple random sampling) and up to ten households are allocated to each SSUs.

BOX 5.7 *Sampling with SPSS and SAS*

SPSS allows selection of random samples, including complex designs. Once you have organized your sampling frame into an SPSS workfile, Simple Random Sampling extraction is straightforward, simply click on DATA\SELECT CASES\RANDOM SAMPLE, then set the sampling size clicking on SAMPLE and draw the sample by clicking on OK. Non-sampled units can be filtered out or deleted. For more complex sampling designs, SPSS has an add-on module (SPSS Complex Samples™) which lets control multi-stage sampling methods and choose design variables, techniques (stratified and cluster sampling plus sampling a number of options on SRS), sample sizes, with room for personalization and an output which includes inclusion probabilities and sample weight for each of the cases. To create complex samples, click on ANALYZE/COMPLEX SAMPLES/SELECT A SAMPLE, then after naming the file corresponding to the sampling procedure, a 'wizard' box will guide you through the steps of sampling.

In SAS, the SAS\STAT component has procedures for the extraction of samples and statistical inference (chapters 11, 61, 62 and 63 in SAS/STAT User's Guide version 8). The procedure SURVEYSELECT allows to extract probability-based samples, while the procedure SURVEYMEANS computes sample statistics taking into account the sample design. It is also possible to estimate sample-based regression relationships with SURVEYREG. Besides simple random sampling, systematic sampling and sampling proportional to size (PSS) as the one described for the China rural household survey, multi-stage sampling is allowed by these procedures, as well as stratified and cluster sampling and sampling methods with unequal inclusion probabilities.

5.3 Non-probability sampling

The previous two sections have shown how the probability foundation of sampling extraction allows an estimation of population parameters together with evaluations of precision and accuracy. While this is not possible for samples that are not extracted according to probability rules, non-probability sampling is a common practice, especially quota sampling. It should also be noted that non-probability sampling is not necessarily biasing or uninformative, in some circumstances – for example when there is no sampling frame – it may be the only viable solution. The key limit is that generally techniques for statistical inference cannot be used to generalize sample results to the population, although there is research showing that under some conditions statistical model can be applied to non-randomly extracted samples (Smith, 1984). Given the frequent use of non-probability samples, it may be helpful to review the characteristics, potential advantages and limits of the most common non-probability samples.

The extreme of non-probability sampling is the so-called **convenience sampling**, where units that are easier to be interviewed are selected by the researcher. There are many examples of studies where the sample representativeness is not important and the aim is to show a methodology, where academics interview students in the class where they are teaching. Clearly this is the cheapest possible method (other than making up the data). As with any other more elegant non-probability sample, inference is not possible. But – even worse – by definition these samples are affected by a *selection bias*. In sampling, selection bias consists in assigning to some units a higher probability of being selected, without acknowledging this within a probabilistic sampling process. If the units with higher inclusion probabilities have specific characteristics that differ from the rest of the population – as it is often the case – and these characteristics are related to the target variable(s), sample measurement will suffer from a significant bias.

There are many sources of selection bias and some depend on the interview method rather than the sampling strategy. Consider this example of a selection bias due to convenience sampling. A researcher wants to measure consumer willingness to pay for a fish meal and decides to interview people in fish restaurants, which will allow him to achieve easily a large sample size. While the researcher will be able to select many fish lovers, it is also true that the sample will miss all those people who like fish but consider existing restaurant fish meals too expensive and prefer to cook it at home. The selection bias will lead to a higher willingness to pay than the actual one. The sample cannot be considered representative of consumers in general, but only of the set of selected sample units.

Even if the researcher leaves behind convenience and tries to select units without any prior criteria, as in **haphazard sampling**, without a proper sampling frame and an extraction method based on probability inference is not valid.

In other circumstances, the use of a prior non-probability criterion for selection of sample units is explicitly acknowledge. For example, in **judgmental sampling**, selection is based on judgment of researchers who exploit their experience to draw a sample that they consider representative of the target population. The subjective element is now apparent.

Quota sampling is possibly the most controversial (and certainly the most adopted) of non-probability technique. It is often mistaken for stratified sampling, but it generally does not guarantee the efficiency and representativeness of its counterpart. As shown in box 5.1, quota sampling does not follow probabilistic rules for extraction. Where there are one or more classification variables for which the percentage of population units is known, quota sampling only requires that the same proportions apply to the sample. It is certainly cheaper than stratified sampling, given that only these percentages are required as compared to the need for a sampling frame for each stratum. The main difference from judgmental sampling is that judgment is based on the choice of variables defining quotas. Sampling error cannot be quantified in quota sampling and there is no way to assess sample representativeness. Furthermore, this sampling method is exposed to *selection biases* as extraction within each quota is still based on haphazard methods if not on convenience or judgment. There are some relevant advantages in applying quota sampling, which may explain why it is common practice for marketing research companies. Quota sampling is often implemented in computer packages for CATI, CAPI and CAWI interviews (see chapter 3), where units are extracted randomly but retained in the sample only if they are within the quota. While this alters the probability design, it offers some control on biases from non-response errors, since non-respondents are substituted with units from the same quota. The point in favor of quota sampling is that a probability sample with a large portion of non-responses is likely to be worse than quota sampling dealing with non-responses. Obviously, probability sampling with appropriate methods for dealing with non-responses as those mentioned in section 5.2.5 are preferable.

If the objective of a research is to exploit a sample to draw conclusion on the population or generalize the results, non-probability sampling methods are clearly inadequate. For confirmatory studies where probability sampling is possible, non-probability methods should be always avoided. However, in many circumstances they are accepted, especially when the aim of research is *qualitative* or simply preliminary to *quantitative* research (like piloting a questionnaire). Qualitative research is exploratory rather than confirmatory, it provides an unstructured and flexible understanding of the problem and relies on small samples (see box 3.2). It can be a valuable tool when operating in contexts where a proper sampling process is unfeasible. A typical situation is when the target population is a very rare population, small and difficult to be singled out. In this circumstance, a useful non-probability technique is **snowball sampling**. As indicated by its name, snowball sampling starts with the selection of a first small sample (possibly randomly). Then, to increase sample size, respondents are asked to identify others who belong to the population of interest, so that the referrals have demographic and psychographic characteristics similar to the referrers. Suppose, for example, that the objective is to interview those people who climbed Mount Everest in a given year (or, say, the readers of this book) and no records are available. After selecting a first small sample, it is very likely that those that have been selected will be able to indicate others who accomplished to the task.

Summing up

Sampling techniques allow one to estimate statistics on large target populations by running a survey on a smaller sub-set of subjects. When subjects are extracted randomly to be included in a sample and the probabilities of extraction are known, the sample is said to be probabilistic. The advantage of probability samples compared to those extracted through non-probability rules is the possibility of estimating the sampling error, which is the portion of error in the estimates which is due to the fact that only a sub-set of the population is surveyed. This does not exhaust the total error committed by running a survey because non-sampling errors like non-response errors also need to be taken into account (see chapters 3 and 4). The basic form of probability sampling is simple random sampling, where all subjects in the target population have the same probability of being inextracted. Other forms of probability sampling include: systematic sampling, where subjects are extracted at a systematic pace and cluster sampling, where the population is first subdivided into clusters that are similar between each other and then a sub-set of clusters is extracted; stratified sampling, where the population is first subdivided into strata that contains homogeneous subjects but are quite different from those in other strata. These techniques can be combined in more complex sampling strategies, as in the multi-stage techniques adopted for official household budget survey. Once the sample has been extracted it is possible to generalize (infer) the sample characteristics like mean and variance to the whole population by exploiting knowledge of the probability distribution (see chapter 6). Estimates of the parameters are accompanied by estimates of their precision, as generally measured by the standard error. Accuracy (that is departure from the true population mean) can also be assessed with some degree of confidence. When planning a sample survey, there is a trade-off between sample size (cost) and precision. The latter also depends on the variability and dimension of the target population. Statistical rules allow to determine the sample size depending on targeted accuracy and vice versa.

Non-probability samples, including the frequently employed quota sampling, are not based on statistical rules and depend on subjective and convenience choices. They can be useful in some circumstances where probability sampling is impossible or unnecessary, otherwise probability alternatives should be chosen.

Appendix

Estimators

	Simple random sampling and systematic sampling	Stratified sampling	One-stage cluster sampling with unknown N
Mean	$$\bar{x} = \frac{\sum\limits_{i=1}^{n} x_i}{n}$$	$$\bar{x}_i = \frac{\sum\limits_{i=1}^{n_j} x_{ji}}{n_j} \text{ (stratum);}$$ $$\bar{x}_{ST} = \frac{\sum\limits_{j=1}^{s} \bar{x}_j N_j}{N}$$	$$\bar{x}_j = \frac{\sum\limits_{i=1}^{N_j} x_{ji}}{N_j} \text{ (cluster);}$$ $$\bar{x}_{CL} = \frac{G \cdot \sum\limits_{j=1}^{g} \sum\limits_{i=1}^{N_j} x_{ji}}{g \cdot \sum\limits_{j=1}^{g} N_j}$$
Variance	$$s^2 = \frac{1}{n-1} \sum_{i=1}^{n} (x_i - \bar{x})^2$$	$$s_j^2 = \frac{1}{n_j - 1} \sum_{i=1}^{n_j} (x_{ji} - \bar{x}_j)^2 \text{ (stratum);}$$ $$s_{ST}^2 = \frac{1}{N-1} \left[\sum_{j=1}^{s} (N_j - 1) s_j^2 + \sum_{j=1}^{s} (\bar{x}_j - \bar{x}_{ST})^2 N_j \right]$$	$$s_w^2 = \frac{1}{n-1} \sum_{j=1}^{g} \sum_{i=1}^{N_j} (x_{ji} - \bar{x}_j)^2 \text{ (within)}$$ $$s_b^2 = \frac{1}{g-1} \sum_{j=1}^{g} (\bar{x}_j - \bar{x}_{CL})^2 \text{ (between)}$$ $$s_{CL}^2 = \frac{G(N-G)s_w^2 + (G-1)Ns_b^2}{G(N-1)}$$
Proportion	$$p = y/n$$	$$p_j = y_j/n_j \text{ (stratum);}$$ $$p_{ST} = \sum_{j=1}^{s} p_j N_j / N$$	$$p_j = y_j/N_j \text{ (cluster);}$$ $$p_{CL} = \frac{\sum\limits_{j=1}^{g} y_j}{n}$$

Precision estimates

	Simple random sampling and systematic sampling	Stratified sampling	One-stage cluster sampling with unknown N
Standard error of the mean	$$s_{\bar{x}} = s_x\sqrt{\frac{N-n}{Nn}}$$	$$s_{\bar{x}_{ST}} = \frac{1}{N}\sqrt{\sum_{i=1}^{S} s_i^2 N_i^2\left(\frac{N_i - n_i}{N_i}\right)}$$	$$s_{\bar{x}_{CL}} = \sqrt{\frac{(G-g)g}{G(g-1)}}\sqrt{\sum_{i=1}^{g}\left(\bar{x}_i - \bar{x}_{CL}\right)^2\left(\frac{N_i}{n}\right)^2}$$
Standard error of the proportion	$$s_p = \sqrt{p(1-p)}\sqrt{\frac{N-n}{N(n-1)}}$$	$$s_{p_{ST}} = \frac{1}{N}\sqrt{\sum_{i=1}^{S}\frac{p_i(1-p_i)N_i^2}{n_i-1}\left(\frac{N_i-n_i}{N_i}\right)}$$	$$s_{p_{CL}} = \frac{1}{N}\sqrt{\frac{G(G-g)}{g(g-1)}\sum_{i=1}^{g}\left(y_i - p_{CL}N_i\right)^2}$$

Post stratification

Mean

$$\bar{x}_{PS} = \frac{\sum\limits_{i=1}^{S}\bar{x}_i N_i}{N}$$

Standard error of the mean

$$s_{\bar{x}_{PS}} = \sqrt{\left(\frac{N-n}{Nn}\right)\sum_{i=1}^{S}\left(\frac{N_i}{N}\right)s_i^2 + \left(\frac{N-n}{n^2 N}\right)\sum_{i=1}^{S}s_i^2\left(\frac{N-N_i}{N-1}\right)}$$

EXERCISES

1. Open the EFS data-set in SPSS
   a. Compute the mean of the variable cc3111 (clocks, watches and jewelery) on the whole data-set
   b. Now, extract a sample of 60 units and compute the sample means using:
      i. Simple random sampling
      ii. Stratified sampling (using INCRANGES - income quartiles as stratification variable)
      iii. Systematic sampling
   c. What is the sampling error? Which one is the most accurate method? And the most precise?
   d. Sort the sample by INCANON and perform systematic sampling again – how do results change? Why?
2. Open the Trust data-set in SPSS
   a. Compute the descriptive statistics of the variable q4kilos on the whole sample (N = 500)
   b. Suppose you need to extract a simple random sample of 50 elements, what is the estimated relative accuracy at a 95% confidence level? (HINT: $t_{\alpha/2} = t_{0.025} = 1.68$)
   c. If the target accuracy is ±15%, what is the sample size?
3. Open the EFS data-set in SPSS
   a. Compute the mean of the variable p609 (recreation expenditure)
   b. If the distribution of age in the population is the following:
      i. Less than 30 = 20%
      ii. 30–55 = 40%
      iii. More than 55 = 40%

Post-stratify (using the AGEBIN variable and the proportion above) and compute the mean again. Does it differ? Why?

Further readings and web-links

❖ **Is random sampling necessary** – This chapter has explained the advantages of random sampling, however, scientists are still debating on the extent of advantages from sampling. A good discussion is provided in Johnstone (1989).

❖ **Good quota sampling** – Many studies have looked into the problems of quota sampling compared to random sampling (see e.g. Moser and Stuart, 1953). King (1985) looks at the combination of quota sampling with probabilistic sampling, with a varying weight of the two methods depending on the budget constraints. See also the review by Marsh and Scarbrough (1990).

❖ **Area sampling** – When a sampling frame of the population is not available, a good way to ensure probabilistic extraction is the application of area sampling. See Hansen and Hauser (1945) and Baker (1991).

❖ **Failure of non-probability sampling and the effect of non-response** – There is an infamous case of survey failure, the prediction of the 1936

US elections (Landon vs. Roosevelt). To know more about this story, see the original results article on the 'History Matters' web-site (historymatters.gmu.edu/d/5168/) or the actual results from the elections (search on wikipedia.org for 'United States presidential election, 1936') and read the paper by Squire (1988).

Hints for more advanced studies

☞ If you are not convinced about the desirability of probabilistic sampling methods, how can one test the potential biases of non-probability sampling? Try and build your population data, extract several non-random and random samples and compare the performances.

☞ What are the sampling algorithms available in the most popular software for CATI interviews?

☞ This chapter discussed inference on means and variability. However, most of the times one may want to infer complex relationships from the sample by using multivariate techniques like regression analysis, cluster analysis, etc... What are the implications for sampling?

Notes

1. See the 'Further Readings' section for a more advanced discussion of the topic of random sampling vs.non-probability sampling.
2. For a review of the essential univariate statistics see the appendix.
3. The expected value of an estimator is the average across all possible values of the statistic and may be interpreted as the average value of the estimator across the sampling space (see the book appendix for an interpretation of 'expected values').
4. A more thorough discussion of the concepts of confidence and significance can be found in chapter 6.

CHAPTER 6

Hypothesis Testing

T HIS CHAPTER introduces the principles of hypothesis testing, describes
how to compute confidence intervals and illustrates the procedures for
testing hypotheses on one or two means. Through hypothesis testing, the
researcher makes an assumption on the population and obtains a statistical
assessment using sample data. These procedures are especially useful to
see whether there are significant differences between two sub-groups of the
population or to evaluate the impact of marketing measures on different
groups.

Section 6.1   describes the probabilistic foundations for confidence intervals
and tests

Section 6.2   illustrates tests on the sample mean for individual samples

Section 6.3   illustrates tests for comparing two means from different samples

Section 6.4   deals with non-parametric tests, which do not require
assumptions on the distribution of the data

Section 6.5   extends parametric tests to frequencies and variances

---

THREE LEARNING OUTCOMES

This chapter enables the reader to:

➡ Understand the probabilistic foundations of hypothesis testing and inference

➡ Distinguish among the wide range of available tests according to the
situations

➡ Appreciate the role of testing in taking statistically-based marketing
decisions

PRELIMINARY KNOWLEDGE: Readers should be familiar with the probability
and statistical concepts reviewed in the appendix, as well as the key notions
of probability sampling.

6.1 Confidence intervals and the principles of hypothesis testing

Chapter 5 stressed the relevance of probabilistic sampling as the only method opening
the way to statistical inference. The fact that only a single sample is observed does not

prevent the researcher from drawing statistical considerations on the target population. Hypothesis testing consists in accepting or rejecting a statement about the target population on the basis of statistics computed on a sample. This is possible if the sampling distribution is known.

To get introduced into the realm of hypothesis testing, a few steps are necessary. First, by exploiting the probabilistic nature of sampling it can be shown how statements on the true population parameters can be based on sample estimates, obviously with some degree of uncertainty (as defined by the *confidence level*). Then, the known probability distribution of sample means can be used to build a *confidence interval*, that is a range of values with known boundaries which is expected to include the true population mean at a given confidence level. The final step, introduced in section 6.1.1, is hypothesis testing, which also exploits the known probability distributions, but with a slightly different purpose than building confidence intervals for parameter estimates. Hypothesis testing aims at computing the probability that a given statement on the population parameters (or on the parameters of different populations) is true, thus allowing the researcher to maintain or discard the hypothesized statement depending on its probability level.

To appreciate how probabilities guide the above steps, suppose we extract a simple random sample of 100 units on a population of 400 regular customers of a local pub to measure the average expenditure in beer. Calculus tells one that the number of selections of 400 units taken 100 at a time is huge,[1] but sampling theory demonstrates that by computing the sample mean for all of these samples, one would obtain a normal distribution centered on the true (population) mean.[2]

The characteristic of the normal distribution is that its central value (corresponding to the mean, median and modal values) is the most likely one, with a probability which depends on how flat the curve is, that is, how large is the variability – specifically the standard error of the mean. For example, we know that 95% of all possible values fall within a range of about two standard errors from the mean,[3] while values outside this range only represent 5% of all possible sample means.

Figure 6.1 represents the normal distribution of sample means across the sample space. If one knew the population mean μ and the exact standard error[4] σ_μ, it would be possible to compute exactly the probability of each of the possible sample means. Following the considerations above, we also know that whatever mean value we got

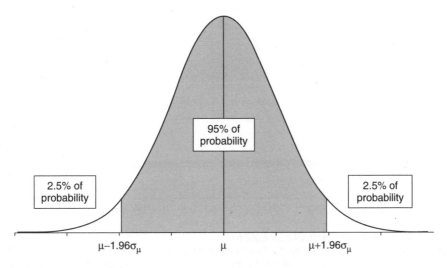

Figure 6.1 *Normal distribution of the sample means*

from our sample (the estimated $\bar{x}$), it will have a 95% probability of falling within the range[5] $[\mu - 1.96\sigma_\mu; \mu + 1.96\sigma_\mu]$.

However, our problem is somehow different, because the population mean μ and the population standard deviation σ (hence the standard error of the mean σ_μ) are unknown.

Luckily, even without knowledge of the population mean μ, some information can be drawn from the single value we have computed from the sample.

To exploit this probabilistic knowledge, another concept is helpful, the *level of confidence*, briefly encountered in chapter 5. If we had complete confidence (that is being 100% sure) that the sample mean value $\bar{x}$ falls within the range $[\mu - 1.96\sigma_\mu; \mu + 1.96\sigma_\mu]$, it would follow that the population mean μ certainly falls within the range $[\bar{x} - 1.96\sigma_\mu; \bar{x} + 1.96\sigma_\mu]$, because in the worst case scenario we would pick one of the extremes of the sampling space, computing a mean of either $\mu - 1.96\sigma_\mu$ or $\mu + 1.96\sigma_\mu$.

While we cannot state this with certainty, we still have a good degree of confidence, since (as shown in figure 6.1) 95% of the possible sample means are included in that interval. Thus, it is still possible to state, *with a 95% confidence level*, that the population mean will fall in the interval $[\bar{x} - 1.96\sigma_\mu; \bar{x} + 1.96\sigma_\mu]$.

Complications are not over, because there is a further problem. In most cases, if μ is unknown σ_μ is also unknown. As shown in chapter 5, one can estimate the standard error of the mean using the sample standard deviation. By doing this, a further element of approximation is introduced. It is possible to deal with this additional source of imprecision by replacing the normal curve with a more conservative distribution. The *Student-t* distribution is very similar to the normal curve, but has 'higher tails,' which means that it attaches lower probabilities to values close to the mean and higher probabilities to those more distant. In other words, the curve is flatter than the normal (see figure 6.2). The actual shape of the Student t distribution depends on the sample size. The larger the sample size, the more similar the two distributions become.[6]

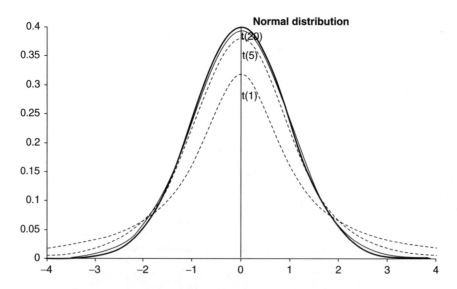

Figure 6.2 *Student-t and normal distributions*
Source: Dotted lines show the *Student-t* distribution with different degrees of freedom (1, 5 and 20), the bold line represents the standard normal distribution (with mean zero and standard deviation equals to 1). The degrees of freedom are equal to the sample size minus one.

As for the normal curve, it is possible to attach confidence levels to intervals around the population mean, as 95% of sample means will fall within the interval $[\mu - t_{\alpha/2}\sigma_\mu; \mu + t_{\alpha/2}\sigma_\mu]$ where $t_{\alpha/2}$ is a known value which depends upon the sample size and the confidence level. As mentioned in chapter 5, α is called *significance level* and is complementary to the confidence level $1 - \alpha$. Since one can commit an error in excess or in defect (the two extremes of the distribution), the actual significance level on each direction is $\alpha/2$. For example, with a confidence level of 95%, we allow for a 5% total probability of getting it wrong, which means a probability of 2.5% on getting it wrong in excess or in defect.

For a confidence level of 95% and a sample of 20 units, the value of $t_{\alpha/2}$ is 2.09. As one may expect, this value is slightly larger than the 1.96 value we used for the normal distribution, because a larger interval is needed to account for 95% of the probabilities as compared to the case where the exact standard error is known, since in this specific situation we have a further element of uncertainty. As the sample size increases, this uncertainty reduces and we go back to the normal case.

Thus, if one extracts a sample of 20 units and estimates the sample mean $\bar{x}$ and the standard error of the mean $s_{\bar{x}}$, with a confidence level of 95% it becomes possible to state that the actual population mean μ will fall within the interval $[\bar{x} - 2.09s_{\bar{x}}; \bar{x} + 2.09s_{\bar{x}}]$.

This is the *95% confidence interval* for the population mean. In summary, these are the steps to estimate a confidence interval:

1. Compute the sample mean $\bar{x}$ and standard deviation s
2. Estimate the standard error of the mean $s_{\bar{x}}$
3. Choose a confidence level α
4. Choose the appropriate coefficient (see table 6.1), using the *Student t* approximation instead of the *normal* (z) value if the sample size is below 50
5. Compute the lower and upper confidence limits as $[\bar{x} - z_{\alpha/2}s_{\bar{x}}; \bar{x} + z_{\alpha/2}s_{\bar{x}}]$ for the normal distribution or $[\bar{x} - t_{\alpha/2}s_{\bar{x}}; \bar{x} + t_{\alpha/2}s_{\bar{x}}]$ for the *Student t* approximation.

Table 6.1 shows the critical value for 90%, 95% and 99% confidence levels for different sample sizes.

6.1.1 Hypothesis testing

The same principles exploited for computing confidence intervals open the way to another key action in statistical inference: testing hypotheses. The rationale is similar. Albeit we only have information on the sample, we can attach probability levels to hypotheses on the population parameters. If a hypothesis has a very low probability

Table 6.1 *Critical values (two-sided) for confidence intervals and hypothesis testing*

| | Student t values | | | | Normal value |
| | t according to sample size | | | | z |
	10	20	30	40	
Level of confidence					
99%	3.17	2.85	2.75	2.70	2.58
95%	2.23	2.09	2.04	2.02	1.96
90%	1.81	1.72	1.70	1.68	1.64

level, below an arbitrarily fixed threshold, then one decides to reject that hypothesis in favor of an alternative.

Consider the example of box 6.1, where the overall average expenditure in chicken is computed at £5.67, while the sample estimate of the mean for those respondents who belong to consumer organizations is £5.04. May we safely conclude that those who belong to consumer organizations spend less than other people? This obviously depends on the precision of these estimates. As a matter of fact, the upper limit of the 99% confidence interval of the sub-sample of respondents associated to consumer organizations is 6.84, while the upper limit of the confidence interval for the overall sample is 6.18. So there is a chance that those who belong to consumer organization actually spend *more* than other people. Statistical tests based on the sample observations help in deciding whether the hypothesis should be rejected, for example, that there is no difference between respondents who belong to a consumer organization and those

---

BOX 6.1 *Confidence interval in SPSS*

To compute the confidence interval for chicken expenditure in the Trust data-set (question q5), use the ANALYZE/DESCRIPTIVE STATISTICS/EXPLORE dialog box. The STATISTICS button allows to choose the desired confidence level.

For the whole sample ($n = 500$), the mean is 5.67, the standard deviation is 4.13, and the estimated standard error of the mean is $4.13/\sqrt{500} = 0.20$. The sample size is large enough to use the normal distribution. For a 99% confidence level, the coefficient (from table 6.1) is $z = 2.58$, so that the interval becomes $[5.67 - 2.58 \cdot 0.2; 5.67 + 2.58 \cdot 0.2]$, that is $[5.16; 6.18]$ as shown in the SPSS output below.

Descriptives

			Statistic	Std. error
In a typical week how much do you spend on fresh or frozen chicken (Euro)?	Mean		5.6677	0.19640
	99% Confidence Interval for Mean	Lower Bound	5.1596	
		Upper Bound	6.1758	
	5% Trimmed Mean		5.2958	
	Median		5.0000	
	Variance		17.0890	
	Std. Deviation		4.1338	
	Minimum		0.0000	
	Maximum		30.0000	
	Range		30.0000	
	Interquartile Range		4.5000	
	Skewness		2.084	0.116
	Kurtosis		8.005	0.231

If we consider the sub-sample of those respondents who belong to a consumer association (q62a= 1), the sample size falls below 50 ($n = 28$) and it becomes necessary to use the *Student t* approximation. The mean is 5.04, the standard deviation is 3.42, the estimated standard error is 0.65. The coefficient value is $t(27) = 2.77$, so the 99% confidence interval becomes $[5.04 - 2.77 \cdot 0.65; 5.04 + 2.77 \cdot 0.65]$, that is $[3.24; 6.84]$ as shown in the SPSS output below, where values slightly differ for rounding errors.

BOX 6.1 *Cont'd*

Descriptives

			Statistic	Std. Error
In a typical week how much do you spend on fresh or frozen chicken (Euro)?	Mean		5.0357	0.64664
	99% Confidence Interval for Mean	Lower Bound	3.2441	
		Upper Bound	6.8274	
	5% Trimmed Mean		4.7024	
	Median		4.0000	
	Variance		11.708	
	Std. Deviation		3.42171	
	Minimum		0.70	
	Maximum		17.00	
	Range		16.30	
	Interquartile Range		3.75	
	Skewness		1.770	0.441
	Kurtosis		4.462	0.858

who don't. As a matter of fact, as it is shown in section 6.2, the difference between the two means follows a known statistical distribution. If the hypothesis of mean equality is true, the difference between the two means should be zero. The actual difference is it is £0.63 (£5.67 minus £5.04), but since this value is based on an individual sample, it is possible that this difference is generated by sampling error. With the help of statistics, one computes the probability that, even when no difference exists at the population level, a sample with a difference of £0.63 is extracted. Suppose this probability is 90%, then it is wise not to reject the hypothesis of equality between the two means. Instead, if the probability is 0.001%, while there is still a slight chance that the difference is due to sampling error, it is such an unlikely event that one should choose to reject the null hypothesis and think that belonging to a consumer organization actually makes a difference on chicken expenditure.

The limit below which we should decide to reject the initial hypothesis is the (already encountered) *significance level* and is usually denoted with α, expressed in percentage terms and set to 5% or 1%.[7] The smaller is the *significance level*, the smaller is the *rejection region*, which is the set of values that lead to rejection of the hypothesis and the larger is the *acceptance region*, which is the set of values that lead to non-rejection of the hypothesis. Note that statisticians prefer to refer to 'non-rejection' rather than 'acceptance' of a hypothesis, since there is always a remaining degree of uncertainty.

The initial hypothesis is called *null hypothesis* and is denoted by H_0. Contextually, the researcher sets an *alternative hypothesis* (H_1) which is complementary to H_0 and remains valid if H_0 is rejected. In the example above, H_0 states that the mean expenditure of respondents in consumer organizations is equal to the mean expenditure of those who do not belong to consumer organizations. The alternative hypothesis, H_1, would include all cases where the two mean expenditures differ, that is, both the situation where the former group of respondents spends more than the latter and the opposite situation. This is said to be a *two-tailed test*, because the alternative hypothesis can go in either directions. Consider a different formulation, where we state a null hypothesis H_0, in which respondents in consumer organizations spend *less* than those outside. In that case the alternative hypothesis is formulated in a unique direction and includes all cases where respondents in consumer organization spend *more* (or exactly the same amount) than those outside. This is a *one-tailed test*. When one makes a decision according to the significance level α, there is obviously the risk of a potential error. The rejection of the hypothesis H_0 when it is true is called *Type I error* and the significance level α represents the probability of committing a Type I error. A second type of error is the non-rejection of the hypothesis H_0 when the alternative hypothesis H_1 is true. The procedure for testing hypotheses is based on the following steps:

1. Formulate the hypotheses H_0 and H_1
2. Determine the distribution of the sample test statistic under the H_0 hypothesis
3. Choose a significance level α
4. Compute the sample test statistic and its probability level (*p-value*)
5. If the *p-value* is lower than α, then reject H_0

Rejection of the null hypothesis opens the way to testing other hypotheses (and the alternative hypothesis is adjusted according to the outcomes of previous tests). For example, if we reject the hypothesis of equality in our example, the next step could be to test whether respondents in consumer organizations spend more than othes, versus an alternative hypothesis that they spend less. Instead, when H_0 cannot be rejected, the testing procedure comes to an end.

A final distinction when comparing two values concerns the possibility that the two values come from *independently extracted samples* versus provenance from *related samples*. A specific case is the situation where the same units appear in both samples, so that they can be matched. In this case, the samples are said to be *paired*. This distinction is relevant to the hypothesis testing technique, as discussed in the rest of this chapter.

6.1.2 Power of a statistical test

As mentioned in the previous section, once a hypothesis has been formulated, there are two ways to obtain the right conclusion through a test and two ways to take a wrong conclusion. Table 6.2 summarizes the range of situations and errors that one could face when testing a hypothesis.

Consider the situation where the null hypothesis is true. Given that a significance level for rejection has been set (or a confidence level for non-rejection), one must accept that there is a probability that the test still leads to rejection of the null hypothesis. This happens if the test returns a probability below the significance level α. Thus, when the null hypothesis is not rejected, one must be aware that the probability that this is the correct outcome is equal to the confidence level (that is $1 - \alpha$). As mentioned,

Table 6.2 *Error types in hypothesis testing, confidence, significance and statistical power*

	Reject H_0	Non-reject H_0
H_0 is true	Type I Error (prob.=*significance* level α)	Correct (prob. = *confidence* level $1 - \alpha$)
H_0 is false, H_1 is true	Correct (prob. = *power* level $1 - \beta$)	Type II Error (prob. = β)

the error of rejecting a true hypothesis is the *Type I error* and obviously has a probability equal to the significance level α.

Consider now the situation where the null hypothesis is false. Here the test could fail to reject it *(Type II error)* and this happens with a probability β, which is not arbitrarily decided and is not mathematically related to the significance level, although it is intuitive that the smaller is α, the more likely it is that we do not reject the null hypothesis and the larger is the *Type II Error*. The probability of 'getting it right' when the null hypothesis is false is obviously $1 - \beta$. This probability represents the *statistical power* of a test and can be estimated on sample data (see Cohen, 1988). It depends on the sample size, on the significance level α and on the *effect size*. The latter (further discussed in chapter 7) is a measure of 'how wrong' is the null hypothesis. An hypothesis which is quite far from the reality leads to a larger effect size and is more easily rejected, increasing the power of a test.

When the null hypothesis is not rejected, one should accompany this result with an estimate of the power. If the power is above 80%, then the conclusion is usually regarded as robust.

6.2 Test on one mean

The first case to be considered concerns a null hypothesis made on the population mean. For example, according the UK Department of Environment, Food and Rural Affairs (DEFRA), the average weekly household chicken consumption was 1.15 kg in 2004. If we consider the sub-sample of UK households within the Trust data-set, we obtain an average of 1.75 kg on 92 respondents. Given that the DEFRA figure is likely to be more representative, we could test the hypothesis that the population mean is actually 1.15. Thus, we want to test

$$H_0 : \mu = 1.15 \quad \text{versus the alternative } H_1 : \mu \neq 1.15$$

As for confidence intervals, we know that the sample means are distributed according to a normal distribution. Under the null hypothesis H_0, the sample mean will be distributed as a normal with mean 1.15 and unknown variance. If we use the sample standard error of the mean (which is equal to 0.28), with a sample size of 92 we can still use the normal curve (with less than 50 observations we should have referred to the *Student-t* distribution). The actual test is made on the standard normal distribution, with mean 0 and variance 1. To perform the test is sufficient to standardize the test statistic. As also shown in the appendix, standardization consists in subtracting the mean and dividing by the standard deviation (in this case the standard error of the mean):

$$\bar{z} = \frac{\bar{x} - \mu}{s_{\bar{x}}}$$

which returns a variable with zero mean and unity standard error. Under H_0 the standardized test statistic $\bar{z}$ is distributed as the standard normal distribution. In our example $\bar{z} = \dfrac{1.75 - 1.15}{0.28} = 2.14.$

This value needs to be compared with the *critical values* which separate the acceptance region from the rejection regions. This is a two-tailed test, since the alternative hypothesis is formulated in both directions, as $\mu \neq 1.15$ holds for either $\mu < 1.15$

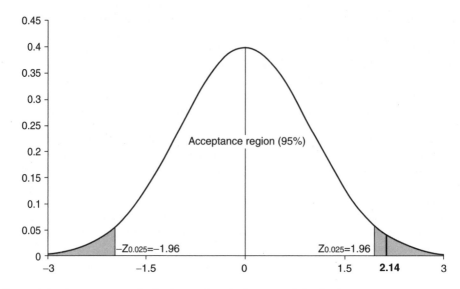

Figure 6.3 *Acceptance and rejection regions in the normal distribution*

or $\mu > 1.15$. Considering a significance level $\alpha = 0.05$, the rejection and acceptance regions are shown in figure 6.3, together with the sample statistics. The two critical values, are indicated as $-z_{\alpha/2}$ and $+z_{\alpha/2}$ as the former defines the left rejection region (excluding those negative value with less than 2.5% of probability) and the latter the right rejection region (excluding those positive values with less than 2.5% of probability). For a 5% significance level (or a 95% confidence level), $-z_{0.025} = -1.96$ and $+z_{0.025} = +1.96$.

The test statistic is outside the acceptance region, as it is larger than the critical value 1.96. As a matter of fact, the exact p-value for $\bar{z} = 2.14$ is 0.032 (for a two-tailed test), which is below the threshold $\alpha = 0.05$. Hence, we reject the null hypothesis that the population mean is 1.15, and there must be some reason why our data-set returns a sample mean of 1.75, which is significantly different from the test value of 1.15. Most software provides critical values and p-values from the Student t distribution even for large sizes. Note, however, that this would not change the outcome as the critical value $t_{0.025}$ with 91 degrees of freedom (that is the sample size minus one) would be 1.99 versus the 1.96 value used with the normal approximation and the p-value for $\bar{z} = 2.14$ would be 0.035 instead of 0.032, still leading to rejection.

A second example shows the procedure for one-tailed tests. Suppose that a meat producer is uncertain about introducing a new product label ensuring conformity with animal welfare standards. The Trust survey measures the importance of animal welfare when purchasing food on a seven point scale, from extremely unimportant to extremely important (q24k). Since the labeling scheme implies additional cost, the producer will introduce it only if – on average – the score is above 4.9. It is convenient to formulate the hypotheses as follows:

$$H_0 : \mu \leq 4.9 \quad \text{versus the alternative} \quad H_1 : \mu > 4.9$$

The above formulation is more relevant to decision making than testing the null hypothesis $H_0: \mu > 4.9$, since non-rejection would be inconclusive, while confidently rejecting the hypothesis that $\mu \leq 4.9$ provides a more robust basis for introduction of the scheme.

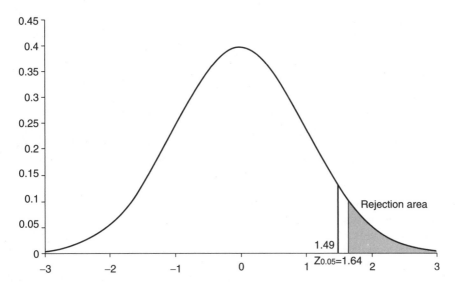

Figure 6.4 *Rejection region in a one-tailed test*

The sample mean is 5.01, with a standard deviation of 1.65 and a standard error of the mean of 0.074. Is it enough to reject the H_0 hypothesis? In this case, the standardized test statistic (on a sample of 497 respondents) is:

$$\bar{z} = \frac{5.01 - 4.90}{0.074} = 1.49$$

Since this is a one-tailed test, it is important to notice that the rejection region lies on the right-hand side of the distribution only, since all values on the left-hand side, even the extreme ones, are consistent with the null hypothesis. Thus, instead of needing two critical values with $\alpha/2 = 0.025$ as in the previous example, we only require a single critical value for $\alpha = 0.05$ (which corresponds to the $\alpha = 0.10$ two-tailed critical values), that is 1.64.

As shown in figure 6.4, the test statistic lies outside the rejection region and within the acceptance region. Thus, we cannot reject the null hypothesis that the average score is below 4.9. A very conservative producer will decide not to implement the labeling scheme, although reverting the null hypothesis to $H_0: \mu > 4.9$ would equally lead to non-rejection. Statistics is not too helpful in this borderline circumstance.

6.3 Test on two means

Tests of equality on two means follow directly from the single-mean test. By computing the difference of the two normally distributed sample means, one still obtains a normally distributed variable. If the means are equal, then their difference would be 0. There are some issues to be taken into consideration. First, the difference follows the normal distribution only if the two samples are unrelated or paired. Secondly, it is possible that the two samples (or sub-samples) have a different standard error, so one needs to compute a joint standard error to proceed with the testing procedure.

Unrelated (independent) samples are those where the sampled units belong to different populations and are randomly extracted. The key condition is that the sampled units should be randomly assigned to the two groups. This would exclude the case where a single sample is drawn and the units are subdivided into two groups according to some variable (gender, living in urban versus rural areas, etc.), because the sampling process might have some selection bias which would imply some dependency between the two groups. Gender or residence areas are not determined by random extraction, but might depend on other factors. For example, in the Trust data-set, about 71% of respondents are female because the targeted population was that of people in charge of food purchases. This means that males and females in the Trust data-set are associated by the fact that they are responsible for food purchases, which excludes those females and males who are not. Hence, any gender comparison would be conditional on that external factor and we could not be conclusive in testing – for example – whether males like chicken more than females.

While truly independent samples are a rare case, most social studies consider the samples to be unrelated if the units are randomly extracted and the groups are mutually exclusive. If one has doubts about the independence of the two samples, both tests for unrelated and related samples could be carried out and if the results differ the issue should be explored in greater detail. An example of two independent samples within the Trust data-set could be the comparison of the mean chicken expenditure in two different countries, since the experimental design was based on independent surveys in each country and the target populations are mutually exclusive.

Related samples are those where the same subjects may belong to both groups. For example, if the same individual is interviewed in two waves, the two samples are said to be related. Similarly, if one compares two measures within the same data-set, the samples are related. The special case where exactly the same units appear in both samples and it is possible to compute the difference for each of the sampled units is denoted as *paired samples*, while sometimes units can be *matched* across the two samples according to some characteristics.

6.3.1 Test for unrelated samples

Suppose, as in the above example for independent samples, that one wants to test whether UK and Italian respondents within the Trust survey show the same appreciation of chicken, as stated in question q9 on a seven point scale ('In my household we like chicken'). The hypothesis to be tested is:

$$H_0 : \mu_{UK} = \mu_{ITA} \quad \text{versus the alternative} \quad H_1 : \mu_{UK} \neq \mu_{ITA}$$

As mentioned, since the two means are normally distributed, under the null hypothesis their difference is also normally distributed. If one considers the variable $D = \mu_{UK} - \mu_{ITA}$ the test becomes identical to the one mean test for $D = 0$ discussed in section 6.2. However, to perform it, a measure of the standard error for D is necessary. If we knew (extremely unlikely) the actual population standard errors we might simply compute the joint standard error as:

$$\sigma_{\mu_1 - \mu_2} = \sqrt{\frac{\sigma_1^2}{n_1} + \frac{\sigma_2^2}{n_2}}$$

and proceed exactly as in section 6.2. More frequently, the standard errors of the mean for each single group are estimated. In this case, there are two alternative situations: (a) the unknown standard errors are equal and; (b) the unknown standard errors are different.

If they are equal, the solution is straightforward and the test statistic (approximated by the t distribution) is provided by:

$$t = \frac{\bar{x}_1 - \bar{x}_2}{s_{\bar{x}}\sqrt{\dfrac{1}{n_1} + \dfrac{1}{n_2}}}$$

The test distribution will be a Student t distribution with a number of degrees of freedom equal to $n_1 + n_2 - 2$.

Instead, when the unknown standard errors are different, there are no straightforward solutions and researchers have investigated the issue for a long time (see for example Reed and Stark, 2004). It is not worth getting into the technicalities, but it is important to know that for large samples one can still proceed by computing

$$z = \frac{\bar{x}_1 - \bar{x}_2}{\sqrt{\dfrac{s_{\bar{x}_1}^2}{n_1} + \dfrac{s_{\bar{x}_2}^2}{n_2}}}$$

which under H_0 follows the standard normal distribution. The issue whether the standard errors are equal or not is usually sorted out by using an appropriate hypothesis test for the equality of two variances. SPSS shows the p-value for the Levene's test (Brown and Forsythe, 1974) and provides the outcomes of both tests, with and without assuming equality of standard errors. In most cases, these two tests provide consistent outcomes.

Getting back to our example, the mean for the 100 UK respondents is 6.12 (standard error 0.15), versus a mean of 5.62 (standard error 0.11) for the 100 Italian respondents. Is a difference of 0.5 *significantly different from 0*? Are the standard errors equal?

The SPSS output is shown in table 6.3.

First, note that the p-value for Levene's test is 0.622. The null hypothesis of equal standard errors is not rejected. At any rate, the two-tailed t-tests (albeit the normal

Table 6.3 *Mean comparison test for independent samples*

Independent Samples Test

		Levene's test for equality of variances		t-test for equality of means					95% Confidence interval of the difference	
		F	Sig.	t	df	Sig. (2-tailed)	Mean difference	Std. error difference	Lower	Upper
In my household we like chicken	Equal variances assumed	0.243	0.622	2.682	198	0.008	0.500	0.186	0.132	0.868
	Equal variances not assumed			2.682	183	0.008	0.500	0.186	0.132	0.868

approximation could be safely assumed), return identical results. The null hypothesis of equal means is strongly rejected (the *p*-value is 0.008) at both the 5% and 1% significance levels.

6.3.2 Paired samples

As a second case study, consider the situation where two measures are taken on the same respondents. For example, all respondents were asked a second question on their general attitude toward chicken, 'A good diet should always include chicken' (q10). Possibly the two questions measure the same item – the general attitude toward chicken. Can we assume that the results are – on average – equal? In this case the samples are paired and it is possible to compute a difference between the response to q9 and q10 for each of the sampled households. It is easy to go back to the one mean test discussed in section 6.2. One may compute a third variable for each of the respondents as the difference between q9 and q10, then compute the mean and the standard error for this variable. SPSS does this automatically (as in box 6.2) and returns a mean of 5.73 for q9 versus a mean of 5.50 for q10, for an average difference of 0.23 and an estimated standard error of 0.06. The *t* test statistic is 3.84, largely above the two-tailed 99% critical value, so that the null hypothesis of mean equality should be rejected. It is not safe to assume that the two questions are measuring the same construct. We leave as an exercise to the reader to compute the difference variable for all households in the data-set and perform the one sample mean test to show that results are identical.

6.4 Qualitative variables and non-parametric tests

When samples are related it becomes quite difficult to make assumptions on the distribution of the test statistics. In this situation a solution is provided by *non-parametric tests*, which do not require knowledge of the underlying distributions (see Gibbons, 1993 for a good introduction to non-parametric testing). Non-parametric tests are also used in situations where the variable to be tested are qualitative.

6.4.1 One-sample tests

Considering one-sample situations, the *Runs test* is useful to test whether a single variable is ordered in a random fashion. This test is useful to assess the sampling procedure (although a non-rejection of randomness does not guarantee that the sample is actually random) and is only useful if the order of observations in the data-set is meaningful, that is the data-set lists the sampled units in the order of their extraction.

For example, if the units in the EFS data-set follow the order of the original sample, it would be surprising to find many consecutive female reference persons and then many consecutive male reference persons. This would cast doubts on the randomness of the sampling process. The *Runs test* can be easily understood by considering a dichotomous variable first, like the gender one. Basically, it compares the empirical (observed) sequence with all potential sequences that could derive from a perfectly random extraction. As usual, if the empirical sequence is very unlikely (below the significance level), we reject the hypothesis that the sequence of observation is random with respect to gender. The test can be also applied to non-dichotomous quantitative variables, provided a cut-point is defined. For example, one might split the sample into two halves through the median, then check whether the sequence of observations below or above the median can be regarded as random. Other cut-points can be the mean, the mode or arbitrarily defined values. It may be easily checked that the *Runs test* does not reject the hypothesis of random order with respect to total consumption expenditure, using the mean as a cut-point.

Another widely used non-parametric test based on a single sample is the *Kolmogorov-Smirnov test*, which tests the null hypothesis that a random sample is drawn from a (user specified) given distribution. The test compares the empirical cumulative distribution functions and the theoretical one and tests whether they differ significantly. For example SPSS allows testing whether a variable is distributed according to a normal distribution, a Poisson curve, a uniform or an exponential distribution. Once the distribution and its parameters are known, it becomes possible to estimate the population parameters.

A similar test is the *Chi-square (Pearson) goodness-of-fit test*, where the observed frequencies for a categorical variable (either a qualitative or a quantitative variable split into groups) are compared to some expected frequencies,which can be freely chosen or determined by a theoretical probability distribution. Under the null hypothesis that the distance between the observed and expected frequencies is zero, the test statistic is compared to critical values provided by a distribution called *chi-square*, which is further discussed in chapter 9.

The *Binomial test* is useful for the specific case of dichotomous variables. For example, we have seen that the Trust data-set contains 71% of female respondents versus 29% of male respondents. Could this proportion be due to sampling error only? If this were the case, the variable should be compatible with a binomial variable where the proportions are given by the proportion of males and females in the population. Assuming a 50% distribution, the null hypothesis is strongly rejected.

Similarly to the one mean t-test seen in section 6.2, the *Wilcoxon signed rank test* allows to test the null hypothesis that the median is equal to some specified value, provided that the distribution is symmetric. It is based on *ranking*, which consists of assigning increasing discrete values 1, 2, etc. to the observations once they have been sorted in ascending order according to the variable to be tested. The observations which differ from the hypothesized mean are ranked according to their distance from the median, and then ranks above and below the median are summated to build the test statistic (Wilcoxon, 1945).

6.4.2 Comparison of two groups

With qualitative variables, it is not possible to apply the standard mean comparison tests. However, one can test whether two populations follow the same probability distribution. The *Wald–Wolfowitz Run test* is the non-parametric form of the *t*-test for two independent samples. It is particularly useful when one is concerned about the independence of two (sub) samples with respect to one variable and it is particularly appropriate when this variable is ordinal, although it can be applied to quantitative variables as well. The null hypothesis which is tested is that two (sub) samples are independent samples extracted from the same population. For example, one might explicitly test whether female and male household reference persons in the EFS data-set are independent samples from the same population, in terms of total consumption expenditure. The alternative hypothesis is that this is not the case, which might mean that they are not independent or they belong to different populations (thus they have different means). The test proceeds as follows – all cases are sorted in ascending order with respect to total consumption expenditure, independently from the gender, then a dichotomous variable is created, assuming value one for males and value two for females. If the two samples are independent of the same population one might expect a random sequence of sparse ones and twos, while if there is some relationship (or different distributions) there will be concentrations and sequences of ones and twos. For relatively large samples (say above 30 observations), all possible sample runs are randomly distributed and it becomes possible to test for independent extraction using a normal population. In our example, the null hypothesis cannot be rejected.

With ordinal variables, the *Mann–Whitney test* (sometimes referred to as the *Wilcoxon-rank-sum test*) assumes that the two samples are random and come from populations that follow the same distribution apart from a translation k. Using the Trust data-set again, one may wish to compare the variable 'trust in the European food safety authority' for French and German respondents. The trust measures are provided in an ordinal scale from 1 to 7. The *Mann–Whitney test* starts by assuming that the probability of Germans answering with a value which is – say – less than 3 is equal to the probability of French answering to the second question with a value which is less than $3 + k$, where k is an integer and the relation is valid for all possible values besides 3. The null hypothesis is that $k = 0$ which would mean that the samples are from the same population and would provide an even stronger support than mean equality. The test statistic is compared with the critical values from the normal distribution. In this example, the value of the test statistics is -2.82, which is below the 95% lower critical value of the two-tailed normal distribution (-1.96). Hence, the assumption that German and French have the same trust level in the European food safety authority should be discarded. The test also provides an evaluation of the 'mean rank,' which allows one to see that Germans show a higher degree of trust. Note that we would get a similar result by treating the variable as quantitative. The Mann-Whitney test is based on *ranking*. All observations are ranked, independently of the groups they belong to and ranks for each group are summated. The U statistics compares these sums, allowing for different group sizes. When the two samples are compared, the U statistic of the Mann-Whitney test is based on the frequency with which the first sample has a higher rank than the second sample. If the two samples come from the same distribution, such frequency should be random, similarly to the rationale behind the *Runs test*. The *Kolmogorov-Smirnov test* and the *Runs test* can also be exploited to compare two distributions, by using either distribution as the benchmark one. SPSS also provides the *Moses extreme reaction test* (Moses, 1952).

When samples are paired, a viable test is the *Wilcoxon paired sample test*, which is simply the Wilcoxon signed rank test on the median difference, provided that the differences in value between the two distributions are symmetrically distributed.

6.5 Tests on proportions and variances

The list of parametric and non-parametric tests provided in this chapter does not cover all the existing tools in hypothesis testing. Some extensions, however, are straightforward and should be mentioned here.

For example, parametric tests can be easily applied to test hypotheses on *proportions*. The mechanism is practically identical to the one exploited for comparing means and for large samples one may simply consider the normal distribution, using the mean and standard error equations seen in chapter 5. Other tests focus on hypotheses on variability measures, as the *F test for equality of variances* mentioned in relation to tests on two means. This test is especially interesting because it is based on a statistical distribution (the *F distribution*) which is central to the analysis of variance explored in chapter 7. Under the null hypothesis of variance equality, if the two samples are extracted from normal distributions or the sample sizes are large enough, the ratio between two variances follows a distribution like the one drawn in figure 6.5.

The F test statistics is simply the ratio between the two estimated variances and the null hypothesis can be tested against the theoretical F distribution, whose shape depends on two values for the degrees of freedom, given by $n_1 - 1$ (the sample size for the first variance minus one) and $n_2 - 1$ (the sample size for the second variance minus one). The notation is $F(n_1 - 1; n_2 - 1)$.

Note that the F distribution is non-symmetrical and for two-tailed test the critical values are defined in a slightly different manner when compared to the t and normal distributions. For a given significance level α the critical value for the right rejection region (the first variance is larger than the second) is denoted by $F_{\alpha/2}$ as usual, since we want to exclude F values larger than $F_{\alpha/2}$ because their probability is below $\alpha/2$. Instead, the critical value for the left rejection area (the first variance is smaller than the second) is denoted as $F_{1-\alpha/2}$, since we set the probability that F is larger than the critical value to $1 - \alpha/2$, which means that the probability that F is smaller than the critical value is actually $\alpha/2$, as desired. This difference is also relevant to one-tailed test where the alternative hypothesis is $\sigma_1 < \sigma_2$, where the critical value will be $F_{1-\alpha}$.

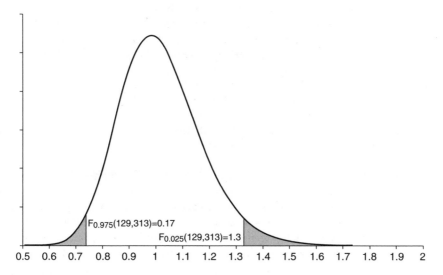

Figure 6.5 *The F distribution*

For example, suppose we want to test whether the variance in chicken consumption for males is equal to the variance in chicken consumption for females. The variances estimated in the sample are 0.87 and 2.60 and the sample sizes 130 and 314, respectively. The null hypothesis is:

$$H_0 : \sigma_M = \sigma_F \quad \text{versus the alternative} \quad H_1 : \sigma_M \neq \sigma_F$$

In this case, the critical value are $F_{0.975}(129, 313) = 0.74$ and $F_{0.025}(129, 313) = 1.33$. The F statistic is given by $0.87/2.60 = 0.33$. Since the test statistic is smaller than the lower critical value, the null hypothesis is rejected in favor of the alternative and we can proceed by assuming that the variance for males is smaller than for females.

Note that the above *F test* is valid under the assumption of normal distribution of the populations. A generalization for non-normal populations as provided in SPSS and SAS is the *Levene's test for homogeneity of variances* which computes a slightly different *F statistic*. Note that the *Levene's test* can be biased by the presence of outliers. An alternative test is the *Brown and Forsythe's test* (Brown and Forsythe, 1974). Further discussion about testing on variances is provided in the next chapter.

Summing up

After data preparation through plots and descriptive statistics, it is time to make some inference, which means drawing statistical conclusions on the population by using data observed on the sample. This is possible because the distributions of variables measured through probabilistic samples are known. Thus, when we observe a particular value on the sample, it becomes possible to compute an interval which will include the true (population) value with some arbitrarily set degree of confidence, which researchers choose according to the risk they are willing to take. The risk is inversely proportional to the width of the confidence interval. Using the same principles, it is possible to test specific assumptions on the population parameters, like means, frequencies or variances. With one mean tests, the researcher can test whether the population mean is equal to a specified value. With two means tests, the population means for two (sub) samples can be compared. When testing hypotheses on a sample, one accepts a given level of risk of getting the wrong conclusions. A true hypothesis can be rejected (type I error) with a probability equal to the significance level, while a different type of error consists in non-rejection of a false hypothesis (type II error). The probability of non-committing a type I error is the level of confidence, while the probability of non committing a type II error is the statistical power of a test. Good tests usually assume a 95% or 99% level of confidence and show a statistical power above 80%.

Proportions and variances can also be tested and compared across two samples through parametric tests. Instead, non-parametric tests allow one to test hypotheses on the distributions of the variables rather than the population parameters. Statistical packages offer a wide range of testing tools, but it is important to know which test is appropriate according to the specific situation (independent samples versus paired samples, quantitative versus qualitative variables).

EXERCISES

1. Open the EFS data-set:
    a. Using a one-mean t-test, test the hypothesis that the average number of cars or vans per household in the UK (variable a124) is exactly one
    b. Is the distribution of the mean of the above variable normal? Plot the histogram and the normal curve, then apply a Kolmogorov-Smirnov test for normality
    c. Test again the hypothesis of one car per household using a non-parametric test. Do results differ? Why?
    d. Compare mean expenditure in restaurants and hotels (variable p611) between the lowest income quartile and the highest income quartile (as classified by variable INCRANGE). Are means significantly different? Does the assumption of equality of variance influence the results?
2. Open the Trust data-set:
    a. Use the Chi-square test to check whether the sample is equally distributed across different categories of town size (q63)
    b. 'About 30% of consumers shop in butcher's shop' – Test this hypothesis using the Binomial test on question q8d
    c. Consider the general attitude toward chicken (q9) – Using the Runs test, check whether the sample is sorted in a random way
    d. Consider that the data-set is sorted according to countries. Now select consumers from the UK only (q64 = 1) and repeat the Runs test. Do results differ? Why?
3. Open the Trust data-set:
    a. Is expenditure in chicken (q5) normally distributed? Use the appropriate test and draw a histogram
    b. Is the frequency of purchases of frozen chicken (q2d) compatible with a Poisson distribution?
    c. Using rank tests, check whether those in the lowest income range like chicken (q9) more than those in the highest income range
    d. Compute the estimated variance of chicken expenditure (q5) for female and male reference persons (q60) and perform an F-test to check for their equality – Hint: the 95% critical values are $F_{0.025}(351,145) = 1.33$ and $F_{0.975}(351,145) = 0.77$
    e. Compare mean expenditures for q5 using the 2-independent samples t-test and check the Levene's F-statistic. Does it lead to a different conclusion?

Further readings and web-links

❖ **All you wanted to know about testing** – The recent book by Andy Field (2005) is a must for those who want to further explore the myriad of situations for hypothesis testing, in both the parametric and non-parametric settings. A visit to Field's web-site, without getting discouraged by the web graphics and the url name, is also recommended: www.statisticshell.com/statisticshell.html

(Continued)

❖ **Does (sample) size matters?** – Central limit theorem, Student-t distribution versus normal distribution, a sample of at least 50 observations… or 30… or 100? And what happens if your sample is *really* small? As usual, it depends from the statistical technique. Wilkerson and Olson (1997) explain very well some misconceptions about significance testing and its relation with sample size. In Sawyer and Ball (1981) there is a discussion on the role of power and size effects in marketing research. For an applied example, see Wiseman and Rabino (2002) study on the effect of sample size on studies for advertising claims.

❖ **Probability distributions and on-line graphical tools** – To familiarize with probability distributions like the normal, the student-t, binomial or Poisson and the parameters needed to determine their shape, a good way is to plot them. There are many helpful web tools which help plotting a statistical distribution. The most complete one is (probably…) the Statistics Online Computational Resource (SOCR) at UCLA (www.socr.ucla.edu/SOCR.html – click on 'distributions,' but other pages may be helpful. Another good graphical introduction is provided by the Java Applets for the Visualization of Statistical Concepts at the University of Leuven (ucs.kuleuven.be/java/).

Hints for more advanced studies

☞ Find out how critical values for non-standard probability distributions can be generated through bootstrapping methods

☞ Write the equation for the confidence interval for a variance

☞ Performing multiple statistical tests may induce problems, when they are not independent from each other. What sort of problems? Why?

Notes

1. There are $2.24 \cdot 10^{96}$ possible different samples.
2. Actually, to guarantee the normal distribution of the sample means, one of the following two conditions should be met: (a) the target variable X is normally distributed; or (b) the sample size is large (generally a sample size of 50 is taken as a threshold). The latter condition is due to one of the fundamental theorems in statistics, the Central Limit Theorem (see e.g. Le Cam, 1986), which states that the sum of independent random variables from the same distribution tends to the normal distribution as the number of variables becomes large (and the approximation becomes acceptable for 50 variables).
3. More precisely, 1.96 times the standard error, as shown in figure 6.1.
4. As defined in chapter 5, the standard error of the mean is a measure of precision of the sample mean statistic and depends on the variability of the target variable and the sample size. The equation to compute the standard error of the mean with simple random sampling is $\sigma_\mu = \sigma/\sqrt{n}$, where σ is the standard deviation of the target variable.

5. The *critical value* 1.96 corresponds to a probability of 95%, but it is possible to know exactly critical values for any probability level, most packages including Microsoft Excel allow computation of critical values.

6. Again, for sample sizes above 50, the differences between the normal distribution and the Student t distribution are negligible.

7. Since α is also used to denote the confidence level in confidence intervals, this generates some confusion between confidence and significance level. The two concepts are closely related, practically they are complementary, so that a confidence level of 95% (99%) is often indicated with $\alpha = 0.05$ (0.01).

CHAPTER 7

Analysis of Variance

T HIS CHAPTER discusses the techniques used to compare more than two means and test whether one or more factors influence the mean of one or more target variables. These techniques belong to a class of methods named analysis of variance (ANOVA), which is particularly useful in cases where the objective is an assessment of the impact of some controllable marketing factor (for example packaging) on consumer response. The type and complexity of the relevant technique varies depending on the characteristics of the sampling design and on the type of variables being analyzed. This chapter aims to shed some light on the basic principles of ANOVA and introduces the reader to more complex designs.

Section 7.1   describes the basic foundation of ANOVA
Section 7.2   illustrates some of the issues and testing procedures
accompanying ANOVA
Section 7.3   extends ANOVA to account for more dependent and
explanatory variables
Section 7.4   provides some initial advice on employing more complex
ANOVA techniques

THREE LEARNING OUTCOMES

This chapter enables the reader to:

➥   Understand the aims and functioning of analysis of variance
➥   Become aware of the statistical requirements and issues in performing ANOVA
➥   Appreciate the difference between the many available ANOVA tools

PRELIMINARY KNOWLEDGE: statistics and probability concepts in the appendix (once more). Mastery of hypothesis testing concepts in chapter 6. It might be quite useful to have some rough idea of what a linear regression model is. While regression is discussed extensively in chapter 9, the notions introduced in the appendix (yes, again…) should suffice.

7.1 Comparing more than two means: analysis of variance

Do employed people spend more or less on alcoholic drinks and tobacco than unemployed people? With the techniques described in chapter 6, it is not difficult to give an answer by applying a mean comparison test. The situation becomes more complex if one wishes to consider other job statuses like retired workers, part-time jobs or students... Obviously one might run a set of two mean tests, but since each of these tests is subject to a source of error (the level of significance α or *Type I Error* as described in chapter 6), running a multitude of test means will result in an overall error larger than the significance level α, because the probability of getting the wrong answer in at least one of the tests increases as the number of tests increase. This is the so-called problem of *inflated family-wise* (or *experimentwise*) *error rate* encountered when multiple and related hypothesis tests are run.

The *analysis of variance (ANOVA)* is a technique which allows to test whether variability in a variable is attributable to one (or more) factors (Scheffé, 1959). Actually, under the term analysis of variance there is a class of techniques, rather than a single one.

The simplest is *one-way ANOVA*, which allows one to test whether a single factor is relevant in explaining variability for a single target variable. The analysis of variance was developed in an experimental milieu, which explains both its terminology and the limitations that need to be taken into account with non-experimental consumer and marketing data. With scientific experiments, it is possible to control the experimental conditions so that one is reasonably sure that the investigated factor is the only source of variability. Thus, the experimental factor is often termed as *treatment effect* and *treatments* are the different categories of the experimental factor. With consumer and marketing data it is not usual to obtain experimental data and it is often difficult to ensure that no other factor intervenes in explaining variability.

To understand the principles of one-way ANOVA consider table 7.1 which refers to the above example using the classification of the EFS data-set for the 'Economic condition of the household reference person' (a093).

Expenditure in alcohol and tobacco varies according to the economic position. Is this a significant variation or does it only depend on sampling error? The null hypothesis to be tested is mean equality for *all* of the treatments (economic conditions). The alternative hypothesis is that *at least* two means differ. To apply ANOVA the first step is the *variation decomposition*. The idea is that if the factor makes a difference in explaining the target variable, then variability *between* the groups defined by the factor levels

Table 7.1 *Expenditure in alcohol and tobacco by economic position (EFS)*

		Economic position of household reference person						
		Self-employed	Fulltime employee	Pt employee	Unempl.	Ret unoc over min ni age	Unoc - under min ni age	TOTAL
EFS: Total		μ_1	μ_2	μ_3	μ_4	μ_5	μ_6	μ
Alcoholic	Mean	18.56	14.64	12.39	19.48	7.34	11.99	12.67
Beverages,		σ_1	σ_2	σ_3	σ_4	σ_5	σ_6	σ
Tobacco	St. Dev.	19.0	18.5	15.0	19.7	14.6	19.1	17.8

Table 7.2 *Classification table for ANOVA*

Economic position of household reference person group (g)					
1	2	3	4	5	6
Self-employed	Fulltime employee	Pt employee	Unempl.	Ret unoc over min ni age	Unoc – under min ni age
Observations					
x_{11}	x_{21}	x_{13}	x_{14}	x_{15}	x_{16}
x_{21}	x_{22}	x_{23}	x_{24}	x_{25}	x_{26}
x_{31}	x_{32}	x_{33}	x_{34}	x_{35}	x_{36}
...	...	...	...	...	...
Number of observations (n)					
n_1	n_2	n_3	n_4	n_5	n_6
Means					
$\bar{x}_1$	$\bar{x}_2$	$\bar{x}_3$	$\bar{x}_4$	$\bar{x}_5$	$\bar{x}_6$
Overall mean	$\bar{x}$				

should be much larger than variability *within* the groups. Thus, by decomposing total variability into two main components, variability within the groups and variability between the group, it becomes possible to evaluate the relevance of the factor. To decompose variation, one should first rearrange the observation in the data-set as shown in table 7.2.

From the data structure of table 7.2 it is possible to compute measures of variability *within* each of the group and *between* the groups, using the usual concept of sampling variance (see chapter 3):

$$s_T^2 = \frac{\sum_{c=1}^{g} \sum_{r=1}^{n_c} (x_{rc} - \bar{x})^2}{n - 1} \qquad \text{(TOTAL VARIANCE)}$$

$$s_{BW}^2 = \frac{\sum_{c=1}^{g} (\bar{x}_c - \bar{x})^2 n_c}{g - 1} \qquad \text{(VARIANCE BETWEEN GROUPS)}$$

$$s_W^2 = \sum_{c=1}^{g} \sum_{r=1}^{n_c} \frac{(x_{rc} - \bar{x}_c)^2}{n_c - 1} \qquad \text{(VARIANCE WITHIN GROUPS)}$$

The equation for *total variance* is the usual quadratic measure of variability around the mean; the only difference from the usual equation is simply notational, since two summation operators are used to reflect the data structure in table 7.2. Instead, *variance between groups* is quantified by measuring the dispersion of the means of the single groups (columns in table 7.2) around the overall mean, using the number of observations in each group as weights. Finally, variance *within groups* is based on the usual equation for sampling variance, this time computed within each group by computing

---

BOX 7.1 *The relationship for variability decomposition*

The relationship linking total variance with variance between and variance within can be explored by considering the numerators only, that is the sums of squared errors. It can be shown that the total sum of squared errors is equal to the sum of squared errors between groups and squared errors within groups:

$$\sum_{c=1}^{g}\sum_{r=1}^{n_c}(x_{rc}-\bar{x})^2 = \sum_{c=1}^{g}\sum_{r=1}^{n_r}(x_{rc}-\bar{x}_c)^2 + \sum_{c=1}^{g}(\bar{x}_c-\bar{x})^2 n_c$$

Thus, the variation expressed in sum of squared errors (SSE_{TOT}) is decomposed in the sum of squared errors *within* groups (SSE_W) plus the sum of squared error *between* the groups (SSE_{BW}) and the relation also holds for the degrees of freedom (the denominators of the variance equations), as $n-1 = n(g-1)+n-1 = g-1+\sum_{c=1}^{g}(n_c-1) = g-1+n-g = n-1$.

---

the variability of observations around the relative group mean and summating the individual group variances. These three measures are all unbiased estimates of variances and can be compared between themselves and are linked by a mathematical relationship (see box 7.1).

At this point it becomes possible to exploit the distribution encountered in section 6.5 as the ratio between two variances follows the F distribution. Under the hypothesis of mean equality, one would expect variance within the groups to be equal to variance between the groups. Since the factor has no effect on the variance between, we expect this variance to reflect the sampling error only, and the same should happen to variance within.

Hence H_0 can be easily tested using a test for equality of the two variances. Using the results of section 6.5, the test statistic is build as $F = s_{BW}^2/s_W^2$ and under the null hypothesis is distributed as a $F(g\text{-}1,n\text{-}g)$.

Table 7.3 shows the SPSS output for the EFS example (use the menu ANALYZE/ COMPARE MEANS/ONE-WAY ANOVA).

The null hypothesis is strongly rejected at the 5% and 1% significance level (the p-value is 0.1%), as the variance between groups is four times the variance within groups. Mean expenditures for different economic status of the household reference person are significantly different, at least for two groups.

Table 7.3 *SPSS ANOVA output*

ANOVA

EFS: Total Alcoholic Beverages, Tobacco

	Sum of squares	df	Mean square	F	Sig.
Between groups	6171.784	5	1234.357	4.024	0.001
Within groups	151535.3	494	306.752		
Total	157707.1	499			

7.2 Further testing issues in one-way ANOVA

7.2.1 Planned comparisons and contrasts

Sometimes one might be interested in something slightly different to perfect mean equality. For example, one might want to test the difference between the average for the unemployed 'treatment' and all other economic conditions. These specific relationships can be expressed as *linear contrasts*, which are linear combinations of the means. By using contrasts, one might test that expenditure in alcohol for unemployed doubles average expenditure for all other economic conditions, or decide which treatments to compare and whether to consider two or more treatments as a single category. A thorough discussion of linear contrasts with precise instruction on how to run them in SPSS is provided in the book by Field (2005) and some examples are given in box 7.2. Besides, when the null-hypothesis of mean equality is rejected, contrasts can be very useful to have a better insight into where the disparity in means lies. As explained, rejecting mean equality means that *at least two* of the means differ. It is usually relevant to understand which particular means are different from each other. This can be pursued through an adequately planned sequence of tests, hence the name of *planned comparisons*. These are especially informative when the treatments follow an ascending or descending order. A specific approach to planned comparisons is provided by the so-called *Helmert contrasts*, where the first treatment is compared with all of the remaining treatments, the second treatment with all the remaining treatments but the first, the third treatment with all of the remaining ones but the first two, and so on. By looking at the results of this battery of tests, it becomes possible to identify those groups whose difference from the others is most relevant. Alternative comparisons between treatments include *polynomial contrasts*, where it is possible to test whether the trend in means follows a linear, quadratic or cubic sequence or any polynomial relationship between the treatments, *repeated contrasts* (each treatment is compared with the one which follows), *reverse Helmert contrasts* (or *difference contrasts*) that are Helmert contrasts going backwards and *simple contrasts* where the user can choose the benchmark treatment between the first and the last category.[1]

7.2.2 Post-hoc testing procedures

There is a second set of testing procedures to identify the treatments which lead to the rejection of the null hypothesis which come under the name of *post-hoc methods*. While contrasts are generally employed as 'a priori' tests in a confirmatory fashion, post-hoc tests are more relevant to data analysis as data-driven exploratory tests. They consist of a set of paired comparisons, although it becomes necessary to deal with the inflated family-wise error problem mentioned at the beginning of the chapter.

Thus, the critical values are corrected to account for the problem of inflating the risk of *Type I Error* (rejecting the null hypothesis when it is true), measured by the *cumulative Type I error* or *family-wise* error. The approach to correcting the critical values determines the type of test being used. SPSS offers quite a few testing procedures. The most widely known are the Scheffe's test, the Bonferroni's test and the Tukey's test. In the *Scheffe test*, comparisons are run simultaneously for all potential pair-wise and linear combinations of treatments using F critical values corrected to account for the number of treatment being compared. Instead the *Bonferroni* post-hoc method starts by running the usual pair-wise t-tests, but to account for the inflated Type I error rate (rejecting the null hypothesis of mean equality when it is true) an adjustment is provided by dividing the family-wise error by the number of tests. The *Tukey's* test, (also known as the honestly significant difference or HSD test) can be used when samples are of equal size, but statistical packages usually provide variants for unequal sizes. With this test significant differences are identified through an adjusted Studentized range distribution, that is an extension of the *Student t* statistic which uses pooled estimates of the standard errors (see Brown, 1974). When a single mean is used as a control or benchmark, the *Dunnett's t-test* is more appropriate. In the *Hsu's MCB* test comparisons are made between the highest treatment level and each of the remaining ones. Stepwise procedures are often employed (see Welsch, 1977), where a standard ANOVA test is performed on all treatments; if the null hypothesis is rejected, one proceeds with all but the last treatment and performs the one-way ANOVA again until the null hypothesis cannot be rejected. Among these stepwise procedures there are *Duncan's* procedure, the *Studentized Newman-Keuls* (SNK) and the *Ryan-Einot-Gabriel-Welsch* (REGW) in two variants, REGWF (based on the F distribution) and REGWQ (based on the Studentized range distribution).

The above tests generally assume equal sample sizes for each of the treatment. When this condition does not hold, it is advisable to explore the results of alternative tests like *Hochberg's GT2* and *Gabriel's* test. Among other tests designed for situations where population variances are different there is the *Dunnett's C* test. SPSS (see box 7.2) and SAS provide a complete toolbox.

When choosing which post-hoc test should be applied, one must check whether the group sizes are equal or not, compare the population variances and consider the trade-off between the test statistical powers (the probability of accepting the alternative hypothesis when it is true) and the probability of committing a *Type I error*. The most powerful tests are also those where the probability of a *Type I error* are higher, while other tests are more conservative, with smaller *Type I errors* (but higher *Type II errors*). In the latter category there are the Scheffé and Bonferroni's test, while Tukey's test is more appropriate when testing large number of means (see Field, 2005).

7.2.3 Effect size and power in ANOVA

A significant F test leads to reject the hypothesis that the factor is irrelevant to explain differences in the target variable, but it is usually important to assess how relevant it is. An obvious statistic to assess how much of the variability in the target variable is explained by the factor is the ratio between SSE_{BW} and SSE_{TOT} (this is the η^2 or Eta-squared statistic), but more conservative diagnostics exist to account for sampling errors (Olejnik and Algina, 2003). The effect size indicator is usually a proportion, hence included between 0 and 1. Note that a larger F statistic does not necessarily mean a higher effect size, since the F statistic also depends on sample size. Another measure which is important to assess the result of a test is based on statistical power (see section 6.1.2). Most packages provide power estimates for ANOVA and for more complex designs.

BOX 7.2 One-way ANOVA, contrasts and post-hoc testing in SPSS

As an example, let us consider the income level (classified into quartiles), as a determinant of alcohol and tobacco expenditure.

Descriptives

EFS: Total Alcoholic Beverages, Tobacco

	N	Mean	Std. Deviation	Std. Error	95% Confidence Interval for Mean		Minimum	Maximum
					Lower Bound	Upper Bound		
Low income	125	7.467	12.8693	1.1511	5.188	9.745	0.0	70.0
Medium-low income	125	11.381	17.9038	1.6014	8.212	14.551	0.0	93.9
Medium-high income	125	13.040	16.9137	1.5128	10.046	16.035	0.0	79.4
High income	125	18.789	20.8025	1.8606	15.106	22.472	0.0	92.5
Total	500	12.669	17.7777	0.7950	11.107	14.231	0.0	93.9

ANOVA

EFS: Total Alcoholic Beverages, Tobacco

			Sum of Squares	df	Mean Square	F	Sig.
Between Groups	(Combined)		8289.5	3	2763.161	9.172	0.000
	Linear Term	Contrast	7932.7	1	7932.717	26.333	0.000
		Deviation	356.8	2	178.382	0.592	0.554
Within Groups			149417.6	496	301.245		
Total			157707.1	496			

The one-way ANOVA F-test strongly rejects the null hypothesis at the 1% significance level ($F = 9.2$). Assuming a polynomial linear contrast (i.e. mean increasing or decreasing) results in two statistics. The contrast test ($F = 26.33$) refers to the null hypothesis that the means does not follow the assumed functional form (linear in this case). The deviation test ($F = 0.59$), instead, refers to the null hypothesis that the means do not follow any of the functional form of higher degree (quadratic, cubic and quartic in this case). Thus, we reject the hypothesis that the means do not follow a linear trend, while we cannot reject the hypothesis that they do not follow any of the other trends. Hence, we can proceed assuming that average expenditure in alcohol and tobacco proceeds linearly through income categories. Note that, alternatively, one can specify the coefficients of the contrasts in order to test specific mean relationships on sub-groups. For example, suppose that one wants to (a) restrict the comparison to the medium-low income group and the medium-high income group and (b) test that the average of low income and medium-low income groups taken together is equal to the average of the medium-high and high income groups averages taken together. These two comparisons can be obtained by assigning specific coefficients (weights) to each mean in the contrast equation, which is written as $w_1 \mu_L + w_2 \mu_{ML} + w_3 \mu_{MH} + w_4 \mu_H = 0$, under the condition that the sum of weights w is 0. For example, hypothesis (a) is reflected by a weight structure $w_1 = w_4 = 0$, $w_2 = 1$, $w_3 = -1$ or [0,1,–1,0], which corresponds to writing $\mu_{ML} = \mu_{MH}$. Instead hypothesis (b) requires a set of weights [1, 1,–1,–1]. When several sets of contrasts are planned (remember that planned comparisons precede the analysis), one should check that they are *orthogonal* (non-redundant), otherwise the type I error is inflated.

BOX 7.2 *Cont'd*

To check this, one should multiply the weights for the same groups across the hypotheses, then sum the products and the total should be 0. In the above case $(0 \times -1) + (1 \times 1) + (-1 \times -1) + (0 \times -1) = 2$, thus the contrasts cannot be regarded as orthogonal. In fact, both hypotheses compare μ_2 with μ_3 in some way. A hypothesis orthogonal to (a) will be certainly the one comparing μ_1 with μ_4, that is $[1,0,0,-1]$.

Contrast Coefficients

Anonymized hhold inc + allowances (Banded)

Contrast	Low income	Medium-low income	Medium-high income	High income
1	0	1	−1	0
2	1	0	0	−1

Contrast Tests

		Contrast	Value of Contrast	Std. Error	t	df	Sig. (2-tailed)
EFS: Total	Assume	1	−1.659	2.1954	−0.756	496	0.450
Alcoholic	equal	2	−11.322	2.1954	−5.157	496	0.000
Beverages,	variances						
Tobacco							
	Does not	1	−1.659	2.2029	−0.753	247.202	0.452
	assume	2	−11.322	2.1879	−5.175	206.788	0.000
	equal						
	variances						

Both with equal and unequal variances assumed, the first comparison is not rejected and the second one is strongly rejected. Note that it is possible to plan non-orthogonal comparisons in SPSS and adjust the significance levels by exploiting simple adjustment rules. For example, the sequential Dunn-Sidák method suggests to use the following threshold: $\alpha^* = 1 - (1 - \alpha)^{1/k}$, where k is the number of planned comparison. Thus, if $\alpha = 0.05$ and $k = 2$, then $\alpha^* = 0.025$. Let us go back to the non-orthogonal hypotheses (a) and (b). The output is the following:

Contrast Coefficients

Anonymized hhold inc + allowances (Banded)

Contrast	Low income	Medium-low income	Medium-high income	High income
1	0	1	−1	0
2	1	1	−1	−1

(Continued)

BOX 7.2 Cont'd

Contrast Tests

		Contrast	Value of Contrast	Std. Error	t	df	Sig. (2-tailed)
EFS: Total Alcoholic Beverages, Tobacco	Assume equal variances	1	−1.659	2.1954	−0.756	496	0.450
		2	−12.981	3.1048	−4.181	496	0.000
	Does not assume equal variances	1	−1.659	2.2029	−0.753	247.202	0.452
		2	−12.981	3.1048	−4.181	450.920	0.000

With $\alpha = 0.05$ (that is $\alpha^* = 0.025$) we can still reject hypothesis (b).

Post-hoc tests (not reported here, we leave their computation in SPSS to the reader) show for which pairs of income classes the difference between mean expenditure is significant. The differences among the post-hoc tests are not large, both the Bonferroni's and Scheffe's tests (the most conservative) show that expenditure for high incomes is significantly different from both the low-income and medium-low income classes, but not from the medium-high income class. Since an F test (Levene's statistic) on the sample variances tell us that population variances are different, we also check the Dunnett's C test, which confirms the results of the Bonferroni's and Scheffe's ones. Stepwise procedures are shown below

EFS: Total Alcoholic Beverages, Tobacco

	Anonymized hhold inc + allowances (Banded)	N	Subset for alpha = 0.05		
			1	2	3
Student-Newman-Keuls[a]	Low income	125	7.467		
	Medium-low income	125	11.381	11.381	
	Medium-high income	125		13.040	
	High income	125			18.789
	Sig.		0.075	0.450	1.000
	Low income	125	7.467		
	Medium-low income	125	11.381	11.381	
Duncan[a]	Medium-high income	125		13.040	
	High income	125			18.789
	Sig.		0.075	0.450	1.000
Ryan-Einot-Gabriel-Welsch Range[b]	Low income	125	7.467		
	Medium-low income	125	11.381	11.381	
	Medium-high income	125		13.040	
	High income	125			18.789
	Sig.		0.145	0.698	1.000

Means for groups in homogeneous sub-sets are displayed.
  a. Uses Harmonic Mean Sample Size = 125.000.
  b. Critical values are not monotonic for these data. Substitutions have been made to
      ensure monotonicity. Type I error is therefore smaller.

Results of the three tests lead to identical conclusions. Three sub-sets can be identified, one exclusive for high income, low income and medium-high income also end up in different sub-sets, while medium-low income is not significantly different from either of these two sub-sets.

7.2.4 Fixed, random and mixed effects

A distinction which is particularly relevant to the extension of the one-way ANOVA discussed in the following sections is the experimental nature of the treatments used in the analysis. Consider the following two examples:

1. explore differences in monthly food expenditure for different geographical regions; and
2. explore differences in monthly food expenditure according to the point of purchase for the last food shopping.

In the first case, we can assume the treatment to be fixed, as geographical regions for each respondent are generally measured with no significant risk of error. Instead, in case 2, the point of purchase for the last food shopping is likely to be affected by some degree of randomness. While in terms of probability most respondents are likely to have done their shopping at their usual point of purchase, it is still possible that for some specific reason someone who always shops in supermarkets did their last shopping in a discount store. Hence, we want to allow for some randomness in the measurement of treatments.

To try and make an uneasy comparison with scientific experiments, the first case represent *fixed factors* because the researcher can actually 'control' the treatment with precision, while the second case consists of *random* factors because the researcher has no full control on the explanatory variable but measures it with some random error. Another example of random effects is the situation where the number of treatments is too large to consider all of them. Suppose that a chair producer wants to launch a newly designed chair, but wants to check whether consumer evaluation of the product varies with the chair color. Instead of looking for difference across the (infinite?) range of all potential colors, one might decide to check whether evaluations vary for a selection of colors, say the rainbow colors. These colors can be viewed as one specific 'sample' of the whole population of colors, but if we notice different evaluation for this sub-set, we could generalize the conclusion.

So far, we have introduced one-way ANOVA in the context of fixed effects. When we observe random treatments, we need to account for an increased degree of uncertainty. Some multi-variable experimental designs include both fixed and random effects, hence they are called *mixed effects* designs.

7.2.5 Assumptions of one-way ANOVA and non-parametric tests

Two key assumptions are needed for running analyses of variance without risks. First that the sub-samples defined by the treatment are independent (see the discussion in chapter 6 for mean comparison tests) and second that no big discrepancies exist in the variances of the different sub-samples. Theory also requires a normal distribution within the sub-sample, but within limits departure from normality is not a serious issue. When different variances exist results are still reliable, if the sizes of sub-samples are equal. Instead, when both variances and sample sizes differ, there is a high risk of biased results although adjustments to get robust tests are still possible by using the *Brown-Forsythe* test and/or the *Welch* test instead of the usual F test.

Instead, when the normality or independence conditions are violated, there are non-parametric counterparts to analysis of variance just like it happens with two-mean comparison tests. When the samples are exclusive, but are drawn from the same population, there are two non-parametric tests that are particularly helpful. The *Kruskal–Wallis* test simply extends the Mann-Whitney test (see chapter 6) to the case of a higher number of (sub) samples. Thus, it tests the null hypothesis that all the sub-populations have the same distribution function. SPSS also provides the *Jonckheere-Terpstra* test, which tests the same null hypothesis, but against the alternative hypothesis that an increase in treatment leads to an increase in the (median of the) dependent variable. If the samples are actually related (which means that the same respondent may appear in several treatment sub-samples, then one should employ the *Friedman* test, the *Kendall* test or the *Cochran Q* test (see Field, p. 557 for a detailed explanation), which extend to the multiple sample case some of the non-parametric tests seen in chapter 6.

At this point, most readers might be confused by the multitude of tests that have been listed, with the additional frustration that they do not cover the whole range of available tests. The worst has yet to come as the above tests only concern one-way univariate ANOVA and consumer research often requires more complex designs. As a partial consolation, this chapter ends with a table (7.5) trying to draw some directions on which test to run and in which situation.

7.3 Multi-way ANOVA, regression and the general linear model (GLM)

One-way ANOVA is more powerful than a combination of t-tests but obviously using a single factor is a strong limitation. In the example about alcohol expenditure, can we actually assume that the economic position is the only factor affecting expenditure? It is very likely that other confounding factors exist, like income or education. Before generalizing ANOVA as being applicable to the *multi-way* case or to other broader approaches, it is useful to change slightly the perspective. While regression is discussed in greater detail in chapter 9, it can be anticipated here that one-way ANOVA is equivalent to a linear model (see appendix), where the target variable becomes the dependent variable and each of the treatments is transformed into a *dummy variable*, which assumes a value of 1 if respondents are subject to that treatment, which means that they belong to that economic condition and are zero otherwise (see box 8.5).

The notation for the usual example of this chapter would be the following:

$$y_i = \beta_0 + \beta_1 SE_i + \beta_2 FT_i + \beta_3 PT_i + \beta_4 UN_i + \beta_5 RE_i + \beta_6 UA_i + \varepsilon_i$$

where y_i is the amount spent in alcohol and tobacco by the i-th respondent, $SE_i = 1$ if the respondent is self-employed, $FT_i = 1$ for full-time employees, $PT_i = 1$ for part-time employees, $UN_i = 1$ for unemployed resependents, $RE_i = 1$ for retired or inactive respondents and $UA_i = 1$ for those under working age.[2] One variable only will assume value 1 for each respondent (since classes are mutually exclusive), all others will be zero. Linear regression allows one to estimate the unknown parameter β, which measures the (average) shift in y that can be imputed to each of the dummy variables, specifically to belonging or not to a specific economic class. The parallel with ANOVA and hypothesis testing in general is provided by the regression diagnostic toolbox.

As a matter of fact, t-tests as those described in chapter 6 can be used to test the null hypothesis that a specific coefficient β is zero. If the general linear model is simplified to have only the relevant explanatory variable, the t-test on the coefficient is perfectly equivalent to a mean comparison test between the two groups defined by the corresponding variable. For example, a t-test on the hypothesis H_0: $\beta_1 = 0$ for the model $y_i = \beta_0 + \beta_1 SE_i + \varepsilon_i$ corresponds to testing the hypothesis H_0: $\mu_1 = \mu_2$ where μ_1 is the average alcohol and tobacco expenditure for those self-employed and μ_2 is the average alcohol and tobacco expenditure for those *not* self-employed part-time.

Instead, if one considers the general linear model in its full specification, another diagnostic tool usually provided with the regression output is an F-test on the hypothesis that *all* of the coefficients β are zero, which corresponds to the one-way ANOVA hypothesis that all of the means are equal. So H_0: $\beta_1 = \beta_2 = \beta_3 = \beta_4 = \beta_5 = \beta_6 = 0$ corresponds to testing $\mu_1 = \mu_2 = \mu_3 = \mu_4 = \mu_5 = \mu_6$.

Thus, t-tests for mean comparisons correspond to t−tests on the coefficient of bivariate regressions, while one-way ANOVA corresponds to an F-test in multiple regression. The advantages of this similarity so far are pretty negligible. However, they become substantial when one considers *Multi-way (Factorial) ANOVA* or *Multivariate ANOVA (or MANOVA)*.

First, let us explain the differences between these two techniques. Factorial ANOVA is the extension of ANOVA to account for *multiple factors*. If several target variables are also considered simultaneously, then multivariate ANOVA (MANOVA) is a further generalization. MANOVA and other complex ANOVA designs are explored in section 7.4.

It is not worth here to get into the technical notation, which can be quite confusing to inexperienced practitioners (a clear introduction can be found in Rutherford, 2001), but it may be helpful to think that all of these complex analyses of variance can be accommodated into a general linear model. After familiarizing with linear regression (chapter 8), using a GLM perspective is likely to be an easier approach to analysis of variance. Using some simple matrix algebra (see appendix), it is possible to write the GLM as follows:

$$\mathbf{Y} = \mathbf{X}\boldsymbol{\beta} + \mathbf{E}$$

where $\mathbf{Y}$ is a matrix where each row correspond to a respondent and each column to a different dependent variable. The dimensions of $\mathbf{Y}$ are $n \times d$, where n is the overall sample size and d is the number of dependent variables. The matrix $\mathbf{X}$ defines the experimental factors and treatments as exemplified in the univariate case above. It has the same number of rows (n) as $\mathbf{Y}$, but the number of columns can be quite large, as there will be one column for each of the possible treatments for each of the factors, plus a number of columns accounting for the *interactions* between the different factors. The matrix $\mathbf{E}$ represents the error component. With this in mind, it is possible to get an idea of what happens with GLM models, but those interested in mathematical rigor and statistical foundation should refer to Fox (1997).

7.3.1 Multi-way (factorial) analysis of variance

Multi-way (factorial) analysis of variance still has a single dependent variable as one-way ANOVA, but more than one factor . Besides the influence of each individual factor, it provides testing of interactions between treatments belonging to different factors. Clearly, as the number of factors increases, also the number of interaction terms rise exponentially. Thus, albeit theoretically possible, factorial ANOVA with more than two factors is rarely employed, as interpretation of results becomes quite complex.

As an example, consider the following one from the EFS data-set. Does soft drink expenditure vary according to income (in four quartiles) and the number of children (in four categories – no children, one child, two children, more than two children)? There are $n = 500$ observations (including missing values) in the EFS data-set, so the Y matrix has 500 rows and one column. Once the income and number of children categories have been transformed into a set of dummy variables and considering that the dummy trap (see box 8.5) implies that one category is discarded from each factor, the number of columns for X is 16. This number of columns derives from the following sum: 1 (a column of ones for the constant in the GLM) plus 3 columns for the included income categories, 3 columns for the included children categories and 9 columns (3×3) to consider interactions across children and income. Application of two-way ANOVA through the GLM design provides the answers to three questions:

a. Does the income level affect expenditure in soft drinks?
b. Does the number of children affect expenditure in soft drinks? and
c. Does the interaction between income and the number of children affect expenditure in soft drinks?

The basic output of ANOVA is shown in table 7.4. The answer to all three question is yes. All F-tests have a significance level below the usual 1% and 5% threshold levels, with the exception of the interaction term which is above the 1% level but below the 5% one. Many testing procedures can be implemented for multi-way ANOVA in a similar way to one-way ANOVA including planned comparisons and post-hoc tests.

7.4 Starting hints for more complex ANOVA designs

If one is not yet satisfied with the complications of multi-way ANOVA, there is still room for more complex analyses and issues. This does not translate into unaffordable burdens in terms of time of computing efforts (for the statistical packages), as the GLM structure is easily extended using SPSS or SAS. But this comes at a cost. What becomes increasingly difficult is interpreting and commenting on the results. Considering social data as those relevant to the purposes of consumer and marketing research, it is not frequent to find (even in the academic literature) examples of these more articulated

Table 7.4 GLM output for factorial ANOVA

Tests of between-subjects effects
Dependent variable: soft drinks

Source	Type I Sum of squares	df	Mean square	F	Sig.
Corrected model	482.593[a]	15	32.173	8.008	0.000
Intercept	1259.253	1	1259.253	313.425	0.000
incranges	166.654	3	55.551	13.827	0.000
Childband	239.440	3	79.813	19.865	0.000
incranges × Childband	76.499	9	8.500	2.116	0.027
Error	1944.573	484	4.018		
Total	3686.419	500			
Corrected total	2427.166	499			

[a] R-Squared $= 0.199$ (Adjusted R-Squared $= 0.174$).

techniques (Niedrich et al., 2001). Exceptions come from those researchers who manage to set up consumer experiments with real people (see for example Lusk et al., 2004), so that the purposes of the analysis and the planning of the experimental design become more similar to those of other disciplines.[3] Instead, consumer researchers interested in exploring reaction to changing factors tend to opt for other techniques which share the complexity of those discussed in this section, but provide a more readable output, like structural equation models (see chapter 15).

7.4.1 Multivariate ANOVA (MANOVA)

Multivariate ANOVA or MANOVA (Bray and Maxwell, 1985) is a generalization of ANOVA to the multivariate case, when two or more dependent variables are treated simultaneously. Within the GLM framework, it implies that **Y** has more than one column. Consider the situation where the researcher wants to explore whether the household income and the number of children have an impact not only on soft drink expenditure but also on a wider set of products, for example ice creams, DVD rental, etc. Running this analysis as a set of separate ANOVAs leads to a problem which is pretty similar to the one encountered when running a set of mean comparison tests instead of a single ANOVA test (see section 7.1). The usual problem is that the probability of rejecting the null hypothesis when it is in fact true (the well-known *Type I Error*) is inflated by running multiple tests. Furthermore, MANOVA allows one to take into account the fact that the dependent variables may be correlated with each other. As a matter of fact, it is possible that the dependent variables (taken individually) are not affected by the factors, while considering their interactions may lead to rejection of the null hypothesis of equality. To put it in different terms, MANOVA allows one to test whether the factors lead to significant differences in a set of variables. To take another marketing example. Suppose that a company producing canned soft drinks is innovating the packaging of its product by changing the color of the can, its size and the font of the brand name. They run a survey asking to respondents how much they like the changes on a scale from 1 to 10, for each specific packaging feature. The research question might be whether socio-demographic characteristics (age, marital status, income range, etc.) affect appreciation of the new package as a whole. Thus, the dependent variables would be the three measurements of consumer appreciation, while the socio-demographic factors enter the **X** matrix. Note that MANOVA also allows to assign a different weight to each of the dependent variables; for example, one may consider the color of the can as more relevant than the other packaging features. MANOVA has close ties with *Discriminant Analyses*; these are discussed in chapter 11.

7.4.2 Repeated measures and mixed design

It has already been emphasized that the fundamental ANOVA (and MANOVA) techniques assume that the respondents belong to independent groups. If the sampled units are not the same across treatments and factors (that is different respondents belong to different treatment groups), the factorial design is said to be an *independent* one and leads to standard ANOVA, sometimes referred to as *between groups ANOVA* to distinguish it from *related measures ANOVA*. As for mean comparison tests (see chapter 6), it is quite useful to relax this assumption and generalize the statistical methods to allow for some dependency across the sub-group of respondents. If the explanatory factors are measured on the same participants, we have a *related factorial design*. *Mixed designs* are possible when some of the factors are independent and others are not. Both cases (related samples and mixed design) can be dealt with in a satisfactory way. The specific case of related samples called *repeated measures* is quite frequent in consumer research. As an example, consider the situation where the objective is assessing the change in consumer attitudes that can be imputed to an advertising campaign. The best way to measure the impact of the advertisements is recording consumer attitudes before and after the campaign, having exactly the same respondents in the sample (as for panel studies). In this case, it is necessary to take into account the fact that the observations across groups are not independent any more, although there are many advantages as it is more likely to observe homogeneity in the two groups being compared. When running ANOVA for repeated measures, it is important to check that some other conditions (which go under the name of *sphericity*) are met, namely that the variance of the differences between two treatment levels is equal across all combinations of treatments. The sphericity condition can be tested to evaluate the reliability of the *F*-test, although – as usual – adjusted statistics to account for its violation have been made available in literature (see Everitt, 1995). Finally, as one can easily gather, *mixed ANOVA* deals with mixed factorial design where some of the variables are measured independently and others are taken as repeated measures (Naes et al., 2001).

7.4.3 Analysis of covariance (ANCOVA)

A final generalization which is quite interesting for consumer research is the *analysis of covariance (ANCOVA)*, which is the appropriate technique when some of the factors are metric and continuous quantitative variables instead of being measured on a nominal or ordinal scale. For example, consider the case when we use the actual measurement of income instead of the income quartiles as in examples throughout this chapter. To appreciate how useful ANCOVA can be for consumer and marketing research, one should remember the main problem highlighted for ANOVA. When running social surveys it is difficult to control for other factors that may affect the final results of an ANOVA exploration. For example, this textbook often refers to examples on the effects of income or socio-demographic variables on expenditure. However, these examples ignore that economic theory states that one major determinant of purchase decisions is price. If different consumers face different prices, how can one believe to isolate the effect of the variable 'number of children' on soft drink expenditure? If the information on prices is actually available, this is likely to be continuous, which would make it problematic to consider prices as an additional factor and exploit *factorial ANOVA*. Instead, *ANCOVA* is a further generalization of the GLM approach, where one or more continuous *covariates* are added to the right-hand side of the equation. If some conditions are met, relating to the requirement that the relationship between the dependent variable and the covariate is similar across the various groups, then ANCOVA allows one to control for the covariate effect and the rest of the output can be interpreted as in the standard ANOVA approach.

7.4.4 ANOVA in short

The compression of so many concepts as those relevant to ANOVA in a short chapter does not do justice to decades of researchers and lakes of ink poured to deal with all

Table 7.5 What to do when and how in SPSS and SAS

Number of target variables	Number of factors	Measurement of factors	Technique	SPSS	SAS (SAS/STAT)
1	1	nominal/ordinal independent samples	One-way ANOVA	Analyze/Mean Comparison/One way ANOVA	Proc ANOVA (equal sample sizes) – Proc GLM
1	2 or more	nominal/ordinal independent samples	Factorial ANOVA	Analyze/General Linear Model/Univariate	Proc ANOVA (equal sample sizes) – Proc GLM
2 or more	1 or more	nominal/ordinal independent samples	MANOVA	Analyze/General Linear Model/Multivariate	Proc ANOVA (equal sample sizes) – Proc GLM
1	2 or more	nominal/ordinal and continuous, independent samples	ANCOVA	Analyze/General Linear Model/Univariate (Insert covariates)	Proc ANOVA (equal sample sizes) – Proc GLM
2 or more	2 or more	nominal/ordinal and continuous, independent samples	MANCOVA	Analyze/General Linear Model/Multivariate (Insert covariates)	Proc ANOVA (equal sample sizes) – Proc GLM
1	1 or more	nominal/ordinal repeated samples	Repeated ANOVA	Analyze/General Linear Model/Repeated measures	Proc ANOVA (equal sample sizes) – Proc CATMOD (categorical responses) – Proc GLM

(Continued)

Table 7.5 Cont'd

Number of target variables	Number of factors	Measurement of factors	Technique	SPSS	SAS (SAS/STAT)
1	1 or more	nominal/ordinal mixed samples	Mixed ANOVA	Analyze/General Linear Model/Repeated measures	Proc ANOVA (equal sample sizes) – Proc GLM – Proc MIXED
1	1 or more	nominal/ordinal random effects	Variance Component Model	Analyze/General Linear Model/Variance Component	Proc VARCOMP
1	1	nominal/ordinal independent samples, non-normal data and/or non-homogeneous independent samples	Non-parametric tests: Kruskal–Wallis test or Jonckheere–Terpstra test	Analyze/Non-parametric Tests/K independent samples	Proc FREQ/Proc NPAR1WAY
1	1	nominal/ordinal independent samples, non-normal data and/or non-homogeneous related samples	Non-parametric tests: Friedman, Cochran Q or Kendall's test	Analyze/Non-parametric Tests/K-related samples	Proc FREQ/Proc NPAR1WAY

of the potential situations (and their interactions…). However, once one is aware of the multitude of options, most statistical packages make life very easy. It is strongly recommended to become more familiar with the basic design before adopting complex ANOVA designs, but table 7.5 provides a quick summary of the potential alternatives.

A last consideration refers to the generalization of the ANOVA approach to the case where the target variable is not a metric (quantitative) one, which makes it impossible to look at the influence of factors strictly in terms of variability. However, there is a route for checking the impact of categorical factors on a categorical target variable, which is explored in the next two chapters. Chapter 9 introduces a measure of association between two categorical variables, and a generalization to the case of more than two categorical variables (*log-linear analysis*).

Summing up

Often the researcher is interested in comparing more than two means, but there are statistical problems in simply running multiple mean comparison tests. However, there is a class of techniques, called analysis of variance (ANOVA) which enables one to compare more than two means simultaneously without loss of statistical power. Thus, it becomes possible to check whether an explanatory factor which appears in different levels (treatments) for sub-groups of the population (or different populations or in different times) is relevant in explaining differences in means. To choose the appropriate technique among those available in the ANOVA class, it is necessary to take into consideration the nature of samples (whether they are independent or related), the type of variables (nominal, ordinal or continuous), the nature of treatments (fixed, random or mixed) and the number of target variables and explanatory factors being considered. One-way ANOVA simply evaluates the effect of several treatments of a single factor on a single dependent variable. More general experimental designs can be accommodated through the GLM (linked to multivariate regression analysis) and include factorial (multi-way) ANOVA, Multivariate ANOVA (MANOVA), analysis of covariance (ANCOVA). When some of the assumptions for the parametric ANOVA tests do not hold, it is still possible to perform non-parametric tests similar to those employed for bivariate mean comparisons.

EXERCISES

1. Open the Trust data-set
   a. Build a table showing average chicken consumption (q4kilos) for different economic conditions, self-assessed as in q61
   b. Perform one-way ANOVA to test the null hypothesis of mean equality
   c. Considering the five response categories of q61 in the original order, use linear contrasts to test the following hypotheses:
      i. $\mu_1 + \mu_2 = \mu_3 + \mu_4$
      ii. $\mu_1 = \mu_2 + \mu_5$
   d. Are the above comparisons orthogonal?

    e.   Consider a third (non-orthogonal) comparison and report results considering the correct significance levels:
        i. $\mu_1 + \mu_2 = \mu_5$
    f.   Report results of post-hoc tests

2. Open the EFS data-set. Using the univariate GLM, do the following steps in a two-way ANOVA:
    a.   Test the effects of income (*incrange*) and age of the reference person (*ageband*) on expenditure for transport (p607). Is the interaction of the two factors significant?
    b.   Compute and comment on the Helmert planned comparisons
    c.   Perform the Bonferroni, Sidak and Tukey post-hoc tests
    d.   Estimate the effect size (OPTIONS menu)
    e.   Estimate the power (OPTIONS menu)
    f.   Comment on the results

3. Open the EFS data-set:
    a.   Using the univariate GLM menu, run a two-way ANOVA to check the influence of sampling month (a055) and income bracket (*incrange*) on mean expenditures for books (c95111)
    b.   Repeat the analysis considering the sampling month and income range as a random rather than fixed factor. Do results change? Why?
    c.   Instead of using income as a categorical variable, include the metric (scale) measure of income (*incanon*) as a covariate and perform ANCOVA. Comment the results

Further readings and web-links

❖ **Multiple testing and family-wise error rates** – For a comprehensive review of the problems (and solutions) in multiple testing, see Shaffer (1995). For a more recent (and advanced) discussion of the family-wise error rate, see Lehmann and Romano (2005).

❖ **The General Linear Model** – Besides the books by Field (2005) and Rutherford (2001) mentioned in the chapter, Horton (1986) discusses the general linear model within a social science background. To get the SAS perspective on the application of the GLM in its full potential (and complexity) see Timm and Mieczkowski (1997).

❖ **Random or fixed effects?** – The use of random effects in ANOVA is discussed amongst others in Jackson and Brashers (1994). To get a better idea of the implications of assuming fixed or random effects, see the application to sensory panellists discussed in Lundahl and McDaniel (1988).

❖ **Measure of effect size** – A good review of the issues in computing effect size for ANOVA and ANCOVA designs can be found in Cortina and Nouri (2000).

❖ **MANOVA** – Wind and Denny (1974) provide a good example on how multivariate analysis of variance can be applied to evaluate the effectiveness of advertising using seven criterion variables, overall liking, intentions to buy, value for money and four product attributes.

Hints for more advanced studies

☞ How is the power of a test estimated?
☞ What are unbalanced designs? What are their implications for ANOVA?
☞ What are multilevel models? What is their relationship with ANOVA?

Notes

1. For a thorough explanation of planned comparisons, linear contrasts and their use in SPSS readers should refer to the excellent treatment in Field (2005).
2. For reasons that become apparent in chapter 8 (see *dummy trap*), the actual model cannot contain all of the modalities, but one needs to be excluded and used as the reference group.
3. Once more, we suggest that those interested in applying these techniques refer to more specific textbooks, where the wide range of theoretical and empirical issues is explained with greater competence. The book by Field (2005) makes it simple to understand the many features of the GLM, while those who prefer SAS should refer to the less reader-friendly but extensive and precise 'treatment' provided by Khattree and Naik (1999).

PART III

Relationships Among Variables

The third part of the book opens the discussion on the joint and simultaneous exploration of several variables and their relationships, the initial route into the vast world of multivariate statistics.

Chapter 8 begins with the discussion of bivariate correlation, the basic measure of relationship between two metric variables, and then extends the concept to the cases of more than two variables and moves to the 'star' of statistical analysis, the regression model. **Chapter 9** explores methods which deal with the relationships among non-metric (qualitative) variables – those that are measured in ordered or nominal categories. With two categorical variables summarized in a frequency table, measures of association and the related Chi-square test allow an evaluation of the intensity of the relationship. These principles can be generalized to more complex frequency tables for more than two variables and an evaluation of associations can be obtained through log-linear analysis. Finally, canonical correlation analysis looks at the correlation between two sets of variables rather than two individual variables and can deal with both metric and non-metric variables. **Chapter 10** builds on the detection of relationships among variables to present the main data-reduction techniques, factor analysis and principal component analysis, which facilitate exploration of large data-sets.

CHAPTER 8

Correlation and Regression

T HIS CHAPTER shows how relationships between two or more metric variables can be quantified, opening the door to the unlimited world of regression analyses. Correlation looks at the relation between two or more metric (scale) variables without assuming a causal link. When two variables are correlated, for example sales and advertising levels, their changes are associated in some fashion. Simple (bivariate) correlation can be extended to control for other influential variables (partial and semi-partial correlation). Regression introduces a causal direction, for example, looking at the impact of advertising on sales. In its simplest bivariate form, regression relates a metric dependent variable to an explanatory variable, assuming that the latter causes the former. The model can be extended to deal with multiple explanatory variables. More complex regression models are also briefly introduced.

Section 8.1   introduces the measuring of covariance, correlation and partial correlation

Section 8.2   provides the main concepts of bivariate regression analysis

Section 8.3   extends the discussion to multiple regression, with SPSS examples

Section 8.4   deals with the method for automatically selecting the explanatory variables

Section 8.5   links the basic regression models to their generalizations

THREE LEARNING OUTCOMES

This chapter enables the reader to:

➥ Become familiar with the concepts of correlation analysis

➥ Understand estimation of regression models considering the underlying assumptions

➥ Interpret the output of correlation and regression using commercial software

PRELIMINARY KNOWLEDGE: Review the definitions of covariance, functions and parameters in the appendix. Recall the key concepts of hypothesis testing, especially t-tests and F-tests.

8.1 Covariance and correlation measures

Correlation is a very intuitive concept in statistics, but also one of the most powerful, since it introduces the tool for analyzing the relationship between two or more variables and opens the way to the realm of regression models.

The idea of measuring the link between two variables is straightforward. For example, the law of demand tells one that when price increases, we expect a reduction in consumption. Quantifying this relationship requires some further steps. Suppose we want to estimate the average relationship between prices and consumption across the whole range of products in a supermarket. First, the fact that a higher price leads to lower consumption is not necessarily true for all goods. For example, some products might reflect a higher quality at higher prices, so that consumers are more inclined to purchase them. What we expect is that the negative relationship between price and consumption is valid *on average*.

Second, what sort of quantification is needed?

Chapter 4 discussed the use of *univariate statistics*, built to describe the frequency, the tendency and the variability of individual variables. With cross-tabulation and contingency tables, the exploration of the relation between two and more variables begins. In chapter 7, with analysis of variance, the general linear model and analysis of covariance, the discussion of quantitative relationships between two or more variables became explicit. The term *correlation* is rather intuitive and thus refers to a shared relation between variables. This chapter formalizes a concept which will be generalized and extended by the more complex methods covered in the rest of this book.

Before defining correlation it is useful to think at the broader object. To find a shared relation between two variables it is necessary to check whether they vary together across our sample. Thus, the first concept is *covariance*, which is introduced in the appendix and measures the co-movement of two variables x and y. In samples, covariance is estimated through the following statistic:

$$COV(x, y) = s_{xy} = \frac{\sum\limits_{i=1}^{n} (x_i - \bar{x})(y_i - \bar{y})}{n - 1}$$

For each observation i, a situation where both x and y are above or below their respective sample means will increase the covariance value, while the situation where one of the variables is above the sample mean and the other is below decreases the total covariance. Contrary to variance, covariance can assume both positive and negative values. If x and y always move in opposite directions, all terms in the summation above will be negative leading to a large negative covariance. On the other hand, if they always move in the same direction there will be a large positive covariance. In most real cases, as the one of prices and consumption across the whole range of supermarket goods, some observation will be in one direction, others in the opposite one. If one direction prevails on the other, then the covariance will have either a positive or a negative sign. If no relationship exists and our sample is representative, the summation terms will neutralize each other and the overall covariance will be close to zero. On average, we expect a negative relationship – hence a negative covariance – between price and consumption.

The step toward bivariate (Pearson's) correlation is straightforward. Covariance, like variance depends on the measurement units. Thus, if one measures prices in dollars and consumption in ounces, a different covariance value is obtained as compared to the use of prices of Euros and consumption in kilograms, although both situations refer exactly to the same goods and observations. This is obviously not desirable; some form of normalization is needed to avoid the measurement unit problem. The usual approach is standardization (see chapters 6 and the appendix), which requires subtracting the mean and dividing by the standard deviation. Considering the covariance expression, where the numerator is based already on differences from the means; all that is required is dividing by the sample standard deviations for both x and y. This generates the *correlation coefficient*, generally indicated with r when referring to samples (and with ρ for populations) – associated with the name of Karl Pearson for the mathematical formalization although the original idea is credited to Sir Francis Galton (1869).

$$CORR(X, Y) = r_{xy} = \frac{s_{xy}}{s_x s_y} = \frac{\dfrac{\sum\limits_{i=1}^{n} (x_i - \bar{x})(y_i - \bar{y})}{n - 1}}{\sqrt{\dfrac{\sum\limits_{i=1}^{n} (x_i - \bar{x})^2}{n - 1}} \sqrt{\dfrac{\sum\limits_{i=1}^{n} (y_i - \bar{y})^2}{n - 1}}} = \frac{\sum\limits_{i=1}^{n} (x_i - \bar{x})(y_i - \bar{y})}{\sqrt{\sum\limits_{i=1}^{n} (x_i - \bar{x})^2 \sum\limits_{i=1}^{n} (y_i - \bar{y})^2}}$$

The standardization of covariance into correlation returns an indicator which is bound to vary between −1 and 1, where:

a. $r = -1$ means *perfect negative correlation*, so that a $p\%$ increase in x corresponds to a $p\%$ decrease in y and vice versa;
b. $r = 0$ means *no correlation*, so that the two variables move with no apparent relation; and
c. $r = 1$ means *perfect positive correlation*, where a $p\%$ increase (decrease) in x corresponds to a $p\%$ increase (decrease) in y.

Note that no assumption or consideration is made on *causality*, that is the existence of a positive correlation of x and y does not mean that is the increase in x which leads to an increase in y, but only that the two variable move together to some extent. Thus, correlation is symmetric, so that $r_{xy} = r_{yx}$. The next section looks closely at alternative ways of computing the correlation coefficient considering causality (as it happens in regression analysis), which is also closely related to analysis of variance.

So far we have explored correlation only as an indicator to be computed on the observed sample values. However, it can be considered as a *sample statistic* (which allows hypothesis testing) by considering the fact that the sample measure of correlation is affected by sampling error. For example, if a small but positive correlation is observed on a sample, this does not rule out the possibility that the true correlation (in the population) is zero or even negative.

To make a proper use of the correlation coefficient, it is necessary to assume (or check) that:

a. the relationship between the two variables is linear (a scatterplot could allow the identification of non-linear relationships);
b. the error variance is similar for different correlation levels; and
c. the two variables come from similar statistical distributions.

If, besides these conditions, the two variables can be assumed to follow normal distributions, it becomes possible to run hypothesis testing. This last condition can be ignored when the sample is large enough (50 or more observations). Then, as widely discussed in chapter 6, it is possible to exploit the probabilistic nature of sampling to run an hypothesis test on the sample correlation coefficient. The null hypothesis to be tested is that the correlation coefficient in the population is zero.

Thus, when looking at bivariate correlation, there are two elements to be considered. The value of the correlation coefficient r, which indicates to what extent the two variables 'move together' and the significance of the correlation (a *p value*). The latter helps to decide whether the hypothesis that $\rho = 0$ (no correlation in the population) should be rejected. For example, one could find a correlation coefficient $r = 0.6$, which suggests a relatively strong relationship, but a *p value* well above 0.05, so that the hypothesis that the actual correlation is zero cannot be rejected at the 95% confidence level. Consider a different situation with $r = 0.1$ and a *p value* below 0.01. In the latter case, thanks to a larger sample, one can be confident (at the 99% level) that there is a positive relationship between the two variables, although the relationship is weak.

The correlation coefficient r allows one to look at the relationship between two variables x and y. This is only meaningful if one can safely assume that there is no other intervening variable which affects the values of x and y. If we go back to the supermarket example, a negative correlation between prices and consumption is expected. However, suppose that one day the government introduces a new tax which reduces the average available income by 10%. Consumers have less money and consume less. The supermarket tries to maintain its customers by cutting all prices, so that the reduction in prices mitigates the impact of the new tax. If we only observe prices and consumption, it is possible that we observe lower prices and lower consumption and the bivariate correlation coefficient might return a positive value. Thus, we can only use the correlation coefficient when the *ceteris paribus* condition (all other relevant variables being constant) holds. This is rarely the case so it is necessary to control other influential variables like income in the price-consumption relationship.

The *partial correlation coefficient* allows one to evaluate the relationship between two variables after controlling for the effects of one or more additional variables. For example, if x is price, y is consumption and z is income the partial correlation coefficient is obtained by correcting the correlation coefficient between x and y after considering the correlation between x and z and the correlation between y and z:

$$r_{xy|z} = \frac{r_{xy} - r_{xz}r_{yz}}{\sqrt{1 - r_{xz}^2}\sqrt{1 - r_{yz}^2}}$$

When the correlations between x and z and between y and z are zero, then the partial correlation coefficient corresponds to the bivariate correlation coefficient.

The equation can be generalized to more complex situation, where it is controlled for more than one variable. For extremely complex models (more than four conditioning variables), it is preferable to exploit structural equation models as discussed in chapter 15.

In some situations it is desirable to control the correlation between the influencing variable z and only one of the two variables x and y. This is achieved by using the *part* (or *semi-partial*) *correlation coefficient*:

$$r_{x(y \cdot z)} = \frac{r_{xy} - r_{yz}}{\sqrt{1 - r_{yz}^2}}$$

Correlations can also be explored using non-parametric statistics, like the *Spearman's rho* and *Kendall's tau-b statistics* (see section 8.1.1). These statistics are based on the rank-order association between two scale or ordinal variables.

Finally, the *multiple correlation coefficient* is explored in greater detail in the sections devoted to regression analysis. Multiple correlation looks at the joint relationship between one variable (the dependent variable) and a set of other variables.

8.1.1 Correlation measures in SPSS and SAS

To explore the implementation of correlation analysis in SPSS, consider the following variables in the Trust data-set:

- Average weekly chicken consumption (*q4kilos*);
- Chicken price (*price*); and
- Monthly income in Euro (*inclevel*)

We can look at the bivariate correlation between these three variables by selecting ANALYZE/CORRELATE/BIVARIATE. Within the initial window it is possible to ask SPSS to compute three correlation statistics – the Pearson's correlation coefficient, the Spearman's rho and the Kendall's tau-b statistic. When the relationship is in one direction only (which is not the case for this example), it is possible to opt for one-tailed testing. The OPTION button allows one to compute some additional descriptive statistic.

Bivariate correlations are shown in table 8.1.

There is an inverse relationship, as expected, between consumption and price. The correlation is not very strong (−0.33), but highly significant, while the correlations between consumption and income and between price and income are not significant.

These results are confirmed by the non-parametric tests in table 8.2, which detect a stronger negative correlation between consumption and price.

Table 8.1 *Bivariate correlations: Pearson's correlation coefficients*

Correlations				
		In a typical week how much fresh or frozen chicken do you buy for your household consumption (kg)?	price	inclevel
In a typical week how much fresh or frozen chicken do you buy for your household consumption (kg)?	Pearson's correlation	1	−0.327**	0.088
	Sig. (2-tailed)		0.000	0.125
	N	446	438	304
price	Pearson's correlation	−0.327**	1	0.091
	Sig. (2-tailed)	0.000		0.117
	N	438	438	300
inclevel	Pearson's correlation	0.088	0.091	1
	Sig. (2-tailed)	0.125	0.117	
	N	304	300	342

** Correlation is significant at the 0.01 level (2-tailed).

Table 8.2 *Bivariate correlations: Kendall's and Spearman's statistics*

Correlations

			In a typical week how much fresh or frozen chicken do you buy for your household consumption (kg)?	price	inclevel
Kendall's tau_b	In a typical week how much fresh or frozen chicken do you buy for your household consumption (kg)?	Correlation coefficient	1.000	−0.469**	0.059
		Sig. (2-tailed)	.	0.000	0.176
		N	446	438	304
	price	Correlation coefficient	−0.469**	1.000	−0.008
		Sig. (2-tailed)	0.000	.	0.854
		N	438	438	300
	inclevel	Correlation coefficient	0.059	−0.008	1.000
		Sig. (2-tailed)	0.176	0.854	.
		N	304	300	342
Spearman's rho	In a typical week how much fresh or frozen chicken do you buy for your household consumption (kg)?	Correlation coefficient	1.000	−0.630**	0.079
		Sig. (2-tailed)	.	0.000	0.171
		N	446	438	304
	price	Correlation coefficient	−0.630**	1.000	−0.009
		Sig. (2-tailed)	0.000	.	0.880
		N	438	438	300
	inclevel	Correlation coefficient	0.079	−0.009	1.000
		Sig. (2-tailed)	0.171	0.880	.
		N	304	300	342

** Correlation is significant at the 0.01 level (2-tailed).

It seems unlikely that consumption is independent on the income level. To check whether this is really the case, we may look at the partial correlation between consumption and income, this time controlling for price. The results are in table 8.3 and show that after controlling for price, the correlation between consumption and income is positive (0.13) and significant at the 5% significance level. The negative correlation between consumption and price is also higher and more significant after controlling for income.

For correlation analysis in SAS, the base procedure CORR produces a variety of statistics. After defining the variables to be included in the analysis through the

Table 8.3 *Partial correlations*

Control variables			In a typical week how much fresh or frozen chicken do you buy for your household consumption (kg)?	inclevel
price	In a typical week how much fresh or frozen chicken do you buy for your household consumption (kg)?	Correlation	1.000	0.129
		Significance (2-tailed)	.	0.026
		df	0	297
	inclevel	Correlation	0.129	1.000
		Significance (2-tailed)	0.026	.
		df	297	0

Control variables			In a typical week how much fresh or frozen chicken do you buy for your household consumption (kg)?	price
inclevel	In a typical week how much fresh or frozen chicken do you buy for your household consumption (kg)?	Correlation	1.000	−0.350
		Significance (2-tailed)	.	−0.000
		df	0	297
	price	Correlation	−0.350	1.000
		Significance (2-tailed)	0.000	.
		df	297	0

VAR statement, there are four types of correlation statistics that can be produced in SAS according to the option. Three of them are the same statistics described for SPSS (PEARSON, KENDALL and SPEARMAN options), the fourth (HOEFFDING) computes the *Hoeffding's D statistics*, a non-parametric measure of association that detects more general departures from independence (see SAS Institute, 2004b, proc CORR on page 283).

The output correlation matrices can be saved into data-sets through the options OUTP, OUTK, OUTS and OUTD. Within the CORR procedure it is also possible to compute the Cronbach's Alpha (see chapter 1, box 1.2) using the ALPHA option and the covariance matrix (through the COV option). Partial correlations can be easily obtained by specifying the PARTIAL statement, followed by the list of variables to be controlled for after the CORR procedure. Besides Pearson's partial correlation coefficients, SAS can also compute the Kendall and Spearman partial correlations using the above options for simple correlations. For Pearson's and Spearman's correlations p-values for the null hypotheses of no correlation are also produced.

The exploration of relationships between more than two non-metric variables can be further pursued through *log-linear analysis*, which parallels the objective of ANOVA for metric variables and is based on a modeling approach which closely resembles the general linear model but with a dependent categorical variable instead of a scale one. The discussion of log-linear analysis requires more familiarity with the regression concept and is postponed to chapter 9.

8.2 Linear regression

Nowadays it is quite unlikely to find a degree course in any topic which does not cover, at least once, the use of regression analysis. For those who are not allergic to mathematical notation, the equation below should be already familiar:

$$y_i = \alpha + \beta x_i + \varepsilon_i$$

The above equation represents bivariate linear *regression*. For those who have read the appendix, y is expressed as a stochastic linear function of x and there are unknown parameters to be estimated. Consider the example used for correlation analysis. In a sample of n individuals there are – among others – two variables, one which measures chicken purchases in kilograms, the other which reports the price paid in Euros. According to economic theory consumers make up their mind on how much chicken to buy on the basis of price (together with many other determinants). Thus, we are assuming a causal relationship between the price (which we indicate here with the variable x) and consumption (noted here as y). The term regression itself is meant to capture the idea of regressing (going backward) from the dependent variable to the independent variables which cause it, which puts the emphasis on the causality assumption. Note that forgetting that causality is assumed rather than tested may lead to serious misinterpretation (see box 8.2). Consumption (y) is said to be the *dependent variable*, as its value depends on the price. On the other hand, we are assuming that price is the determining factor in this relationship, so that x is usually termed as the *independent variable*, sometime also known as *explanatory variable*. More sophisticated users can choose other terminologies; sometimes y is indicated as the *endogenous variable* (since it is determined inside the regression model), while x is called (not always properly) *exogenous variable*, because we assume (albeit this is not necessarily true) that its value is determined outside the regression model. For example, the consumption decision is taken by the household (thus it is endogenous), whereas the price is not chosen by the household but it is determined outside (which makes it exogenous).

The regression model for a bivariate linear relationships is portrayed by the equation which opened this section. If we take the generic observation i, we can express the i-th observation of the dependent variable y as a linear function of the explanatory variable x, where β is the regression coefficient which measures the impact of the explanatory variable on the dependent variable. Most of the time, the model also includes an *intercept term* α, which represents the average (expected) value of the dependent variable when the explanatory variable is zero. Presumably, even if chicken is freely distributed for some promotional reason (especially if the objective is to clear stocks after some scary bird flu news), consumers will not fill up their fridge and freezers with chicken, but rather stop at a reasonable amount, measured by α.

Assuming that the purchased quantity only depends on the price level is very simplistic. There are many variables, some of which are probably more important than price, which guide consumers in deciding how much chicken to purchase. Therefore a simplification of this behavior through a bivariate relationship, will lead to an error in most cases. The final term of the equation (ε) guarantees that the actual chicken quantity purchased by the i-th respondent equals the right-hand side term of the equation. Thus, the *residual term* (or *error term*) ε embodies anything which is not accounted for by the linear relationship. As with other methods, the regression equation *parameters* α and β are usually estimated on a sample of observations rather than on the whole target population and these estimates are generally indicated with lower case Latin letters, a and b. There are several ways for estimating a and b and the usual one is very intuitive. We want to find those values of a and b which make the error ε as small as possible – at least on average. For any given set of observations **y** and **x** (note that the bold value indicate that these are vectors containing all of the observations in the sample) and estimated parameters a and b, the sample residual term e for the i-th observation is computed as:

$$e_i = y_i - a - bx_i$$

Since e_i can assume both positive and negative values and minimizing the above function poses some problems, a straightforward solution consists in minimizing the squared residuals e^2. This is the notorious *ordinary least squares* method, so that linear regression is sometime referred to as *least square regression* to distinguish it from the multitude of other methods of the regression family.

Before getting into the technical hitches, it is probably wise to list some conditions which (a) open the way to the estimation of parameters a and b and (b) allow to perform a series of statistical diagnostics on the results of the regression analysis:

1. The error term has a mean of zero (otherwise there would be some sort of systematic bias which should be captured by the intercept).
2. The variance of the error term does not vary across cases. For example, the error variability should not become larger for cases with very large values of the independent variable, so that the linear function is adequate to describe consumption for very large and very small prices. This assumption has a name which can make the distinction between those who can pretend to know statistics and those who cannot – *homoskedasticity*. *Heteroskedasticity* is the opposite condition.
3. The error term for each case is independent of the error term for other cases. In other words, there should not be a relationship between errors of different consumers. Usually this translates in the condition that the errors are non-correlated. The omission of relevant explanatory variables would break this assumption, since the omitted independent variable (which is correlated across cases by definition) ends up in the residual term and induces correlation.
4. The error term is also independent of the values of the explanatory (independent) variable. If this condition does not hold, it would mean that the variable is not truly independent and is affected by changes in the dependent variable. A frequent problem with this assumption is the *sample selection bias*, which occurs when non-probabilistic sample is used (Heckman, 1979), that is, the sample only includes units from a specific group. For example, if one measures the response to advertising by sampling those who bought a deodorant after seeing an advert, those who decided not to buy the deodorant are not taken into account even if

they saw the advert. Thus, observations that are likely to be correlated with the dependent variable do not enter the analysis and this leads to a correlated error term.

5. The error term is normally distributed, which corresponds to the assumption that the dependent variable is normally distributed for any value of the independent variable(s). Normality of the residuals makes life easier with hypothesis testing on the regression results, but there are ways to overcome the problem if the distribution is not normal.

While these conditions can be sometime relaxed, in general they are required to hold. Otherwise the estimates of the parameters will be biased and hypothesis testing incorrect. Further problems arise when the data is not continuous or they are truncated.

When the above conditions hold, it is possible to obtain the *least square estimates* for a and b. With these estimates, we can use the linear regression model to *predict* values as follows:

$$\hat{y}_j = a + bx_j$$

where j might refer to an observation within the sample (*within sample forecasts*) or to other observations outside the sample (*out-of-sample forecasts*). In the latter case x_j does not represent a value which appears in the sample and allows to predict the value of the dependent variable, provided the regression relationship holds outside the sample as well.

For example, we might evaluate the effect on chicken consumption of a 10% price increase. Before employing the regression model for predictions it is advisable to evaluate how well it works. This can be accomplished by using some of the standard output of a regression analysis.

The estimated *residual variance* (the estimated variance of ε, usually denoted by s_ε) is a first diagnostic on the *goodness-of-fit* of the regression. If the residuals have a large variance then the error term is relevant and the regression model does not fit the data too well. Since the parameters a and b are estimated in the sample just like a mean, they are accompanied by the *standard error of the parameters*, which measures the precision of these estimates and depends on the sampling size. Knowledge of the standard errors opens the way to run hypothesis testing and compute *confidence intervals* for the regression coefficients (see chapter 6).

There are two standard tests run on regression output. The first is a *t*-test on each of the individual parameters (see chapter 6) for the null hypothesis that the corresponding population parameter is zero. The t statistic is simply obtained by dividing the estimate (for example a, the estimate of α) by its standard error (s_α). A p-value will help us to decide whether to reject or not the null hypothesis that $\alpha = 0$, depending on the confidence level we set. The second test is an F-test (which is more useful when working with multiple independent variables as discussed in section 8.3) run jointly on all coefficients of the regression model to test the null hypothesis that all coefficients are zero. Those with a good memory might recall the discussion about the general linear model of chapter 7 where the F-test was used for factorial ANOVA.

Another typical output of regression analysis is the *coefficient of determination*, also known as the R^2 statistic. The R^2 measures how well the model fits the data (within the sample) by looking at the key measure for least squares, the *sum of squared errors (SSE)* which is the criterion to be minimized to obtain least-squares estimates. The sum of squared residuals (errors) is defined as:

$$SSE = \sum_{i=1}^{n} (y_i - a - bx_i)^2 = \sum_{i=1}^{n} e_i^2$$

Consider that total variability of the dependent variable is related to the total sum of squared deviation from the mean (SST):

$$SST = \sum_{i=1}^{n} (y_i - \bar{y})^2$$

It is quite easy to show that the total sum of squared deviations can be decomposed (in a similar fashion to the variability decomposition of chapter 7) into:

$$SST = SSE + SSR$$

where $SSR = \sum_{i=1}^{n} (a + bx_i - \bar{y})^2$ is the variation accounted for by the regression model. The larger is SSR as compared to SSE, the better the model fits the data. Thus, an obvious indicator of goodness-of-fit is the *coefficient of determination*:

$$R^2 = \frac{SSR}{SST} = 1 - \frac{SSE}{SST}$$

which varies between zero (when the regression model does not explain any of the variability of the dependent variable) and one (when the model perfectly fits the data) Box 8.1 shows how the coefficient estimates are also related to the variability measures and the relation between the coefficient of determination and the correlation coefficient. With multiple explanatory variables (see section 8.3), the square root of the coefficient of determination is the *multiple correlation coefficient* and measures the correlation between the dependent variable and the set of explanatory variables.

An assessment of the validity of the model outside the sample is generally obtained by trying to predict values outside the sample (that is not used for estimation) and comparing predictions with the actual values. A common measure of *forecast accuracy* is the *root mean standard error* (RMSE) which corresponds, for in-sample diagnostics, to the *standard error of residuals* (or *standard error of estimates* in SPSS), that is the square root of SSE divided by $n - 1$.

To show an application of bivariate regression in SPSS, let us consider the EFS dataset. For example, let us look at the relationship between the weekly expenditure in eggs (c11471) and the household size (a049), where one can reasonably assume that the latter causes the former and not the other way round. To run the regression, we click on ANALYZE/REGRESSION/LINEAR, we select household size as the independent variable and eggs as the dependent one. Select ENTER as the method for estimation (the role of alternative methods is described in section 8.4) and optionally ask for more STATISTICS, like confidence intervals for the coefficients (see chapter 6 for details). Most of the other options are only relevant to multiple regression (next section), although the OPTIONS box allows to exclude the intercept from the model (which is not our case).[1]

Residual diagnostics are shown in table 8.4. First, the correlation between the two variables is around 0.23 (which means that the R^2 is $0.23^2 = 0.05$). Obviously the

---

BOX 8.1   *Some 'curious' relationships in regression analysis*

With a few mathematical passages it can be shown that the regression coefficient b is directly related to the Pearson's correlation coefficient r. Furthermore, while the coefficient b obtained when Y is the dependent variable and X is the independent one (we can write it as b_{YX}) differs from the coefficient obtained by regressing X on Y (the coefficient b_{XY}), these two parameters are also related. Finally, the intercept a can be obtained as a last step after the estimation of b.

1. The coefficient b_{YX} can be estimated as $b_{YX} = \dfrac{Cov(X, Y)}{Var(X)} = \dfrac{s_{xy}}{s_x^2}$

2. The intercept a can be then computed as $a_{YX} = \overline{Y} - b_{YX}\overline{X}$

3. The *standardized coefficient* $\hat{b}_{YX}$ is the regression coefficient when both Y and X are standardized to eliminate the effect of the measurement unit. In this case, Y and X have zero mean, thus $a = 0$. The bivariate regression of Y on X corresponds to the regression of X on Y, since $s_x = s_y = 1$. Thus $b_{XY} = b_{YX} = s_{XY} = r$. In facts, the covariance of standardized variables is the correlation coefficient r. It is also easy to pass from the raw regression coefficient to the standardized one, since $\hat{b}_{YX} = b_{YX}\dfrac{s_x}{s_y} = b_{XY}\dfrac{s_y}{s_x}$

4. Finally, it is worth noting why the coefficient of determination is termed as R^2. It can be shown mathematically, for a bivariate regression, that R^2 corresponds to the square of the Pearson's correlation coefficient (which we term r). More generally, it is always true (also for multiple regression), that the R^2 is the squared correlation between the original variable Y and the variable containing its values as predicted by the regression model.

---

correlation would be the same if we had eggs as the independent variable and household size as the dependent one (although this makes little sense in a causal relationship). The *standard error of the estimate* is computed as the ratio between the sum of squared residuals and the number of observations minus one. It is an indicator of the precision of predictions (within the sample) of the dependent variable. Considering that the mean expenditure for eggs is about 0.47, a standard error of 0.53 is not very encouraging, but we might have expected that, since the R^2 tells us that only 5% of the variability in the dependent variable is explained by the regression model. Usually, in cross-section

Table 8.4   *SPSS summary diagnostics for regression analysis*

Model summary				
Model	R	R-square	Adjusted R-square	Std. error of the estimate
1	0.232(a)	0.054	0.052	0.532591

[a]Predictors: (Constant), Household size.

ANOVA(b)					
Model		Sum of squares	df	Mean square	F	Sig.
1	Regression	8.036	1	8.036	28.329	0.000(a)
	Residual	141.259	498	0.284		
	Total	149.295	499			

[a]Predictors: (Constant), Household size.
[b]Dependent Variable: Eggs.

Table 8.5 *Estimates of regression coefficients*

coefficients[a]

Model		Unstandardized coefficients		Standardized coefficients	*t*	Sig.
		B	Std. error	Beta	B	Std. error
1	(Constant)	0.235	0.049		4.834	0.000
	Household size	0.095	0.018	0.232	5.323	0.000

[a]Dependent variable: Eggs.

BOX 8.2   *Spurious regression*

While the power and the fame of regression is undisputable, there is always a risk in relying too much on the outcomes of a regression model. In regression, the causal link is *assumed* (or founded on some theory or evidence), but the method by itself does not allow to test whether it is real causality what we find through our estimates. This is the so-called problem of *spurious regression* (or correlation), especially relevant with time series data (observations in a time sequence), where a significant regression coefficient is found simply because the data show the same pattern, without any real connection between the two variables. Suppose that we have data on the number of commercial flights over the last century and we use it as an independent variable to explain the number of traffic lights in London over the same period. Probably both variables have increased, so a strong correlation would emerge leading to a significant regression relationship, although it would be difficult to argue that the traffic lights are a consequence of commercial flights.

---

studies, an acceptable regression model should explain at least 35–40% of the original variability.

The ANOVA table is the *F*-test on the hypothesis that all the coefficients are zero and is based on the ratio between the mean square error of regression and the mean residual square error. A *p*-value of 0.000 (which means that some non-zero decimal number comes at a later digit) means that the null hypothesis is rejected at the 99% confidence level, so at least one coefficient is relevant to explaining the variability of the dependent variable. This is confirmed by the *t*-test on the coefficients which are shown in table 8.5 together with the raw coefficients' estimates, their standard errors and the standardized coefficient.

The *t*-statistic is the ratio between the raw coefficient and the standard error and is highly significant for both the constant and the independent variable. The standardized coefficient, as expected, is equal to the correlation between the two original variables. The model (which as seen does not fit the data very well) tells us that, on average, even in households with no members (hard to define), there is still an expenditure of £0.235 in eggs. This might seem a bit odd, because there is no-one to eat eggs in a household (and noone who pays the eggs vendor) with no members. In practice, it is not uncommon to find poor estimates for the intercept, because data-sets often have no data points with *x* values close or equal to zero, which suggests that one should be careful in the interpretation of intercept terms. Considering the explanatory variables, household size, and according to the model, the expenditure in eggs increases by about £0.10 for each additional household member.

8.3 Multiple regression

The use of a bivariate regression model does not add much (merely an intercept) to correlation analysis, although it opens the way to predict values of the dependent variable. Regression becomes much more interesting when there is more than one independent variable, as in the multiple regression model. In this case it becomes possible to separate the individual contribution of each regressor (or explanatory or independent variable) – a quite useful result for situations where there are many determinants of the dependent variable. Going back to the example of section 8.1, if chicken consumption depends both on its price and the income of the household, it is possible to separate the contribution of these two explanatory factors through a multiple regression model which also allows one to test the significance of each of the regression coefficients. Without the need of replicating the notation and the explanations of the previous section, it is easy to understand that the multiple regression model with k explanatory variables $(x_1, x_2, ..., x_k)$ is written as follows:

$$y_i = \alpha_0 + \alpha_1 x_{1i} + \alpha_2 x_{2i} + ... + \alpha_k x_{ki} + \varepsilon_i$$

There are not many additional considerations with respect to the bivariate regression model, but it is necessary to add a sixth assumption for inference purposes:

6. The independent variables are also independent from each other. Otherwise we could run into some double counting problems and it would become very difficult to associate a separate meaning to each coefficient.

Assumption six refers to the so-called *collinearity* (or multicollinearity) problem (as *collinearity* exists when two explanatory variables are correlated). *Perfect collinearity* means that one of the variables has a perfect correlation with another variable or with a linear combination of more than one variable. In this case it is impossible to estimate the regression coefficient. Even with non-perfect but strong collinearity, estimates of the coefficients become unstable and *inefficient*, where inefficiency means that the standard errors of the estimates are inflated as compared to the best possible solution. Furthermore, the solution becomes very sensitive to the choice of which variables to include in the model. In a nutshell, when there is multicollinearity the model might look very good at a first glance but is quite likely to produce poor forecasts.

Another element to be considered when dealing with multiple explanatory variable is that the goodness-of-fit to the original data, as measured by the coefficient of determination R^2 (still computed as the ratio between SSR and SST), always increases with the inclusion of additional regressors. This is not desirable because of the 'parsimony' principle, since a model with many explanatory variables is more demanding in terms of data (which usually means higher costs) and computations. If the R^2 index is used to select across models, those with more explanatory variables will regularly result in a better fit. Thus, a proper indicator is the *adjusted R^2* which accounts for the number of explanatory variables:

$$\overline{R}^2 = 1 - (1 - R^2)\frac{n-1}{n-k-1}$$

where n is the number of observations in the sample and k the number of explanatory variables. Thus, an increase in R^2 does not necessarily lead to an increase in $\overline{R}^2$, because increasing k has a negative contribution on the value of the adjusted R^2 which might offset the positive contribution of the increase in R^2.

BOX 8.3 *Collinearity diagnostics*

Consider the output of the multiple regression model from table 8.6. Is it possible that the small number of significant variables could be due to the presence of multi-collinearity, which inflates the standard errors? Here are some diagnostics. The *Tolerance value* (*T*) is computed by regressing each of the independent variables on the remaining independent variables and computing how much of the variability of that independent variable is *not* explained by the other independent variables (that is one minus the R^2 of this auxiliary regression model). The *Variance inflation factor (VIF)* is simply computed as $1/T$. Collinearity symptoms are low values of *T* and high values of the VIF. In our case, the *T* values are all above 0.9 and the *VIF* all below 1.11, suggesting that multi-collinearity is not a big issue. A value of 0.9 means that the R^2 of the auxiliary regression is about 0.1, hence a correlation with the other independent variables of 0.32. When this correlation rises above 0.5 (which corresponds to a tolerance value of 0.75), collinearity could become a problem as variance is inflated by about 33%. The table also shows zero-order correlation, partial correlation and semi-partial (part) correlation (see chapter 8). *Zero order correlation* is simply another name for Pearson's bivariate correlation (or product-moment correlation) between the dependent variable and the explanatory variable. *Partial correlation* controls for all other independent variables and their influence on both the dependent and independent variable, while *part* (or *semi-partial*) *correlation* removes the effect of the other explanatory variables on the independent variable alone, thus is useful to evaluate the unique contribution of the independent variable to the dependent variable. SPSS also produces (upon request) a specific table for collinearity diagnostics.

Collinearity Diagnostics

Model	Dimension	Eigenvalue	Condition Index	(Constant)	price	inclevel	In my household we like chicken	Chicken is a safe food	Age	Number of people currently living in your household (including yourself)
							Variance Proportions			
1	1	6.044	1.000	0.00	0.01	0.01	0.00	0.00	0.00	0.00
	2	0.346	4.179	0.00	0.00	0.82	0.00	0.01	0.02	0.00
	3	0.310	4.419	0.00	0.72	0.00	0.00	0.00	0.01	0.09
	4	0.149	6.374	0.00	0.11	0.15	0.00	0.02	0.15	0.71
	5	0.082	8.601	0.00	0.04	0.00	0.03	0.32	0.66	0.09
	6	0.058	10.234	0.02	0.03	0.01	0.30	0.52	0.03	0.09
	7	0.013	21.827	0.97	0.09	0.01	0.66	0.11	0.14	0.02

[a]Dependent Variable: In a typical week how much fresh or frozen chicken do you buy for your household consumption (kg)?

In the ideal situation of no multicollinearity at all (fully independent explanatory variable), all eigenvalues are one, as eigenvalues larger and smaller than one (the sum of eigenvalues is equal to $k + 1$ where k is the number of explanatory variables excluding the intercept) suggest departure from the independence. If there are very large and very small eigenvalues collinearity is likely to be a problem. The *condition index* for each dimension is the ratio between the corresponding eigenvalue and the largest eigenvalue in the table (the first). Generally condition indices above 10 or 15 indicate some weak problem and as the condition index increases toward 30–40 the problem becomes more serious. When the condition index is too high (like for the 7th dimension in our table), one can look at the variance proportion to assess the seriousness of the problem. In our case, the 7th dimension accounts for a large proportion of variance only for one independent variable ('In my household we like chicken'), so there is no need to worry. Had the variance proportion been large for more than one variable, multi-collinearity would be an issue.

As an example of multiple regression let us consider the following case, this time taken from the Trust data-set. Suppose that we want to explain average weekly chicken consumption (q4kilos) as a function of:

- Chicken price (price variable);
- Monthly income in Euro (inclevel);
- Whether people in the household like chicken (q9);
- The perception of chicken as a safe food (q12d);
- Age of the respondent (q51); and
- Household size (q56).

As before, we select the ENTER method for estimation. On the STATISTICS dialog box we now ask for *part and partial correlations* and *collinearity diagnostics*.

Table 8.6 shows that the correlation between the actual values of q4kilos and those predicted by the model is 0.44 and that the regression model explains about 19% of the original variability of q4kilos. However, to compare this model with another one with a different number of variables, the *adjusted* R^2 is the indicator to be used.

The ANOVA *F*-test reassures that the regression model is not useless, since at least one of the coefficients is significantly different from zero at a 99.9% confidence level.

Have a look at the individual coefficients. If we set a significance level of 5%, only two variables are relevant to explain chicken consumption, namely price and household size.

If price increases by £1, consumption falls by 109 grams, whereas each additional member of the household brings – on average – an increase in consumption of 280 grams. Potential collinearity issues are dealt with in box 8.3, while box 8.4 introduces SAS regression procedures.

8.4 Stepwise regression

The specification of a regression model might lead to two undesirable situations of incorrect specification – the problem of *omitted variables*, where some important independent variable is not included among the regressors, and the problem of *redundant variables*, where some non-significant variables are included even if non-necessary. The first problem is particularly dangerous because it is likely to generate correlation in the residuals thus it violates one key assumption of the model and leads to inefficient estimates (with inflated standard errors). Instead, the problem of redundancy is related to the estimation of a non-parsimonious regression model.

The selection of those independent variables to be included in the model can be entrusted to the researcher or left to a sequential procedure which chooses which variables should be retained in the model out of a given set of potential explanatory variables.

Three types of procedures can be followed for model selection – stepwise, backward and forward.

Stepwise regression considers each single variable before entering it in the model. It starts by adding the variable which shows the highest bivariate correlation with the dependent variable. Then the partial correlation of all remaining potential independent variables (after controlling for the independent variable already included in the model) is explored and the explanatory variable with the highest partial correlation coefficients enters the model.

Table 8.6 *SPSS output for multiple regression analysis*

Model Summary

Model	R	R-Square	Adjusted R-Square	Std. Error of the Estimate
1	0.439[a]	0.193	0.176	1.54945

[a]Predictors: (Constant), Number of people currently living in your household (including yourself), Chicken is a safe food, In my household we like chicken, Age, price, inclevel

ANOVA[b]

Model		Sum of Squares	df	Mean Square	F	Sig.
1	Regression	166.407	6	27.734	11.552	0.000[a]
	Residual	696.233	290	2.401		
	Total	862.639	296			

[a]Predictors: (Constant), Number of people currently living in your household (including yourself), Chicken is a safe food, In my household we like chicken, Age, price, inclevel.
[b]Dependent Variable: In a typical week how much fresh or frozen chicken do you buy for your household consumption (kg)?

Coefficients[a]

Model		Unstandardized coefficients		Standardized coefficients			Correlations			Collinearity Statistics	
		B	Std. Error	Beta	t	Sig.	Zero-order	Partial	Part	Tolerance	VIF
1	(Constant)	−0.386	0.686		−0.562	0.574					
	In my household we like chicken	0.124	0.078	0.084	1.575	0.116	0.134	0.092	0.083	0.974	1.027
	Chicken is a safe food	0.091	0.062	0.078	1.457	0.146	0.097	0.085	0.077	0.972	1.028
	price	−0.109	0.020	−0.301	−5.513	0.000	−0.343	−0.308	−0.291	0.934	1.071
	inclevel	5.31E-006	0.000	0.080	1.445	0.149	0.087	0.085	0.076	0.919	1.089
	Age	0.005	0.006	0.047	0.876	0.382	0.069	0.051	0.046	0.955	1.047
	Number of people currently living in your household (including yourself)	0.280	0.073	0.212	3.815	0.000	0.286	0.219	0.201	0.902	1.109

[a]Dependent Variable: In a typical week how much fresh or frozen chicken do you buy for your household consumption (kg)?

After the model is re-estimated with two explanatory variables, the decision whether to keep the second one is based on the increase of the *F*-value (that is the variability explained by the regression model with two variables compared to the one with a single explanatory variable) or other goodness-of-fit statistics like the Adjusted *R*-Square, Information Criteria, etc. If such increase is not significant, then the second variable is not included in the model. Otherwise the second variable stays in the model and the process continues for the inclusion of a third variable, again selected on the basis of partial correlations, controlling for all of the variables already included in the model. Furthermore, at each step, the procedure may drop one of the variables already included in the model if the model without that variable

BOX 8.4 *SAS regressions*

There are several procedures in SAS which help running regression analyses with a large degree of flexibility. The base procedure is *proc REG* in SAS/STAT, which allows to use different estimation methods for estimating a linear regression model. When the data are collected in the form of contingency tables (as frequencies), it is possible to use *proc CATMOD*, which allows to estimate logistic regression models. Other procedures for estimating logistic, probit and ordered probit and logistic models (see chapter 16) are *proc LOGISTIC* and *proc PROBIT*. Alternatively, one might use a broader specification through *proc GENMOD*, which allows to specify a generalized linear model by choosing different link functions (as explained in section 8.5). Note again the distinction between GENMOD (for the generalized linear model) and *proc GLM*, for the general linear model which focuses on analysis of variance. Non-linear models can be fitted through *proc NLIN*.

does not show a significant decrease in the *F*-value (or any other targeted stepwise criterion).

Forward regression works exactly like stepwise regression, but variables are only entered and not dropped. The process stops when there is no further significant increase in the *F*-value.

Backward regression starts by including all the independent variable and works backward according to the stepwise approach, so that at each step the procedure drops the variable which causes the minimum decreases in the *F*-value and stops when such decrease is not significant.

Let us test the SPSS stepwise method on the above model. This only implies changing the value in the drop-down menu of the regression dialog box to STEPWISE.

Table 8.7 shows the model being considered at each step. The first model only includes *price* as an explanatory variable and the second model adds the household size without removing price. No other variable is considered for inclusion because it would not significantly increase the *F* criterion. Note that the *F* criterion for exclusion and inclusion or the relative *p*-values for determining significance can be adjusted by clicking on the OPTIONS button.

The first model has an *Adjusted R²* of 0.11, the second one rises to 0.167. Recall that when we included all of the independent variables (see section 16.2), the Adjusted R^2 was 0.176, which means that the remaining four variables barely change the goodness-of-fit.

The SPSS output continues by showing the usual results for both model one and model two, including estimates of coefficients and *t*-values for those explanatory variables not included in the model. Table 8.8 shows the estimated coefficients for the two models.

The *t*-tests are all very significant and the collinearity statistics are satisfactory for model two. Note an additional household member leads to a 314 grams increase in consumption, while an extra £1 reduces consumption by 111 grams.

8.5 Extending the regression model

Regression is not a method – it is a whole world with many potential applications in consumer research (see box 8.6). In this short summary of regression analysis (and in the remaining chapters of this book), there are some situations that require the regression

Table 8.7 *Stepwise regression output*

Variables Entered/Removed[a]

Model	Variables entered	Variables removed	Method
1	price	.	Stepwise (Criteria: Probability-of-F-to-enter <= 0.050, Probability-of-F-to-remove >= 0.100).
2	Number of people currently living in your household (including yourself)	.	Stepwise (Criteria: Probability-of-F-to-enter <= 0.050, Probability-of-F-to-remove >= 0.100).

[a]Dependent Variable: In a typical week how much fresh or frozen chicken do you buy for your household consumption (kg)?

Model Summary

Model	R	R-Square	Adjusted R-Square	Std. error of the estimate
1	0.343[a]	0.118	0.115	1.60634
2	0.416[b]	0.173	0.167	1.55799

[a]Predictors: (Constant), price.
[b]Predictors: (Constant), price, number of people currently living in your household (including yourself).

Table 8.8 *Stepwise regression: coefficient estimates*

Coefficients[a]

Model		Unstandardized coefficients B	Std. Error	Standardized coefficients Beta	t	Sig.	Collinearity statistics Tolerance	VIF
1	(Constant)	2.067	0.170		12.190	0.000		
	price	-0.124	0.020	-0.343	-6.270	0.000	1.000	1.000
2	(Constant)	1.128	0.268		4.204	0.000		
	price	-0.111	0.019	-0.305	-5.685	0.000	0.975	1.026
	Number of people currently living in your household (including yourself)	0.314	0.071	0.238	4.427	0.000	0.975	1.026

[a]Dependent Variable: In a typical week how much fresh or frozen chicken do you buy for your household consumption (kg)?

BOX 8.5 *Dummy explanatory variables and regression*

In regression analysis it is quite frequent to include *dummy explanatory variables*, that is binary variables that can only assume two values (usually 0 and 1) depending on whether the case belongs to one situation (or group) or not. For example, in the EFS or Trust data one could include a dummy variable which assumes a value of one if the household has kids and zero otherwise.

The use of dummy variables in regression analysis has several implications. First, one might recall from section 7.3 that a regression with only dummy explanatory variables corresponds to a *multi-way analysis of variance* run through the *general linear model*, where the joint F-test on the dummy coefficients is the ANOVA test of the null hypothesis that the discriminating factors taken altogether do not influence the dependent variable and each t-test on individual coefficients test the specific contribution of that factor, after accounting for all other factors.

Second, the use of dummy variables opens the way to the inclusion of nominal (categorical) variables in a regression model, since each of the categories can be regarded as a binary variable with a value of one if the case belongs to that category. However, when the model includes an intercept, one of the categories should be omitted to avoid the so-called *dummy trap*, which means perfect collinearity due to the fact that if one case does not belong to any of the other categories, necessarily belongs to the one left (to put it mathematically, the value of each of the dummy variables obtained from a categorical variable is equal to 1 minus the sum of all other dummies). In SPSS, it is relatively simple to move from a categorical variable to a set of dummies, by using the TRANSFORM/RECODE INTO DIFFERENT VARIABLES menu. To save time, the best way is to do it for the first category, and click on the PASTE button to obtain the syntax for recoding. Extending the syntax to create a dummy for each of the categories is then straightforward. Otherwise, macros and syntaxes to recode categorical variables into dummies can be easily found on the web.

Finally, it is inappropriate to run standard regression when the dependent variable is a dummy, as the dependent variable (and the residuals) are not normally distributed conditionally to the explanatory variables, as required by the regression assumptions. In these case, one should refer to *discrete choice models*, regression models for discrete and truncated dependent variables. These are discussed in chapter 16. Another option to deal with a target variable which is binary is *discriminant analysis* explored in chapter 11.

---

model to be extended. Without pretending to build an exhaustive list (sometimes is even difficult to decide whether an alternative model is an extension of the regression model or simply a different method), this section provides a list of cases where alternative types of regression model might be useful.

In chapter 7 it was shown through the GLM how *factorial ANOVA* or *MANOVA* can be pursued through a multiple regression with only dummy variables on the right-hand side (see box 8.5 to learn more on dummy variables).

Analysis of Covariance (ANCOVA) exploits both dummy and metric dependent variables on the right-hand side. Some similarities also emerge with methods still to be discussed, like *discriminant analysis* in chapter 11, especially the use of stepwise methods for the choice of predictors.

It is also important to consider what happens when some of the six assumptions listed in this chapter do not hold. When the dependent variable is a binary one (which happens in discriminant analysis) standard regression analysis does not work properly, as the residuals are not *normally distributed*. The technique described for discriminant analysis is an alternative in this situation, but not the only one. *Logistic regression* (see chapter 16) deals with binary dependent variables and can be extended to deal with categorical dependent variables (*multinomial logistic regression*) and ordered variables (*ordered logistic regression*). The right-hand side variables can be either metric or non-metric.

BOX 8.6 *Examples of regression models in consumer and marketing research*

When the idea and the data are good, a well-specified regression model can do the job better than complex models. A good example is the article by Moschis and Churchill (1978) on adolescent consumer socialization, which refers to the youth learning process of consumer-related knowledge, attitudes and other 'consumption skills.' They estimate a set of (independent) regression equations where selected consumer skills (for example knowledge or consumer activism) are related to a set of independent variables like age, social class, newspaper reading, etc. Another example is the study by Liao and Cheung (2001) on the willingness of Singaporeans to e-shop on the Internet, which is the dependent variable in a regression analysis where the explanatory factors are the perceived risks associated with transactions security, the level of education and training in computer applications, the e-market retail price, the perception of the relative life content of Internet-based e-shopping, the perceived quality of Internet e-vendors, the level of Internet usage and the network speed. If we extend the use of regression models to time series data, some complications may arise due to the particular issue of residuals not being independent over consecutive periods of time. Any introductory econometrics textbook covers these issues in sufficient details and they are not so complex as they might look (see for example the classic book by Greene, 2003). Regression models on time series data open the way to a number of applications in advertising research, like the models discussed in Parsons (1976), that are only an early example within a really vast literature. In these models, sales for a specific brand are expressed as a function of the current and past expenditure in advertising.

More generally, a model with a dichotomous dependent variable can be transformed in a model with a continuous dependent variable through a transformation of the binary variable. While a more complete discussion of these models is provided in chapter 16, here it may be useful to highlight the main feature of *discrete choice models*, for example the *logit model* and the *probit model*. Discrete choice models can be defined in terms of a *generalized linear model* (not to be confused with the general linear model of chapter 7) where the expectation (average value) of the dependent variable is related to the set of explanatory variables through a *link function*, a function which translates the dichotomous variable into a continuous one by transforming its expected value. The idea is to obtain a continuous variable which can be related to the regressor as in a standard regression, but with a cut-off point which determines whether an observation ends up with a value of zero or one. The link function depends on the probability distribution chosen. The different link functions are shown in chapter 16. Here it is sufficient to anticipate that the *logit* and *probit* models simply correspond to different assumptions on the link function.

To continue with our list of broken assumptions, *heteroskedastic errors* are situations where the residuals become larger for some values (usually the extremes) and the independent variables require the use of *weighted least squares (WLS)* instead of ordinary least squares. The objective of WLS is essentially to give less relevance to those observations with a larger variance to avoid biased results. To do this each observation is weighted according to a given *weighting variable*, usually the inverse of its variance, when this can be computed.

Finally, the regression analysis discussed in this chapter has focused on linear relationships. When the relationship is non-linear, different estimation methods are used, generally iterative procedures, like *non-linear least squares*.

Summing up

Correlation and regression both work with metric variables. The former assumes no causal relationships, while the latter does. The concept of correlation is introduced through the covariance statistic. Covariance looks at whether there is an association between the variance of two individual metric variables. Since covariance depends on measurement units, a normalization exists which produces a correlation measure (the Pearson's correlation coefficient) which varies between −1 (perfect negative correlation) and +1 (perfect positive correlation), where zero means the absence of correlation. Positive correlation means that the two variables increase (or decrease) together; negative correlation indicates that an increase in either variable is associated with a decrease in the other one. Other correlation indicators exist, like the non-parametric Spearman's and Kendall's Tau b statistic. Provided that the sample is large and probabilistic these statistics can be accompanied by an hypothesis test for checking that correlation in the overall population is not zero. If a third (or more) variable(s) are likely to influence the relationship between the two variables being explored, then the appropriate technique is partial correlation which purges the bivariate correlation from the effects of these control variables.

From correlation to regression the step is small. Take one dependent variable, for example, sales, and one variable which is expected to determine it, for example, advertising. Put them in a linear relationship where sales are depending (on average) on the values of advertising through a (regression) coefficient after correcting for an intercept. This is bivariate linear regression analysis and the parameters to be estimated (the intercept and the regression coefficient) are those which minimize the squared distance between the actual values of sales and those predicted by the model (through the advertising variables). This is the ordinary least squares method. All that follows are simply generalizations, like multiple regression analysis, which allow for more than one explanatory variables. The choice of the explanatory variables might be cumbersome but there are methods which help choosing among a list of potential candidates through stepwise procedures which compare the individual contribution of the independent variables to the explanation of the variability of the dependent variable. Regression analysis relies on some important assumptions, like independence and normality of the residuals, normality and constant variance of the dependent variable for any value of the explanatory ones and non-correlation of the explanatory variables between themselves (absence of collinearity). While it is important to check that these conditions hold, in case they don't, it is still possible to exploit extended regression models.

EXERCISES

1. Open the EFS data-set
    a. Explore the bivariate correlation matrix for the following variables
        i.   Expenditure in petrol (c72211)
        ii. Expenditure in railway and tube fares (exluding season; c73112)

    iii. Expenditure on taxis (c73212)

    iv. Anonymized household income (incanon)

  b. Compute the Spearman's Rho and the Kendall's Tau b – are they very different from Pearson's bivariate correlations?

  c. Income is probably influencing the relationship between expenditure in petrol and expenditure on taxis – use partial correlations to account for such an effect

  d. Now look at the partial correlation between taxis and petrol after accounting for both income and railway/tube fares

  e. What do the above corrrelation values mean? Which ones are significant?

2. On the EFS data-set

  a. Run a bivariate regression where expenditure on petrol (c72211) is the dependent variable and income (incanon) the explanatory variable

  b. Compare the results with the bivariate correlations of previous exercise

  c. Generate the residuals from the regression (hint – click on SAVE in the regression dialog box and ask for *unstandardized residuals* – a new variable is created)

  d. Plot the residuals into a histogram. Do they look normally distributed with mean zero? Perform a test for normality among those suggested in chapter 6.

3. Open the Trust data-set

  a. Run a multiple regression where expenditure on chicken (q5) is the dependent variable and the explanatory variables are:

    i. Price (*price*)

    ii. General attitude toward chicken (*q9*)

    iii. Perceived risk from chicken consumption (q27d)

  b. What is the goodness-of-fit?

  c. Estimate the model again and add the income variable (*income*) and the number of household components (q56). What is the goodness-of-fit now? Which goodness-of-fit indicator allows comparison between the two models?

  d. Estimate the regression equation again with the stepwise method. Which variables are kept in the model?

  e. Interpret the coefficients, relating unitary changes in the explanatory variables to changes in the dependent variables.

Further readings and web-links

❖ **Assumptions of the regression model** – A good and simple discussion of the theory behind regression, with examples, can be found in Lewis-Beck (1980). A more critical and effective review of regression methods is provided by Berk and Berk (2004).

❖ **Time series regression** – When data are observations over time, rather than from a single survey, regression analysis faces many specific problems. The literature on time series regression models tends to be infinite; however for a complete coverage see the classic textbook by Hamilton (1994). A more concise treatment can be found in Ostrom (1990).

❖ **Model comparison and information criteria** – One could write an entire book on the process of choosing among alternative models. Someone did write an entire book – for example McQuarrie and Tsai (1998). A recent review of the use of information criteria is provided by Burnham and Anderson (2004).

❖ **Forecasting with regression models** – Regression is quite useful for forecasting and planning, but before a model can be safely used to this purpose there are many issues to be taken into account. See chapter 4 of the book by Schleifer and Bell (1995) for a review with reference to managerial decisions. The web-site of CAPDM (a distance learning company, www.capdm.com/demos/software/html/capdm/qm/forecasting/) and has a Java applet to explore the role of forecasting with regression models in management decisions.

Hints for more advanced studies

☞ What happens if the homoskedasticity condition is not met?
☞ Find the meaning of the following properties of estimates: unbiasedness, consistency, efficiency
☞ What are systems of regression equations?

Notes

1. Excluding the intercept from a regression model has consequences on the estimates and inference. It is advisable to opt for a no-intercept model only when there are grounds for such choice (i.e. a zero value of x necessarily means a zero value for y).

CHAPTER 9

Association, Log-linear Analysis and Canonical Correlation Analysis

T HIS CHAPTER discusses methods which explore associations among non-metric (categorical) variables and relationships between sets of variables of any type. Where two categorical variables are concerned, measures of correlation are substituted by measures and tests for associations based on frequency tables. When more than two variables are organized in complex contingency tables, log-linear analysis allows one to explore their association in a simplified and parsimonious way. Finally, Canonical Correlation Analysis (CCA) looks at the correlation between two sets of variables rather than two individual variables.

Section 9.1   explains association measures
Section 9.2   introduces log-linear analysis
Section 9.3   describes the theory and application of canonical correlation analysis

THREE LEARNING OUTCOMES

This chapter enables the reader to:

➡ Learn methods looking at the relationships among two or more non-metric variables
➡ Appreciate the potential use of CCA
➡ Understand SPSS output for log-linear analysis and CCA

PRELIMINARY KNOWLEDGE: Review frequency tables and cross-tabulation of categorical variables in chapter 4, the essential elements of hypothesis testing in chapter 6 and the GLM in chapter 7.

9.1 Contingency tables and association statistics

When two qualitative and ordinal variables are represented in a frequency table through cross-tabulation (see chapter 4 and table 4.4), there are several statistics aimed at measuring the degree of *association*. Two variables are said to be associated when they cannot be regarded as statistically independent. As for correlation, this means that changes in the value of one variable are reflected by changes in the values of the other variable. For example, education is generally associated with job position.

Association measures for ordinal variables are based on tables of frequencies, also termed *contingency tables*. Let one consider two categorical variables, with j and k categories, respectively. The *marginal totals*, that is the row and column totals of the contingency table, represent the univariate frequency distribution for each of the two variables. If these variables are independent, one would expect that the distribution of frequencies across the internal cells of the contingency table only depends on the marginal totals and the sample size.

In probability terms two events are regarded as independent when the probability that they happen jointly is exactly the product of the probabilities of the two individual events, that is $Prob(X = a, Y = b) = Prob(X = a)Prob(Y = b)$. Similarly, when two categorical variables are independent, the joint probability of two categorical outcomes is equal to the product of the probabilities of the individual categorical outcomes. Thus, the frequencies within the contingency table should be not to different to these expected values:

$$f_{ij}^* = \frac{n_{i0} \cdot n_{0j}}{n_{00}} = \frac{f_{i0} \cdot f_{0j}}{f_{00}} = f_{i0} \cdot f_{0j}$$

where n_{ij} and f_{ij} are the absolute and relative frequencies respectively, n_{i0} and n_{0j} (or f_{i0} and f_{0j}) are the marginal totals for row i and column j respectively and n_{00} is the sample size (hence the 'total' relative frequency f_{00} equals 1). The distance between the actual (observed) frequencies and the expected ones is processed into a single value, which is the chi-square (χ^2) statistic, through the following equation:

$$\chi^2 = \sum_{i,j} \frac{(f_{ij} - f_{ij}^*)^2}{f_{ij}^*}$$

The more distant the actual joint frequencies are from the expected ones, the larger is the chi-square statistic. Under the independence assumption, the chi-square statistic has a known probability distribution, so that its empirical value can be associated with a probability value to test independence similarly to other hypothesis testing situations described in chapter 6. Obviously the observed frequency values may differ from the expected values f_{ij}^* because of random errors, so that the discrepancy can be tested using a statistical tool, the *chi-square distribution*. As widely discussed in chapters 6 and 7 for tests on means and variances, here the basic principle is also to measure the probability that the discrepancy between the expected and observed value is due to randomness only. If this probability value (taken from the chi-square theoretical distribution) is very low (below the usual 5 or 1% thresholds), then one should reject the null hypothesis of independence between the two variables and proceed assuming some association.

An alternative measure of association is the *contingency coefficient*, which ranges from zero (independence) to values close to 1 for strong association. However, this measure is dependent on the shape (number of rows and columns) of the contingency table, so that a more accurate transformation is the *Cramer's V*, which is also bound between zero and 1 and does not suffer from the above shortcoming, although strong associations may often translate in relatively low (below 0.5) values. When the two variables are strictly nominal it is possible to refer the *Goodman and Kruskal's Lambda*, which compares predictions obtained for one of the variables using two different methods – one which only considers the marginal frequency distribution for that variable and the other which picks up the most likely values after considering the distribution of the second variable. The more one gains in precision by using the second method the closer to 1 will be the Lambda value. If the value is one, it means that each category of the explanatory variable corresponds to a single category of the variable to be predicted. A mirror approach considers the reduction in the prediction error rather than the rate of correct predictions. This is called *uncertainty coefficient*. It is appropriate to compute the Lambda value (or uncertainty coefficient) for both the variables, and then take a weighted average. In this case these measures of association are said to be *symmetric*, while if only one variable is taken into account, they are *asymmetric* measures and depend on the number of categories and marginal totals for the chosen variable. When the categories of the two variables follow a specific order (thus they can be classified as ordinal), more measures are available. SPSS provides output for the *Gamma* statistic, the *Somer's d* statistic and the *Kendall's Tau b* and *Tau c* statistics. Basically these statistics check all pairs of values assumed by the two variables to see if:

(a) a category increase in one variable leads to a category increase in the second one (positive association); or
(b) whether the opposite happens (negative association); or
(c) the ordering of one variable is independent from the ordering of the other.

The Gamma statistic ranges between -1 (negative association) and 1 (positive association), with values close to zero indicating independence. The Gamma statistic does not account for different sizes of the contingency tables, while the Tau statistics do. With square tables (number of rows equals the number of columns), the *Tau b* statistic should be chosen, with rectangular tables the *Tau c* statistic is more appropriate. The *Somer's d* statistic is an adjustment of the Gamma statistic which should be preferred when one variable is believed to be dependent on the other, so that the direction of that relationship is taken into account.

In SPSS all association statistics can be computed by using the ANALYZE/ DESCRIPTIVE STATISTICS/CROSSTABS menu and selecting the appropriate statistics.

Let one consider an example from the Trust sample data-set. Suppose we want to look at the association between the type of purchased chicken, a categorical nominal variable (q6) and the marital status (q52), another nominal variable with three categories. This results are in the contingency table 9.1.

The stronger the association, the more concentrated are the frequencies in a few cells. The main association measures are reported in table 9.2.

Clearly there is no evidence of association. Measures are low and probability values are above the usual threshold of 0.05. Let one now consider the association between

Table 9.1 *Contingency table between q6 and q52 in the Trust data-set (% frequencies)*

		Marital status			
		Single	Married	Other	Total
	'Value' chicken	3.4	7.5	1.2	12.1
In a typical week, what type	'Standard' chicken	15.5	32.1	8.7	56.3
of fresh or frozen chicken do	'Organic' chicken	1.9	8.5	1.2	11.6
you buy for your household's	'Luxury' chicken	3.6	14.3	2.2	20.0
home consumption?	Total	24.4	62.3	13.3	100.0

Table 9.2 *Association measures*

Statistic	Value	Probability
Chi-square	8.51	0.203
Contingency	0.14	
Cramer's V	0.10	
Lambda (sym)		
Uncertainty (sym)	0.10	0.183
Gamma	0.10	0.195
Somer's d (sym)	0.05	0.195
Tau c	0.40	0.195

q6 and q61 (perceived financial status), as reported in table 9.3. In this second example there are several cells with relatively low frequencies, so that some association seems to emerge. This is confirmed by the association measures. For most of the statistics the probabilities are well below both the 0.05 and 0.01 thresholds, although the contingency values are not very high, with the contingency and Gamma measures above 0.25.

Contingency tables can relate more than two categorical variables, and the procedure for testing the significance of the relationship is generalized from the *chi-square* test to *log-linear analysis*, which is the next section. To build a contingency table for more than two variables (like table 4.5 in chapter 4), the response categories of one variable define the rows, while the columns are response categories of the factors, opportunely combined to represent all possible pairs of combinations. Table 9.4 shows an example of a three-way contingency table from the Trust data-set, where trust in information released by the Government (q43j) is crossed with gender (q50) and country of the respondent (q64).

9.2 Log-linear analysis

Log-linear analysis can be better understood by referring to a few concepts presented earlier in this book. The objective of log-linear analysis is to explore the association between more than two categorical variables (like in table 9.4), check whether their association is significant and explore how the variables are associated.

The first known concepts we can exploit are those used for defining the general linear model (chapter 7), where a multivariate regression model was specified as a comprehensive equation which embeds a set of specific ANOVA methods. In the GLM approach, a dependent metric variable is related to a set of variables of any type which represents the various factors and treatments, recoded as dummy variables. The right-hand side of the regression equation can also include interaction terms, represented by the product of different treatment dummies. Tests on the significance of the regression coefficients correspond to factorial ANOVA (or ANCOVA if some of the explanatory variables are metric), with the possibility of including multiple dependent variables for MANOVA and MANCOVA. The limitation of the GLM is the nature of the dependent variable(s), strictly required to be metric (scale) variables (beware, SPSS does not signal nominal dependent variables and treats them as metric).

On the basis of contingency tables, it becomes possible to follow a similar strategy to model the associations and interactions among qualitative (non-metric) variables, through a model which – to emphasize similarity with the GLM – can be named

Table 9.3 Contingency table and associations between q6 and q61

		How would you describe the financial situation of your household?					
		Not very well off	Difficult	Modest	Reasonable	Well off	Total
In a typical week, what type of fresh or frozen chicken do you buy for your household's home consumption?	'Value' chicken	1.9	1.7	4.9	2.7	1.0	12.2
	'Standard' chicken	3.4	4.9	18.2	19.7	9.7	56.0
	'Organic' chicken		1.5	2.9	4.6	2.7	11.7
	'Luxury' chicken	0.2	1.7	5.1	9.7	3.4	20.2
	Total	5.6	9.7	31.1	36.7	16.8	100.0

Statistic	Value	Probability
Chi-square	29.53	0.003
Contingency	0.26	
Cramer's V	0.15	
Lambda (sym)	0.02	0.105
Uncertainty (sym)	0.29	0.001
Gamma	0.24	0.001
Somer's d (sym)	0.16	0.001
Tau c	0.14	0.001

Table 9.4 *Three-way contingency table*

		Government							
		Completely distrust count	2.00 count	3.00 count	Neither count	5.00 count	6.00 count	Completely trust count	Total count
Female	UK	8	6	11	22	12	15	7	81
	Italy	9	4	9	8	8	8	5	51
	Germany	5	3	3	17	14	20	9	71
	Netherlands	2	3	4	13	12	21	18	73
	France	11	4	6	16	13	4	9	63
	Total	35	20	33	76	59	68	48	339
Male	UK	1	1	1	1	1	1	2	8
	Italy	7	3	3	6	11	9	8	47
	Germany	0	4	1	7	4	6	6	28
	Netherlands	0	0	1	5	6	8	6	26
	France	9	2	4	6	6	2	5	34
	Total	17	10	10	25	28	26	27	143
Total		52	30	43	101	87	94	75	482

general log-linear model (GLLM), although in broader terms this type of analysis is also referred to as *log-linear analysis* or *multi-way frequency analysis*.

To understand the principles of log-linear analysis, the starting basis is the relationship between the frequencies in a contingency table and the categories of the interested variables. Consider the example of the three-way contingency tables 9.4, where trust information provided by the Government was related by gender and country of the respondent. Denote with n_{ijk} the absolute frequencies from the combination of the three categorical variables trust, gender and country. For example, n_{713} refers to the 7th trust category (complete trust), the 1st gender category (female) and the 3rd country (Germany) and has a value of 5 (or 1.9%), as shown in table 9.4.

Each joint frequency can be rewritten as:

$$n_{ijk} = \theta u_G u_C u_T u_{GC} u_{GT} u_{CT} u_{GCT}$$

Where u_G is the main effect of gender, u_C is the main effect of country, u_T is the main effect of trust, u_{GC} is the effect of the interaction between gender and country and so on, including a term u_{GCT} for the three-way interaction between gender, country and trust level. Finally, θ is a scale parameter which depends on the total number of observations. We have used the matrix notation and have written them in bold, because-as explained later-the effects can be estimated through a regression model, as coefficients of vectors of dummy variables. Similarly to ANOVA and the general linear model (see chapter 7) each of the categorical variables on the right-hand side can be rewritten as a set of dummy variables so that the estimated coefficients are an estimate of the main and interaction effects. It is worth noting a key difference between the GLM used for ANOVA and the GLLM discussed here. The former has a clearly defined dependent variable (the target metric variable), while the latter simply explore associations among all of the non-metric variables without selecting a dependent one, as the dependent variable of the model are the table frequencies. It is essential to understand that the one described above is an exact relationship which always holds with no error. If all the possible main and interaction effects are considered simultaneously they can fully explain the joint frequency n_{ijk}. This is why the exact model considering all main effects and their interactions is termed *saturated model*.

The *log-linear model* allows one to compute the parameters (effects) **u**. However, when several categorical variables and their interactions are involved, the interpretation of the parameters becomes quite cumbersome. Thus, one of the objectives of *log-linear analysis* is to find a simplified model as compared to the saturated model which can replicate the observed frequencies in a satisfactory way and can be thus regarded as statistically similar to the saturated model, as discussed below.

For this purpose the log-linear model first keeps faith in its name and moves to the logarithmic transformation of all of the elements of the above equation. This is necessary to work with an additive model, which is more similar to the types of models explored throughout this book. Thus, in our three variables example, the saturated log-linear model can be written as:

$$\log(n_{ijk}) = \log(\theta) + \log(\mathbf{u_G}) + \log(\mathbf{u_C}) + \log(\mathbf{u_T}) + \log(\mathbf{u_{GC}}) + \log(\mathbf{u_{GT}})$$
$$+ \log(\mathbf{u_{CT}}) + \log(\mathbf{u_{GCT}})$$

It is straightforward to compute the effects of the saturated model with SPSS; simply click on ANALYZE/LOGLINEAR/GENERAL and be sure to tick *estimates* in the OPTIONS menu. We omit the output here and leave it as an exercise, otherwise the tables could fill the rest of the book. This is why estimating a saturated model is practically useless, because it is virtually impossible to identify which marginal effects are truly important among so many values. The saturated model simply represents the benchmark for simplified models.

Thus, the next step is to start a simplification procedure, which recalls the stepwise procedures of model selection in regression analysis (see chapter 8). A typical approach is *hierarchical log-linear analysis*, which consists in proceeding hierarchically backward, starting with the term corresponding to the highest-order interaction, in our case the term $\log(\mathbf{u_{GCT}})$. However, deleting terms has an important implication: the marginal effects cannot be computed exactly any more, they need to be estimated and there will be an error. Thus, the model becomes:

$$\log(n_{ijk}) = \log(\theta) + \log(\mathbf{u_G}) + \log(\mathbf{u_C}) + \log(\mathbf{u_T}) + \log(\mathbf{u_{GC}}) + \log(\mathbf{u_{GT}})$$
$$+ \log(\mathbf{u_{CT}}) + \varepsilon_{ijt}$$

where the deleted term has been substituted by an error term. Now the effects are not computed exactly, but can be estimated through a regression-like model, where the dependent variable is the (logarithm of) cell frequency and the explanatory variables are a set of dummy variables with value 1 when a main effect or interaction is relevant to that cell of the contingency table with 0 otherwise. Thus, the estimated model coefficients are the (logarithm of) the corresponding main or interaction effects. As for other models now assumptions on the error are needed together with an estimation procedure. For technical details readers should refer to specialized references (as the one by Knoke and Burke, 1980); here it may suffice to know that SPSS exploits an iterative procedure to find the marginal effects which reproduce the observed frequencies in the best possible way. In SPSS click on ANALYZE/LOGLINEAR/MODEL SELECTION and select the backward option. Table 9.5, taken from the SPSS output, summarizes the steps of the backward elimination process.

The first table reports the so-called *k-way effects*. The upper section tests whether the deletion of an effect of order k and all the interaction effects of higher order (for example $k = 3$ is the three-way interaction, $k = 2$ implies deletion of the two-way and

Table 9.5 *Backward elimination steps in hierarchical log-linear analysis*

K-Way and Higher-Order Effects

	k	df	Likelihood ratio Chi-square	Sig.	Pearson Chi-square	Sig.	Number of iterations
K-way and higher	1	69	277.095	0.000	276.963	0.000	0
Order effects[a]	2	58	124.416	0.000	115.059	0.000	2
	3	24	17.155	0.842	14.783	0.927	3
K-way effects[b]	1	11	152.678	0.000	161.904	0.000	0
	2	34	107.262	0.000	100.276	0.000	0
	3	24	17.155	0.842	14.783	0.927	0

[a] Tests that k-way and higher order effects are zero.
[b] Tests that k-way effects are zero.

Partial Associations

Effect	df	Partial Chi-square	Sig.	Number of iterations
q64*q50	4	37.915	0.000	2
q64*q43j	24	63.880	0.000	2
q50*q43j	6	3.225	0.780	2
q64	4	0.752	0.945	2
q50	1	82.057	0.000	2
q43j	6	69.869	0.000	2

Step Summary

Step[b]			Effects	Chi-square[a]	df	Sig.	Number of iterations
0	Generating class[c]		q64*q50* q43j	0.000	0		
	Deleted effect	1	q64*q50* q43j	17.155	24	0.842	3
1	Generating class[c]		q64*q50, q64*q43j, q50*q43j	17.155	24	0.842	
	Deleted effect	1	q64*q50	37.915	4	0.000	2
		2	q64*q43j	63.880	24	0.000	2
		3	q50*q43j	3.225	6	0.780	2
2	Generating class[c]		q64*q50, q64*q43j	20.379	30	0.906	
	Deleted effect	1	q64*q50	39.036	4	0.000	2
		2	q64*q43j	65.001	24	0.000	2
3	Generating class[c]		q64*q50, q64*q43j	20.379	30	0.906	

[a] For 'Deleted Effect,' this is the change in the chi-square after the effect is deleted from the model.
[b] At each step, the effect with the largest significance level for the Likelihood Ratio Change is deleted, provided the significance level is larger than 0.050.
[c] Statistics are displayed for the best model at each step after step 0.

three-way interactions, etc.) lead to a significant departure from the saturated model. The test is based on a *Pearson Chi-square* statistic and a *likelihood-ratio* statistic, and both follow a Chi-square distribution. The output shows that the term corresponding to the three-way interaction can be deleted, since the model still reproduces the observed frequencies in a satisfactory way. This is not the case for the 2nd order interaction terms (and obviously not for the 1st order ones, since the test also considers deletion of higher order terms). The lower half of the table checks whether deletion of effects of order k, while retaining all other effects, leads to problems. Again, there are no issues in deleting the 3rd order interaction term, but the effects of lower order need to be retained.

The *partial association table* (which needs to be explicitly requested in the OPTIONS menu) looks within those effects which should be retained and shows that among the 2nd order effects, deletion of the interaction between gender and trust (q50*q43j) would not worsen the model. The country main effect (q64) can also be deleted without significant consequences.

The third table (*step summary*) reports on the backward elimination procedure. The initial step (step zero) requires the deletion of the interaction of order three. Two chi-square statistics are shown. The first tests the significance of the departure from the observed frequencies (that is from the saturated model) and one hopes to find it non-significant, so that the simplified model has a similar performance to the saturated one. The second tests whether deletion of each of the individual effects of a given order leads to a significant change in the chi-square statistic and it becomes relevant when several terms of the same order are competing to survive in the model.

Thus, the first line (step zero) of the step summary table shows that there is no problem in deleting the 3rd order interaction term. The following step (step 1) explores whether a 2nd order interaction term can be deleted among the three existing ones. As one would expect considering the output of the partial association table, the term which does not change the *Chi-square* statistic is the interaction between trust (q43j) and gender (q50), thus it is removed before step two. In this step, the two remaining 2nd order interaction are explored, because the simplified model is still acceptable (according to the first chi-square statistic of step two). However, elimination of either interaction term would induce a significant change in the *chi-square* statistic, thus the backward elimination process stops and does not consider elimination of lower order terms (in this case, the main effects).[1]

An obvious limitation of proceeding hierarchically is that lower-order effects which could be non-relevant are kept in the model. Thus, a good way is to start from the configuration found with the hierarchical log-linear analysis to fit a general log-linear model. To this purpose, SPSS provides the ANALYZE/LOGLINEAR/GENERAL procedure.

This time, the starting model is the one which came out from the hierarchical analysis (or even a simpler one if theory helps deleting some interactions). The specification of the model to be estimated can be defined by clicking on the MODEL button and selecting the main effects and the interaction terms which will appear in the model.

Then the researcher may try and drop other terms, possibly with the help of theory or the output from the hierarchical analysis, using the *likelihood ratio test* to check whether deletion has significantly worsened the model. As the most experienced readers would gather from the suggested test, estimations within the general log-linear model framework exploit probability assumptions. More precisely one can choose between the Poisson distribution and the multinomial distribution for the frequencies (and the error term). Hopefully, results should be similar.

To obtain estimates of the marginal effects, it is necessary to select *estimates* in the OPTIONS box. Note that the output (as it also happens in the hierarchical approach) informs the user that a 0.5 value (the *delta*) has been added to all cell frequencies to avoid cells with zero counts. This should not affect estimates significantly, but without cells with zero counts one may set the delta to zero. However, it is desirable to always have cells with absolute frequencies higher than 1, otherwise the tests have a much lower power (see chapter 8 for the definition of power).

Table 9.6 shows the goodness-of-fit tests (likelihood ratio and chi-square) and the parameter estimates. The model is acceptable because it is statistically equivalent to the saturated one according to both statistical tests.

Given the high probability values of the statistical tests, we may try to delete other effects lower than the 2nd order ones (since these were already tested in the hierarchical procedure). For example, one might try and delete the main effect of country (q64) which emerged as non-relevant in the association table of the hierarchical analysis. We leave the reader to check that the deletion of this main effect returns a model which is still comparable with the saturated one according to both statistics. Once the process has been completed, one can turn to the output which matters, the parameter estimates. These can be interpreted as *size effects* (see chapter 7), which means that the value of the parameters reflects the relevance of the corresponding effect. Estimates are also accompanied by standard errors, significance levels and confidence intervals, so that conclusions also take into account precision measures. Note that at least one category for each effect is omitted from the analysis and has a coefficient set to zero. This is the reference category. One may compute the odds of being in a given category (or interaction of categories) compared to the reference category by computing the exponent of the parameter estimate. For example, according to the parameter estimates, the odds ratio for being in the lowest category of trust (q43j $= 1$) is about 1.43 (the exponent of 0.357), which means that it is about 43% more likely than being in the top level of trust, the reference category.

Significant parameter estimates indicate that the corresponding effect contributes to explain frequencies in the three-way contingency table, in the sense that it leads to departure from a flat distribution of the frequencies across categories (which would be the case in a no association situation). To identify the most relevant effect one should refer to the Z statistic. Table 9.6 shows that the most relevant significant effect (Z $= 3.97$) is associated with the 2nd order interaction effect $u_{GC}(0,1)$, that is the interaction between the female gender (q50 $= 0$) and the 1st country (United Kingdom). To make sense of the numbers one should refer to the ratios of the Z statistic for two cells.[2] For example, compare UK females (Z $= 3.97$) with German females (Z $= 1.02$) – the ratio is about four – which means that the interaction between being female and from the UK is about four times more important than the interaction between being female and from Germany in explaining departure from a flat distribution, and the effect is positive (it increase frequencies). This would suggest that in a contingency table it is more likely to find UK females than German females, after accounting for all other effects. If one considers the interaction between the lowest level of trust (q43j $= 1$) and country, it is possible to compute that the interaction between Dutch and non-Trusters is about 1.5 times ($-3.49/-2.34$) more influential than the interaction between Germans and non-Trusters. Negative ratios mean that the two effects being compared act in opposite ways. For example, the effect of being female and from the UK is more than twice as strong than the effect of being female and from Italy and the two effects act in opposite directions ($3.97/-1.82 = -2.18$).

SPSS also produces a (usually large) matrix with the correlations of the parameter estimates, which allows one to check whether the correlations between the estimated effects are low. It is desirable to find relatively low correlations between different effects, because this implies that the estimated effects in the analysis are able to explain frequencies independently of each other, making interpretation more robust. Box 9.1 shows some application of log-linear models in consumer and marketing research.

Table 9.6   *Goodness-of-fit tests and estimates of the general log-linear model*

Goodness-of-Fit Tests[a,b]

	Value	df	Sig.
Likelihood ratio	20.379	30	0.906
Pearson Chi-square	18.410	30	0.952

[a] Model: Poisson.
[b] Design: Constant + q43j + q50 + q64 + q43j * q64 + q50 * q64.

Parameter Estimate[b,c]

Parameter	Estimate	Std. Error	Z	Sig.	95% Confidence interval Lower bound	Upper bound
Constant	1.591	0.301	5.287	0.000	1.001	2.180
[q43j = 1.00]	0.357	0.348	1.024	0.306	−0.326	1.040
[q43j = 2.00]	−0.847	0.488	−1.736	0.082	−1.804	0.109
[q43j = 3.00]	−0.336	0.414	−0.813	0.416	−1.148	0.475
[q43j = 4.00]	0.452	0.342	1.322	0.186	−0.218	1.122
[q43j = 5.00]	0.305	0.352	0.867	0.386	−0.385	0.996
[q43j = 6.00]	−0.847	0.488	−1.736	0.082	−1.804	0.109
[q43j = 7.00]	0[a]					
[q50 = 0]	0.617	0.213	2.898	0.004	0.200	1.034
[q50 = 1]	0[a]					
[q64 = 1]	−1.803	0.562	−3.210	0.001	−2.903	−0.702
[q64 = 2]	0.239	0.423	0.567	0.571	−0.589	1.068
[q64 = 3]	−0.146	0.428	−0.340	0.734	−0.984	0.692
[q64 = 4]	0.250	0.401	0.625	0.532	−0.535	1.036
[q64 = 5]	0[a]					
[q43j = 1.00]*[q64 = 1]	−0.357	0.586	−0.608	0.543	−1.506	0.792
[q43j = 1.00]*[q64 = 2]	−0.149	0.511	−0.292	0.770	−1.150	0.852
[q43j = 1.00]*[q64 = 3]	−1.455	0.623	−2.336	0.019	−2.676	−0.234
[q43j = 1.00]*[q64 = 4]	−2.842	0.814	−3.490	0.000	−4.438	−1.246
[q43j = 1.00]*[q64 = 5]	0[a]					
[q43j = 2.00]*[q64 = 1]	0.596	0.701	0.850	0.396	−0.779	1.971
[q43j = 2.00]*[q64 = 2]	0.228	0.677	0.337	0.736	−1.098	1.554
[q43j = 2.00]*[q64 = 3]	0.085	0.669	0.127	0.899	−1.226	1.396
[q43j = 2.00]*[q64 = 4]	−1.232	0.783	−1.574	0.116	−2.767	0.303
[q43j = 2.00]*[q64 = 5]	0[a]					
[q43j = 3.00]*[q64 = 1]	0.624	0.605	1.032	0.302	−0.561	1.810
[q43j = 3.00]*[q64 = 2]	0.256	0.576	0.445	0.656	−0.872	1.385
[q43j = 3.00]*[q64 = 3]	−0.985	0.699	−1.410	0.158	−2.355	0.384
[q43j = 3.00]*[q64 = 4]	−1.232	0.643	−1.917	0.055	−2.492	0.028
[q43j = 3.00]*[q64 = 5]	0[a]					
[q43j = 4.00]*[q64 = 1]	0.486	0.521	0.933	0.351	−0.535	1.507
[q43j = 4.00]*[q64 = 2]	−0.378	0.515	−0.734	0.463	−1.387	0.632
[q43j = 4.00]*[q64 = 3]	0.018	0.475	0.038	0.970	−0.912	0.948

Table 9.6 *Cont'd*

[q43j = 4.00]*[q64 = 4]	−0.740	0.463	−1.599	0.110	−1.647	0.167
[q43j = 4.00]*[q64 = 5]	0ᵃ					
[q43j = 5.00]*[q64 = 1]	0.062	0.559	0.112	0.911	−1.033	1.157
[q43j = 5.00]*[q64 = 2]	0.074	0.504	0.147	0.883	−0.913	1.061
[q43j = 5.00]*[q64 = 3]	−0.123	0.496	−0.248	0.804	−1.096	0.850
[q43j = 5.00]*[q64 = 4]	−0.593	0.470	−1.261	0.207	−1.515	0.329
[q43j = 5.00]*[q64 = 5]	0ᵃ					
[q43j = 6.00]*[q64 = 1]	1.423	0.642	2.217	0.027	0.165	2.680
[q43j = 6.00]*[q64 = 2]	1.116	0.611	1.825	0.068	−0.083	2.314
[q43j = 6.00]*[q64 = 3]	1.397	0.586	2.385	0.017	0.249	2.546
[q43j = 6.00]*[q64 = 4]	1.037	0.561	1.849	0.064	−0.062	2.135
[q43j = 6.00]*[q64 = 5]	0ᵃ					
[q43j = 7.00]*[q64 = 1]	0ᵃ					
[q43j = 7.00]*[q64 = 2]	0ᵃ					
[q43j = 7.00]*[q64 = 3]	0ᵃ					
[q43j = 7.00]*[q64 = 4]	0ᵃ					
[q43j = 7.00]*[q64 = 5]	0ᵃ					
[q50 = 0]*[q64 = 1]	1.698	0.427	3.974	0.000	0.861	2.536
[q50 = 0] * [q64 = 2]	−0.535	0.294	−1.823	0.068	−1.110	0.040
[q50 = 0] * [q64 = 3]	0.314	0.308	1.017	0.309	−0.291	0.918
[q50 = 0] * [q64 = 4]	0.416	0.312	1.331	0.183	−0.196	1.027
[q50 = 0] * [q64 = 5]	0ᵃ					
[q50 = 1]*[q64 = 1]	0ᵃ					
[q50 = 1]*[q64 = 2]	0ᵃ					
[q50 = 1]*[q64 = 3]	0ᵃ					
[q50 = 1]*[q64 = 4]	0ᵃ					
[q50 = 1]*[q64 = 5]	0ᵃ					

ᵃ This parameter is set to zero because it is redundant.
ᵇ Model: Poisson.
ᶜ Design: Constant + q43j + q50 + q64 + q43j * q64 + q50 * q64.

BOX 9.1 *Examples of marketing applications for log-linear analysis*

Log-linear analysis and the log-linear model have been frequently applied for marketing and consumer research. DeSarbo and Hildebrand (1980) review the use of log-linear models in marketing and discuss potential problem areas for marketing applications. Here are some applied examples. Blattberg and Dolan (1981) compare log-linear models with alternative approaches to support several firm marketing strategies, as the selection of target market segments and the development of a direct mail plan (that is the selection of targeted recipients). Danaher (1988) exploits log-linear analysis to model magazine audience, using a theoretical model which aims at measuring the distribution of exposures to magazine adverts. Iacobucci and Henderson (1997) make a slightly different use of log-linear analysis to test for brand switching behavior in buying cars. They consider a 2–way log-linear analysis on cars owned in two different time periods and check for the independence of the two variables and asymmetries in the switching behavior. Finally, Mehdizadeh (1990) explores the student evaluations of economic courses according to a set of criteria.

9.3 Canonical correlation analysis

Canonical Correlation Analysis (CCA) is a very general multivariate technique (see Alpert and Peterson, 1972 and Thompson, 1984), so general in fact that most of the multivariate statistical methods can be interpreted as special case of CCA. Basically, this technique allows one to explore the relationship between a set of dependent variables and a set of explanatory variables. Multiple regression analysis (chapter 8) can be seen as a special case of CCA where there is a single dependent variable. Another feature of CCA is that it is applicable to both metric and non-metric variables.

The link with correlation analysis is immediate; one should think of canonical correlation as the method which maximizes the correlation between two sets of variables rather than individual variables. For example, suppose that you want to explore how attitudes toward chicken relate to general food lifestyles. Considering the Trust dataset, attitudes toward chicken are measured through a set of variables which include taste, perceived safety, value for money, etc. (the items in question q12), while lifestyle measurement is based on agreement with statements like 'I purchase the best quality food I can afford' or 'I am afraid of things that I have never eaten before' (items in q25).

Given that we are dealing with two sets of variables, this technique also needs to combine variables within each set to obtain two composite measures which can be correlated. In standard CCA this synthesis consists in a linear combination of the original variables for each set, leading to the estimation of *canonical variates* or *linear composites*. The bivariate correlation between the two canonical variates is the *canonical correlation*.

Suppose we have a set of m dependent variables, y_1, y_2, ..., y_m and a set of k independent variables, x_1, x_2, ..., x_k. As shown below, canonical variates are linear combinations of the original variables.

The objective is to estimate several (say c) canonical variates as follows:

$$^YS_1 = \alpha_{11}y_1 + \alpha_{12}y_2 + \ldots + \alpha_{1m}y_m$$
$$^XS_1 = \beta_{11}x_1 + \beta_{12}x_2 + \ldots + \beta_{1k}x_k$$
$$^YS_2 = \alpha_{21}y_1 + \alpha_{22}y_2 + \ldots + \alpha_{2m}y_m$$
$$^XS_2 = \beta_{21}x_1 + \beta_{22}x_2 + \ldots + \beta_{2k}x_k$$
$$\ldots$$
$$^YS_c = \alpha_{c1}y_1 + \alpha_{c2}y_2 + \ldots + \alpha_{cm}y_m$$
$$^XS_c = \beta_{c1}x_1 + \beta_{c2}x_2 + \ldots + \beta_{ck}x_k$$

where the (canonical) correlation between the canonical variates $^YS_1$ and $^XS_1$ is the highest possible, followed by the correlation between $^YS_2$ and $^XS_2$ and so on. Furthermore, the extracted canonical variates are not correlated between each other, so that $CORR(^YS_i, ^YS_j) = 0$ and $CORR(^XS_i, ^XS_j) = 0$ for any $i \neq j$, which also implies $CORR(^XS_i, ^YS_j) = 0$.

The linear relationship between variates, $^YS_i = f(^XS_i)$ is the i-th canonical function and the linear coefficient is obviously related to the canonical correlation. This should be clearer by recalling bivariate regression from the previous chapter. The maximum number of canonical functions c (canonical variates) is equal to m or k, whichever the smaller.

A reader familiar with factor (or principal component) analysis (discussed in chapter 10) will recognize the resemblance with these methods although CCA adds an element, since the objective is to estimate the *canonical coefficients* α and β in a way that maximize the *canonical correlation* between the two covariates. The technique starts by maximizing the first canonical correlation, then proceeds by finding the (second) maximum canonical correlation based on the leftover variability and so on. The coefficients are usually normalized in a way that each canonical variable has a variance of one. The method can be generalized to deal with *partial canonical correlation* (which consists in controlling for other sets of variables) and *non-linear canonical correlation* (where the canonical variates show a non-linear relationship).

Canonical correlation analysis produces a variety of outputs, besides the canonical coefficients and the canonical correlations. The *canonical loadings* are the simple linear correlations between each of the original variables and their respective canonical variates, while the *cross-loadings* are the correlations with the opposite canonical variates. As it will be clearer when dealing with principal component analysis, the technique returns *eigenvalues* (or *canonical roots*) which are the squared canonical correlations and represent how much of the original variability is shared by the two canonical variates of each canonical correlation. Furthermore, it is possible to compute *canonical scores* for each of the observations based on the canonical variates. These open the way to plotting cases on a reduced dimensional space, as it happens with factor, principal component and correspondence analyses. Finally, an important indicator is the *canonical redundancy index* which measures how much of the variance in one of the canonical variates is explained by the other canonical variates.[3] If all this terminology is too confusing, hopefully the applied example in the next section will shed some light.

9.3.1 SPSS application

Unfortunately, SPSS does not provide a menu-driven routine for CCA, so that a macro routine through the command editor is necessary, although a very simple one as shown below. To complete the Trust example with an SPSS application, it is sufficient to write the following lines in the Syntax window:

```
INCLUDE '[SPSS installdir] \ Canonical correlation.sps'.
CANCORR SET1=q12a q12b q12c q12d q12e q12f q12g q12h q12i q12j q12k /
        SET2=q25a q25b q25c q25d q25e q25f q25g q25h q25i q25j.
```

The first line of the syntax opens the canonical correlation macro, the other two lines simply define the set of variables. Be sure to write the proper install directory for SPSS and do not forget slashes or full stops to make the routine work. SAS also has a procedure for CCA (see box 9.2). Since there are 11 variables in the first set and 10 in the second one it is possible to extract 10 canonical functions at most. The corresponding canonical correlations are shown in table 9.7.

Since the first canonical correlations are not very high, this suggests that the relationship between our set of dependent variables and the set of independent variables is not too strong. The Wilks' Lambda test (explained in chapter 11) allows to test, for each number of canonical correlation functions, the null hypothesis that the remaining canonical correlations are zero (thus they can be omitted by subsequent analysis). The null hypothesis is rejected at the 95% confidence level for the first three canonical correlation functions (and at the 90% confidence level for the fourth) and provides a first (but not unique) clue about how many dimensions to maintain in the analysis. Table 9.8 reports the estimated standardized canonical coefficients (the raw ones are also available in the SPSS output).

These coefficients are then translated into loadings and cross-loadings, which are the correlations between the original variables and the canonical correlation

Table 9.7   *Canonical correlations*

1	0.440
2	0.359
3	0.321
4	0.277
5	0.232
6	0.210
7	0.149
8	0.130
9	0.081
10	0.039

Test that remaining correlations are zero:

	Wilks	Chi-SQ	DF	Sig.
1	0.502	238.234	120	0
2	0.622	164.049	99	0
3	0.714	116.272	80	0.005
4	0.796	78.689	63	0.088
5	0.862	51.206	48	0.349
6	0.911	32.083	35	0.61
7	0.953	16.462	24	0.871
8	0.975	8.733	15	0.891
9	0.992	2.821	8	0.945
10	0.998	0.52	3	0.915

scores for the same set and for the other set respectively. They are shown in table 9.9.

For example, the first canonical variate is positively related to q12h (chicken is low in cholesterol) and to q12j (buying chicken helps the local economy), but negatively to q12c (chicken is not easy to prepare). While a high score on the first canonical variate probably reflects a positive attitude, note an important point which marks the difference from principal component analysis (chapter 10): the loadings obtained from CCA are not meant to be easily interpreted (as it happens for principal component analysis), since the objective is to obtain the maximum correlation between the two variates and not to synthesize the correlations within each set. The cross-loadings can be useful to explore causality links. For example, there is a positive correlation between q25j ('I am afraid of things I have never eaten before') and the first canonical variate, suggesting that those scared by new food tend to have a better attitude toward chicken. Canonical loadings for Set 2 and cross-loadings between the variables in Set 1 and the second canonical variate can also be explored and it is probably worth looking also at the second and third canonical correlation function, but we leave this to the reader.

Instead, it might be interesting to evaluate the redundancy indices from table 9.10, which shows the proportion of variance explained by each set with respect to itself or to the other. As one might have expected, given the low canonical correlations, the redundancy indices are quite low. For the first function, the first canonical variate only explains 9% of the original variability of the same set of variables and less than 2% for set 2, while the second canonical variate explains about 20% of the second set of variables and something less than 4% of set 1. If one takes the first three canonical functions

Table 9.8 Standardized canonical coefficients

Set 1

	1	2	3	4	5	6	7	8	9	10
q12a	-0.293	0.445	0.392	-0.261	-0.379	0.354	-0.226	0.048	-0.264	0.748
q12b	0.305	-0.362	-0.071	-0.591	0.237	0.267	-0.778	0.276	0.215	-0.25
q12c	-0.365	0.431	-0.267	-0.24	-0.187	0.209	-0.075	0.13	0.639	0.023
q12d	0.244	-0.211	-0.5	0.301	0.026	0.194	0.183	0.039	-0.201	0.026
q12e	0.217	-0.436	0.025	0.708	-0.453	0.398	0.257	-0.464	0.191	0.395
q12f	-0.407	-0.292	-0.005	-0.76	0.354	-0.041	0.706	-0.056	0.304	-0.704
q12g	-0.246	0.343	-0.049	-0.018	-0.452	-0.176	-0.363	-0.505	-0.957	-0.245
q12h	0.633	0.11	0.322	-0.083	0.406	-0.103	0.692	0.639	0.521	0.558
q12i	0.338	0.156	-0.211	-0.252	0.179	0.164	0.045	-0.837	0.072	0.437
q12j	0.349	0.469	0.286	0.099	-0.438	-0.2	-0.183	-0.211	0.313	-0.526
q12k	0.187	0.327	-0.039	0.079	0.23	0.721	0.306	0.117	-0.377	-0.346
q12l	-0.245	-0.23	0.607	0.278	0.344	0.055	-0.341	-0.141	0.165	0.071

Set 2

	1	2	3	4	5	6	7	8	9	10
q25a	-0.55	0.283	-0.018	-0.215	-0.183	0.651	-0.759	0.116	0.431	0.526
q25b	0.14	0.222	0.038	0.135	-0.889	-0.084	-0.277	0.224	-0.533	0.59
q25c	0.323	0.103	-0.127	0.087	0.674	-0.465	-0.09	-0.536	-0.619	0.484
q25d	0.277	0.154	0.614	-0.315	-0.062	-0.258	0.212	-0.219	0.467	0.42
q25e	-0.142	-0.673	0.769	0.665	-0.089	-0.047	-0.487	0.15	-0.217	-0.417
q25f	0.295	0.682	-0.168	-0.425	-0.38	0.391	0.232	-0.258	-0.207	-0.796
q25g	0.322	0.057	-0.12	-0.438	0.39	-0.306	-0.979	0.058	0.421	-0.495
q25h	0.087	-0.504	-0.12	-0.725	0.022	0.326	0.083	0.315	-0.308	0.148
q25i	-0.175	0.329	0.444	0.135	0.476	0.284	0.092	0.766	-0.375	-0.032
q25j	0.5	-0.305	-0.115	0.428	0.025	0.847	0.206	-0.449	0.187	0.09

Table 9.9 *Canonical loadings*

Set 1

	1	2	3	4	5	6	7	8	9	10
q12a	-0.06	-0.1	0.382	-0.384	-0.564	0.371	0.073	0.126	-0.174	0.305
q12b	0.443	-0.404	0.096	-0.512	-0.213	0.339	-0.371	0.177	0.001	-0.077
q12c	-0.354	0.496	-0.363	0.034	-0.018	0.262	-0.049	0.115	0.505	0.056
q12d	0.373	-0.209	-0.284	0.08	-0.326	0.265	0.188	0.215	-0.143	0.001
q12e	0.118	-0.499	0.232	-0.007	-0.604	0.404	0.24	-0.143	0.132	0.04
q12f	-0.085	-0.372	0.308	-0.528	-0.305	0.168	0.502	-0.136	0	-0.234
q12g	0.216	0.096	0.258	-0.357	-0.277	-0.06	0.167	-0.013	-0.452	-0.011
q12h	0.516	0.119	0.375	-0.281	-0.042	-0.09	0.372	0.294	-0.055	0.228
q12i	0.217	0.258	-0.217	-0.174	0.465	-0.017	-0.061	-0.714	0.113	0.246
q12j	0.494	0.219	0.362	0.031	-0.494	-0.026	-0.011	-0.134	0.248	-0.432
q12k	0.025	0.405	0.069	0.216	0.41	0.672	0.078	0.03	-0.18	-0.316
q12l	-0.175	0.038	0.592	0.325	0.506	0.211	-0.157	-0.169	0.115	-0.037

Set 2

	1	2	3	4	5	6	7	8	9	10
q25a	-0.616	0.164	0.178	-0.204	0.046	0.402	-0.372	-0.421	-0.038	0.197
q25b	0.577	0.148	-0.126	0.205	-0.494	-0.111	-0.233	0.409	-0.172	0.284
q25c	-0.108	0.114	0.153	-0.119	0.38	0.003	-0.221	-0.621	-0.567	0.193
q25d	0.286	0.109	0.77	-0.338	-0.048	-0.074	0.154	-0.181	0.283	0.233
q25e	-0.268	-0.317	0.624	0.001	-0.102	0.09	-0.278	-0.397	-0.33	-0.277
q25f	-0.1	0.346	0.305	-0.328	-0.16	0.271	-0.029	-0.492	-0.382	-0.426
q25g	0.678	0.055	-0.183	-0.005	0.128	-0.129	-0.478	0.392	0.271	-0.122
q25h	0.038	-0.56	0.087	-0.7	0.009	0.234	0.026	0.059	-0.348	0.079
q25i	0.305	0.395	0.229	0.184	0.395	0.235	0.071	0.663	-0.081	0.059
q25j	0.719	-0.09	-0.113	0.367	0.089	0.503	-0.02	0.071	0.205	0.141

Cross 1-2

	1	2	3	4	5	6	7	8	9	10
q12a	-0.027	-0.036	0.123	-0.106	-0.131	0.078	0.011	0.016	-0.014	0.012
q12b	0.195	-0.145	0.031	-0.142	-0.05	0.071	-0.055	0.023	0	-0.003
q12c	-0.156	0.178	-0.116	0.01	-0.004	0.055	-0.007	0.015	0.041	0.002
q12d	0.164	-0.075	-0.091	0.022	-0.076	0.056	0.028	0.028	-0.012	0
q12e	0.052	-0.179	0.074	-0.002	-0.14	0.085	0.036	-0.019	0.011	0.002
q12f	-0.037	-0.134	0.099	-0.146	-0.071	0.035	0.075	-0.018	0	-0.009
q12g	0.095	0.035	0.083	-0.099	-0.064	-0.013	0.025	-0.002	-0.037	0
q12h	0.227	0.043	0.12	-0.078	-0.01	-0.019	0.055	0.038	-0.004	0.009
q12i	0.095	0.093	-0.07	-0.048	0.108	-0.004	-0.009	-0.093	0.009	0.01
q12j	0.217	0.079	0.116	0.008	-0.115	-0.005	-0.002	-0.017	0.02	-0.017
q12k	0.011	0.146	0.022	0.06	0.095	0.141	0.012	0.004	-0.015	-0.012
q12l	-0.077	0.014	0.19	0.09	0.117	0.044	-0.023	-0.022	0.009	-0.001

Cross 2-1

	1	2	3	4	5	6	7	8	9	10
q25a	-0.271	0.059	0.057	-0.056	0.011	0.085	-0.055	-0.055	-0.003	0.008
q25b	0.253	0.053	-0.041	0.057	-0.115	-0.023	-0.035	0.053	-0.014	0.011
q25c	-0.047	0.041	0.049	-0.033	0.088	0.001	-0.033	-0.081	-0.046	0.007
q25d	0.126	0.039	0.247	-0.093	-0.011	-0.016	0.023	-0.024	0.023	0.009
q25e	-0.118	-0.114	0.2	0	-0.024	0.019	-0.041	-0.052	-0.027	-0.011
q25f	-0.044	0.124	0.098	-0.091	-0.037	0.057	-0.004	-0.064	-0.031	-0.017
q25g	0.298	0.02	-0.059	-0.001	0.03	-0.027	-0.071	0.051	0.022	-0.005
q25h	0.017	-0.201	0.028	-0.194	0.002	0.049	0.004	0.008	-0.028	0.003
q25i	0.134	0.142	0.073	0.051	0.092	0.049	0.011	0.086	-0.007	0.002
q25j	0.316	-0.032	-0.036	0.101	0.021	0.106	-0.003	0.009	0.017	0.005

Table 9.10 *Redundancy analysis*

	Set 1 by Set 1	Set 1 by Set 2	Set 2 by Set 2	Set 2 by Set 1
CV1-1	0.093	0.018	0.196	0.038
CV1-2	0.096	0.012	0.077	0.01
CV1-3	0.105	0.011	0.125	0.013
CV1-4	0.091	0.007	0.098	0.007
CV1-5	0.158	0.008	0.061	0.003
CV1-6	0.091	0.004	0.064	0.003
CV1-7	0.058	0.001	0.058	0.001
CV1-8	0.066	0.001	0.176	0.003
CV1-9	0.054	0	0.093	0.001
CV1-10	0.047	0	0.051	0

Proportion of Variance Explained by...

BOX 9.2 *Canonical correlation analysis in SAS*

SAS is more flexible than SPSS in terms of CCA, thanks to the CANCORR procedure in SAS/STAT. Besides producing the standard CCA output, the procedure allows to perform partial canonical correlation through the statement PARTIAL, followed by the list of variables to be controlled for. The variables for the first data-set are listed after the statement VAR, while the statement WITH indicates the variables for the second data-set.

---

together, the amount of variance explained is still low, ranging from 4.1% between the first variate and the second set to 39.8% within set 2. This suggests that more than three canonical functions should be considered to explain a relationship which does not look too strong.

Summing up

The concept of correlation with metric (scale) variables is paralleled by measures of association between non-metric (categorical) variables. Association measures are computed from a two-way frequency table (known as contingency table) and the hypothesis of independence between the two categorical variables can be tested through the Chi-square statistic, which compares the empirical distribution of the table frequencies with the distribution one would expect under the null hypothesis of no association. Alternative measures of association are the Cramer's V statistic (useful when the two variables have a different number of categories), the Goodman and Kruskal's Lambda (for nominal variables), the Somer's d statistic and the Kendall's Tau b (for square tables) and Tau c statistics (for rectangular tables). Through multi-way (more than two variables) contingency tables, one is interested in exploring associations and interactions among several non-metric variables, which makes the analysis quite complex. In a similar fashion to the general linear model, the general log-linear model relates the logarithm of the frequencies in each cell of the contingency table to the complete set of main and interaction effects. Log-linear analysis provides two procedures (hierarchical log-linear analysis and general log-linear analysis) to simplify the model and maintain only those effects which enable to produce a contingency table which is not significantly different from the original one according to the Chi-square statistic and the log-likelihood ratio. Finally, canonical correlation analysis is a very comprehensive technique which relates two sets of variables. The first canonical correlation is the highest possible bivariate correlation between a linear combination of the variables within the first set and a second linear combination of the variables within the second set. This is especially useful when many variables are used to measure the same concept, like exploring the joint response of a set of market outcomes (for example sales, market shares and repeated purchases) to a set of advertising impact measures (for example reach, recognition and recall).

EXERCISES

1. Open the Trust data-set
   a. Build a two-way contingency table showing absolute and relative frequencies between the variable q55 (job) and the variable q61 (perceived household wealth)
   b. Evaluate the statistical significance of their association using the range of statistics illustrated in this chapter
   c. Log-linear analysis can also be applied on two-way contingency tables, look at the role of the interaction between job and household wealth through a hierarchical analysis, looking at the significance of the interaction term
   d. What is the difference between the Chi-square statistic computed through CROSSTABS and the one corresponding to the two-way interaction in hierarchical log-linear analysis?
2. Open the EFS data-set
   a. Create a new variable from tenure type (a121) with two categories, rented (a121 from 1 to 4) and owned (a121 from 5 to 7) – HINT: use the TRANSFORM / RECODE INTO DIFFERENT VARIABLES menu
   b. Using log-linear analysis, explore the interactions among the above variables, Government office region (gor) and the number of children considering the first three categories only (*childband* variable from 1 to 3)
   c. Using hierarchical log-linear analysis, compute effects in the saturated model then estimate a simplified model
   d. Check the partial association table – are there lower-order terms that could be discarded?
   e. Re-estimate the simplified model with the general log-linear model and find the most parsimonious model which is still acceptable according to the chi-square statistic (HINT: first exclude cases where childband=4 using the DATA / SELECT menu)
   f. Compare results using alternatively the Poisson and Multinomial distribution assumption. Do estimated effects differ?
   g. What are the odds of having a rented rather than owned house in government region 2?

3. Open the Trust data-set, run a canonical correlation analysis on the following two sets of variables, the first referring to safety and risk aspects of chicken, the second related to trust in information on chicken safety:

    ```
    SET1 = q12d q21d q21e q27d /
    SET2 = q43a q43b q43e q43f.
    ```

    a. What are the values of the canonical correlations?
    b. How many canonical correlations should be retained according to the Wilks' Lambda test?
    c. Look at the canonical loadings for each set and comment on those for the first canonical correlation
    d. Explore the cross-loadings between the first two canonical correlations and comment on them
    e. Is the first canonical correlation relevant in explaining variability of the second one?
    f. Compute the sample averages of the four canonical scores

Further readings and web-links

❖ **Association measures and appropriate tests** – The specific page on the web-site of Emeritus Professor Richard B. Darlington at Cornell University is worth a visit, since it allows one to find the appropriate test for any situation through a useful classification:www.psych.cornell.edu/Darlington/crosstab/table0.htm

❖ **Categorical variables and log-linear analysis** – A classic textbook for the analysis is the one by Agresti (2002). The book also sheds light on the links and differences between log-linear analysis and some of the methods discussed in chapter 16 including the generalized linear model.

❖ **Canonical correlation analysis** – For a more detailed discussion of the method of CCA and the meaning of the output, see the introductory article by Rencher (1992) or the books by Thompson (1984) and Levine (1977). Alpert and Peterson (1972) review the interpretation of canonical correlation in relation to marketing research.

Hints for more advanced studies

☞ What are odds ratio? How can they be computed from contingency tables? What is the relation between parameter estimates from log-linear analysis and odds ratio?

☞ Log-linear analysis can also be exploited to model the relationship between a categorical dependent variable and a set of categorical and metric variables. Find out how logit log-linear analysis works in SPSS.

☞ Canonical correlation analysis can be seen as a general technique which embodies many others, including ANOVA. Find out why.

Notes

1. The algorithm still proceeds to test for specific lower-order terms when all the higher-order interactions for the variables being evaluated have been discarded in previous steps. For example, suppose the second step of the analysis had discarded both q43j*q50 and q60*q50, then the analysis would proceed to test the significance of the main individual effect for q50.
2. For more detailed explanations of odds ratios from frequency tables and log-linear analysis see Rudas (1998).
3. Those who recall regression analysis from chapter 8 will recognize the R^2 index of regression analysis, computed on the canonical function.

CHAPTER 10

Factor Analysis and Principal Component Analysis

T HIS CHAPTER introduces the two most widely employed techniques for summarizing large data-sets into smaller ones, making analysis and interpretation easier. The same techniques are also useful when the researcher wants to measure some directly unobservable or latent quantity (like product quality) through a set of measurable variables (stated satisfaction, perceived quality, compliance to standards, etc.).

Section 10.1   provides an overview of data reduction techniques
Section 10.2   explains the technical steps and the options for carrying out factor analysis
Section 10.3   illustrates principal component analysis
Section 10.4   discusses a practical example to be implemented in SPSS or SAS

THREE LEARNING OUTCOMES

This chapter enables the reader to:

➡ Become familiar with data reduction techniques
➡ Appreciate the different uses and possibilities of principal component analysis and factor analysis
➡ Understand and master the main options for their SPSS and SAS application

PRELIMINARY KNOWLEDGE: Readers should recall the explanation of latent variables in chapter 1 and should be familiar with the methods of correlation and regression in chapter 8. This chapter uses quite a bit of matrix notation and some algebra, too. All the relevant algebraic concepts are explained in the appendix, but the key functioning of the techniques should be equally clear without the support of algebra (in which case, one should probably avoid section 10.2.1).

10.1 Principles and applications of data reduction techniques

A temptation of marketing research is to exploit surveys to collect as much information as possible on target or potential consumers. This might be a waste of time and resources, if one is unable to process this vast amount of information. Even worse, one might get confused by highlighting minor aspects emerging from the data and ignore the key measures and relationships. Data reduction methods can help to avoid these problems.

The idea behind these methods is to avoid 'double counting' of the same information by distinguishing between the individual information content of each variable and the amount of shared information. Take for example household income, household size and the number of cars. On average richer households will own more cars than poorer households. However, the number of cars also depends on household size, whether the family lives in a city or in a rural area and many other household and individual characteristics not necessarily reflected by income or household size. Out of the three original variables, we might use a single variable which efficiently combines the relevant information provided by the original ones.

This chapter explores the characteristics, similarities and differences between the two most commonly used methods for data reduction – factor analysis and principal component analysis. The two methodologies share the same objective which is to summarize information.

Factor analysis and principal component analysis are multivariate statistical methods designed to exploit the correlations between the original variables and create a smaller set of new artificial (or latent) variables that can be expressed as a combination of the original ones. The higher the correlation across the original variables, the smaller is the number of artificial variables (factors or components) needed to describe adequately the same phenomenon.

There are many potential applications for these techniques, both as self-contained methods and as a preliminary tool to other data processing approaches. The obvious use of *data reduction* techniques is to identify a reduced set of dimensions out of a large amount of information. Take the case (from Angelis et al., 2005) where customers' perceived value for private versus public banks is measured through a seventeen-items questionnaire. Out of these variables and using PCA, it is possible to identify and label three main dimensions for perceived value – professional service, marketing efficiency and effective communication. Another popular use of these methods is for perceptual mapping (see chapter 13) and the identification of *new market opportunities* and product space. For example, PCA (combined with cluster analysis) is exploited in Bogue and Ritson (2006) to measure end-users' preferences with respect to light dairy products and identify opportunities for new product development in lighter strawberry yoghurt as opposed to lighter Cheddar-type cheeses. PCA and factor analysis are also widely studied as a preliminary data reduction tool in studies aimed at *segmenting markets*, for example a market segmentation of computer gamers as in Ip and Jacobs (2005). With factor analysis they determine three main drivers of gaming behavior – attitudes (and knowledge), playing habits and buying habits. They eventually identify two types of gamers – hardcore and casual gamers.

10.1.1 Data reduction techniques

Technically, a data-set containing m correlated variables is condensed into a reduced set of $p \leq m$ new variables (respectively factors or principal components according to the method used) that are uncorrelated between each other and explain a large proportion of the original variability.

Data reduction is a necessary step to simplify data treatment when the number of variables is very large or when correlation between the original variables is regarded as an issue for subsequent analysis. The description of complex inter-relations between the original variables is made easier by looking at the extracted factors (components), although the price is the loss of some of the original information and the introduction of an additional error component.

To explain factor analysis, let us consider the Trust data-set once more. The questionnaire includes a set of 23 questions measuring respondent trust in food safety information provided by a variety of sources (q43). Intuitively a large amount of correlation exists across the 23 trust variables. For example, it is likely that those who place a high level of trust in information provided by food processors will also trust information by retailers. Thus, it may be useful to exploit a data reduction techniques for (a) working on a reduced set of variables and (b) identifying latent structures in trust in sources of food safety information. The data-set contains observations on the 23 variables for 500 respondents in five different countries. Suppose the objective of the researcher is to provide a summary data description, for example a cross-tabulation focusing on the relationship across the Trust variables and differences across the five countries. The number of tables necessary to this task would be extremely high, with the likely outcome that the description would be confusing rather than clarifying.

Factor analysis starts from considering how the original variables are correlated between each other. For instance, it is plausible that trust in different types of mass media (television, radio, newspapers) exhibits a (strong) positive correlation. Suppose that Italians place a higher trust in mass media, then a description on the original data-set will return that trust in radio, television and newspapers is relatively higher in Italy. But it would be much simpler if one had a 'mass media' variable. But how can one group variables? One may subjectively include only television in the descriptive analysis, but this would imply a loss of information on the other variables.

Factor analysis, and in a similar way principal component analysis, allow a non-subjective estimation of latent underlying variables that are linear combinations of the original ones. By looking at the relative weight of the original variables in determining the new factors, the researcher can interpret the latter as latent constructs and numerical values for these constructs will be available for each of the consumer in the data-set.

In simple terms, the methodology exploits the fact that for each of the original variables, only a certain amount of variability is specific to that variable, while there is a fair amount of variability which is shared with others. Such shared variability can be fully summarized by the common factors. Note that the method is potentially able to restructure into factors *all* of the original variability, but this would be a useless exercise, as in order to do that the number of extracted factors should be equal to the number of original variables. The researcher faces a trade-off between the objective of reducing the dimensions of the analysis and the amount of information lost by doing so.

Also note that these techniques are more effective and reliable when there are high levels of correlation across variables, but when the links between the original variables are weak it will not be possible to perform a significant data reduction without losing a considerable amount of information.

Finally, it might be useful to point out the main difference between factor analysis and principal component analysis before getting into the technical details. Factor analysis *estimates p* factors, where the number of factors p is lower than the number of variables n and needs to be assumed prior to start estimation. While – as explained later in this chapter – one may compare results with different number of factors, for each of the

BOX 10.1 *Looking for the 'g factor': the origins of Factor Analysis*

Charles Edward Spearman (1904), recognized with Francis Galton as the forefather of factor analysis, laid its foundations by looking at the performance of intelligence testing and trying to look at the 'types' of intelligence in man. Given the multiplicity of tests available to measure intelligence, Spearman assumed that all of them contribute to explain a form of 'general' intelligence (the so-called g factor), and each of them measures a 'specific' type of intelligence (the s factor), hence introducing the key distinction of factor analysis. The claim of a single 'general' intelligence factor instigated a wide debate and many argued that more than one intelligence factor should be taken into account. In his early applications of factor analysis, Thurstone (1946), besides contributing to major methodological advancements, found nine factors (primary mental abilities), such as visual, perceptual, inductive, deductive, numerical, spatial, verbal, memory and word fluency. The debate is still on-going and far from consensus. However, it is interesting to notice that the (shared) drive for this research stream is the search for some underlying intelligence types that are actually believed to exist and not merely the purpose of reducing the number of tests.

analyses it is necessary to set preliminarily the number of factors p. Instead, in PCA there is no need to run multiple analyses to choose the number of components. As a matter of fact, PCA starts by setting $p = m$, so that all of the variability is explained by the new factors (principal components). These are actually not estimated with error, but *computed* as the principal component problem and estimation method has an exact solution. A reduced number of components are then selected at a later stage according to some criterion.

Factor analysis can be used as either an *exploratory* or *confirmatory* technique, depending on the final objective of research (Schervish, 1987). Exploratory factor analysis starts from observed data to identify unobservable and underlying factors, unknown to the researcher but likely to exist. In his research, Thurstone (see box 10.1) went in search of the latent dimensions of intelligence, without prior knowledge of their numbers and types. Under this connotation, nothing excludes that no underlying factors are found and if any is detected, their meaning depends on interpretation. However, there is an alternative use of factor analysis (discussed in chapter 15) when the researcher wants to test a specific hypothesized underlying structure. This is frequently the case in psychometric studies.

10.2 Factor analysis

Factor analysis starts from a very simple principle – the total variability of the original data-set can be split into two parts, shared variability (common to two or more variables) and specific variability (exclusive to each individual original variable). Factor analysis exploits the former part of variability, common to two or more variables, to synthesize and summarize the initial amount of information into a small set of factors.

These factors are a weighted combination of the original variables. The objectives of factor analysis are:

(1) to estimate the weights (*factor loadings*) that provide the most effective summarization of the original variability;

(2) to interpret the sets of factor loadings and derive a meaningful 'label' for each of the factors; and

(3) to estimate the values of the factors (*factor scores*) so that these can be used in subsequent analyses instead of the original variables.

The technical and mathematical aspects of factor analysis are covered in some details in section 10.2.1. While a thorough technical treatment of the estimation methods available for factor analysis is beyond the scopes of this textbook, knowledge of the technicalities may be extremely useful to appreciate the potential and uses of factor analysis. The appendix provides all of the necessary terminology and concepts for those less familiar with matrix algebra. However, it is possible to understand the basic functioning of factor analysis without getting into the mathematical details of this section and jump directly to the main options provided by statistical packages in section 10.2.2.

10.2.1 Factor analysis and variance decomposition

Suppose that a sample of n individuals has been extracted and for each of those individuals, a set of p variables has been observed. The original data-set is written down as:

$$X = \{x_j\}_{j=1,\dots,p} = \{x_{ij}\}_{i=1,\dots,n; \, j=1,\dots,p}$$

so that X is the matrix containing all the data, x_j is the data vector containing all observations for the j-th variable and x_{ij} is the observation of the j-th variable on the n-th individual. Considering the Trust case, each line of the matrix X refers to one of the 500 individual respondents of the survey, while the responses to the individual trust questions are stored in each of the 23 columns.

The overall variability of the data matrix is measured by the covariance matrix (see appendix and chapter 8) $Var(X)=\Sigma$. Factor analysis explains the relationship among these p variables through m latent factors f_k, with $k = 1, \dots, m$. In synthesis:

$$X = \mu + \Gamma F + E = \begin{cases} x_1 = \mu_1 + \gamma_{11} f_1 + \gamma_{12} f_2 + \dots + \gamma_{1m} f_m + \varepsilon_1 \\ x_2 = \mu_2 + \gamma_{21} f_1 + \gamma_{22} f_2 + \dots + \gamma_{2m} f_m + \varepsilon_2 \\ \dots \\ x_p = \mu_p + \gamma_{p1} f_1 + \gamma_{p2} f_2 + \dots + \gamma_{pm} f_m + \varepsilon_p \end{cases}$$

which means that the data matrix X can be expressed as a function of a vector μ, of the matrix F containing the unknown factor values and of a residual matrix E, while the right part of the above equation explicits the comprehensive matrix notation for each of the variables. As we start the analysis we generally don't know how many of these factors are necessary (the value m), nor what will be the meaning of each of them.

Each of the original variables represented by the vector of observations x_j is explained by a specific factor (the constant μ_j which is the mean of vector x_j), by the vectors $f_1, \dots, f_m$, which contains estimates of the latent factors for each observation, and the vectors of random errors ε_j which are all assumed to have a zero mean and a covariance matrix ψ. The errors are assumed to be uncorrelated across equations, which means that the ψ matrix is diagonal, and to be uncorrelated with the factors. These assumptions are the same as those of regression analysis (see chapter 8). Actually, the above model would be a system of regression equations if the factor $f_1, \dots, f_m$, were observable and measured variables. Instead, we have no prior information about the factors apart from the assumption that they are not correlated with each other. Without loss of generality, the factors can be treated as random variables standardized to have zero mean and variance equal to one, so that $Var(\mathbf{F}) = \mathbf{I}$, where $\mathbf{I}$ is the identity matrix (see appendix). The link between the original variables and the unknown factor is measured by the unknown constant coefficients contained in the matrix $\mathbf{\Gamma}$.

An intuitive meaning can be attached to the factor coefficient matrix $\mathbf{\Gamma}$ (sometime referred to as the *factor matrix*). Its elements are called *factor loadings* and represent the link between the latent factors and the original variables, like correlations (see chapter 8). For example, the generic element γ_{ij} measures the importance of the latent common factor $\mathbf{f}_j$ in determining the value of $\mathbf{x}_i$. Furthermore, the higher the absolute value of γ_{ij}, the higher is the correlation between the variable $\mathbf{x}_i$ and the other variables with high loadings with respect to the same factor.

If we move the vector of constants $\boldsymbol{\mu}$ to the left-hand side of the factor analysis equation (or more simply we express the original variables as distances from their mean), we obtain the following matrix relationship:

$$\mathbf{X} - \boldsymbol{\mu} = \mathbf{\Gamma F} + \mathbf{E}$$

From this, some matrix algebra allows one to derive the key relationship between variances, as:

$$\mathbf{Var(X - \boldsymbol{\mu})} = \mathbf{\Gamma Var(F)\Gamma'} + \mathbf{Var(E)}$$

The above equation is probably the most helpful in order to understand the mechanism of factor analysis. The left-hand side $\mathbf{Var(X-\boldsymbol{\mu})}$ corresponds to the original data variability $\mathbf{\Sigma}$ (subtracting a vector of constant values does not make any difference), while the right-hand side shows that this variability is the sum of a portion of variability depending on the common factors $\mathbf{Var(F)}$ at the cost of committing an error, quantified by the residual covariance matrix $\mathbf{Var(E)}$.

Now it becomes possible to exploit the assumption that $\mathbf{Var(F)=I}$, which means that

$$\mathbf{\Sigma} = \mathbf{\Gamma \Gamma'} + \mathbf{\Psi}$$

If the matrix notation is too confusing, one can look at the same relationship for each of the variables:

$$Var(\mathbf{x}_i) = \sum_{j=1}^{m} \gamma_{ij}^2 + \psi_{ii}^2$$

which introduces an important term of factor analysis, the *communality* $\sum_{j=1}^{m} \gamma_{ij}^2$.

Clearly, a high value of the communality for variable i with respect to the *specific variance* ψ_{ii}^2 means that the i-th variable is well summarized by the m common factors estimated in the analysis. Thus, the typical diagnostic to assess the effectiveness of the factor analysis in summarizing each of the variables is given by the *relative communalities*, bounded between zero and 1 and computed as:

$$c_i = \frac{\sum_{j=1}^{m} \gamma_{ij}^2}{Var(\mathbf{x}_i)}$$

where $c_i = 1$ means that the i-th variable is completely explained by the m factors and $c_i = 0$ means that the common factors are unable to explain any of its variability. The relative communality can be interpreted similarly to the R^2 index in regression analysis (see chapter 8), as it measures the goodness-of-fit of the common factors model.

10.2.2 Estimation of factor loadings and factor scores

Estimation in factor analysis involves two objects: the *factor loadings* (and jointly the *specific variances*) and the *factor scores*. The two sets of estimates cannot be obtained simultaneously and estimation of factor loadings always precedes estimation of factor scores.

Assuming that the number of factors is known, a naïve summarization of the estimation process can be provided by the following steps:

1) Correlations among the original variables are exploited to obtain an initial guess of the factor loadings
2) On the basis of this initial guess, specific variances are estimated
3) New estimates of the factor loadings are obtained according to some (variable) criterion, conditional on estimates of the specific variances obtained in step 2
4) New estimates of the specific variances are obtained (again different methods can be used) conditional on estimates of the factor loadings obtained in step 3
5) Return to step 3 and proceed iteratively until estimates of factor loadings and specific variances are stable

Clearly a key choice involves the method or criterion to obtain the estimates in step 3 and 4, as different methods may lead to more or less substantial differences in the results. A thorough researcher may wish to explore the outcomes of different estimation methods and choose the most satisfactory one in terms of goodness-of-fit (communalities).

Box 10.2 lists the estimation methods available in SPSS and SAS and provides a reference to articles discussing technical details.[1] Among these methods, it is crucial to know that the *principal factoring* and the *image factoring* methods are not independent from the measurement scales of the original variables. In fact, one may choose to apply them either to the correlation matrix or to the covariance matrix of the original data. Applying them to the correlation matrix corresponds to standardizing[2] the original variables and eliminates the effects of different measurement units or scales. However, this also implies that all the original variables have the same variability, which is not always a desirable assumption. For instance, if the measurement scales (and units) are the same for different variables, there is no reason why variability should be forced to be equal across variables. In such case it may be more appropriate to apply the method to the covariance matrix which maintains the original measurement scales. As a rule of thumb, if one adopts the principal factoring or image factoring estimation methods, factor analysis should be applied to the covariance matrix if all variables have the same measurement scale and to the correlation matrix otherwise. The other methods, and maximum likelihood should be preferred, are independent from such choice. To assess the influence of the scale of measurement on the results, it may be useful to compare maximum likelihood results with those obtained through principal factoring or image factoring.

BOX 10.2 *Estimation of factor loadings*

The commonly used methods for estimating the *factor loadings* are shown in this table, together with some key references. All of these methods are available as estimation options both in SAS and SPSS. The methods start from different assumption and may lead to different results.

Method	Description	References
Maximum likelihood	Estimates of factor loadings and specific variances are those who maximize a likelihood function conditional on the estimated sample covariance matrix. Multivariate normality is assumed for the data. The method is invariant to the choice of working on the correlation or covariance matrix.	Fuller (1987); Lawley and Maxwell (1962)
Least-squares methods	Estimates of factor loadings and specific variances are those that minimize the sum of the squared differences between the actual and estimated correlation matrices	Joreskog (1962)
Principal factoring	After the initial estimate on the specific variance (usually by using residuals from regression analysis between the original variables), principal components are extracted from the difference between the sample covariance matrix and the estimated specific variances. New factor loadings and specific variances become available and the procedure is repeated iteratively.	Gorsuch (1983)
Alpha factoring	After performing an initial principal component analysis, estimates of the factor loadings and specific variance are obtained through a criterion which maximizes the Cronbach's alpha-reliability index (see box 1.2).	Kaiser and Caffrey (1965)
Image factoring	Each of the original variables is regressed on the remaining ones and the model is used to obtain within-sample predictions. These predictions (images) allow to compute the specific and residual variances and a unique solution exists whether the correlation or covariance matrix is used.	Guttman (1953); Harris, (1962)

The factor analysis equation exploited in previous sections has an undesirable characteristic – none of the variables on the right-hand side is actually known to the researcher. Even after estimation of the factor loadings Γ and specific variances Ψ, there is no unique solution to obtain values for the factor scores F. As a matter of fact it can be shown that the relationship of factor analysis holds for any orthogonal rotation[3] of the factor loadings and factor scores matrices. This is the so-called problem of *indeterminacy* of factor analysis, well-known and highly debated among statisticians (Mulaik and MacDonald, 1978; Acito and Anderson, 1986). Furthermore, the deterministic computation of factor scores using estimates of Γ and Ψ does not

BOX 10.3 *Estimation of factor scores*

Once the factor loadings have been estimated, it becomes possible to estimate the *factor scores*, that is the value of the factors for each of the statistical units. There is a variety of techniques which may lead to different results.

Method	Equation	References
Regression (SPSS, SAS)	$\hat{\mathbf{F}} = \hat{\Gamma}'(\hat{\Gamma}\hat{\Gamma}' + \hat{\Psi})^{-1}(\mathbf{X} - \mu)$ If multivariate normality is assumed for the factors and for the specific component (as necessary in the maximum-likelihood factoring method), a basic method of estimating the factor scores consists in reverting the factor analysis equation to express the factor scores as a function of the observed data (in this way they can be seen as regression predictions). A drawback of this method is that the estimated factors frequently show correlation between each other, contrarily to the original assumption and objective of factor analysis.	Thurstone (1935), pp. 226–231
Bartlett scores (SPSS)	$\hat{\mathbf{F}} = (\hat{\Gamma}'\hat{\Psi}\hat{\Gamma})^{-1}\hat{\Gamma}'\hat{\Psi}(\mathbf{X} - \overline{\mathbf{X}})$ This method considers the factor analysis equation as a system of regression equations where the original variables are the dependent variables, the factor loadings are the explanatory variables and the factor scores are the unknown parameters. As the specific variances differ across the original variables (heteroskedasticity), the estimator for the factor scores is given by weighted least squares. Again, estimated factor scores will result to be correlated even if the theoretical factors aren't.	Bartlett (1937)
Anderson-Rubin method (SPSS)	$\hat{\mathbf{F}} = \mathbf{X}\hat{\Psi}^{-1}\hat{\Gamma}(\hat{\Gamma}'\hat{\Psi}^{-1}\mathbf{R}\hat{\Psi}^{-1}\hat{\Gamma})^{-\frac{1}{2}}$ This method is appropriate when the theoretical factors are uncorrelated and produces perfectly uncorrelated estimates.	Anderson and Rubin (1956)

yield satisfactory results. It is not simply a mathematical problem, because it means that one could get two very different scores for the same factor and the same observation. Again, statistical packages rely on several methods for estimating factor scores. Box 10.3 summarizes the main factor scores estimation methods and provides the technical details.[4] Finally, note that by construction $Var(\mathbf{F})=\mathbf{I}$, which means that the variance for each factor is 1. Sometimes one might want to maintain the relative weight of each factor in terms of variability. This can be achieved by multiplying each factor score by the square root of the corresponding communality.

10.2.3 Choice of the number of factors

A key difference between factor analysis and principal component analysis is that the former requires knowledge (or some prior assumption) on the number of common factors to be extracted prior to estimation. All results will be heavily influenced by this choice and under-extraction (less factors than in the underlying structure) or over-extraction (too many factors as compared to the underlying structure) produce a bias on the results. Over-extraction has less serious consequences than under-extraction and when uncertainty arises it is advisable to risk an excess of factors rather than an inadequate number of them. Note that the need for the number of factors prior to analysis does not mean that the researcher cannot evaluate alternative choices by running several factor analyses and comparing the results on the basis of specific diagnostics, some of them (as the *scree diagram*) discussed in this chapter in relation to principal component analysis. A *confirmatory* approach can be applied after assuming that the original data follow a multivariate distribution. Together with estimates one gets a *likelihood ratio statistic* to compare alternative choices of factor numbers. The *Akaike Information Criterion* (AIC) can be also computed to find the factor structure which best fits the data, corresponding to the lowest AIC value. In all situations where uncertainty is an issue, it is advisable to run a factor analysis for different number of factors and compare the solutions according to some goodness-of-fit statistics. On the one hand, this allows us to find the best 'statistical' solution. On the other hand, it provides an indication of the stability of the solution and the sensitivity to the initial assumption on the number of factors. The *Bartlett's test of sphericity* is another useful statistic which can be computed when multivariate normality of the original variables can be safely assumed and the maximum likelihood estimation method is applied. The test is applied to the correlation matrix. If the null hypothesis is not rejected, the original correlation matrix is not significantly different from an identity matrix, which means that there are no grounds for searching common factors. A second test named after Bartlett is a corrected *likelihood ratio test* (Bartlett, 1950; Rao, 1956) which should not be confused with the sphericity test, as it tests whether the selected number of factors explains a significant portion of the original variability, although there are some dangers in using it routinely as it is sensitive to sample size.

10.2.4 *Factor interpretation and rotation*

The preceding discussion has illustrated the main features of factor analysis and the procedures for extracting the common factors. However, it might still be unclear what these factors actually are, whether a mere mathematical tool to reduce the dimension of the analysis or indeed a procedure to identify some underlying latent factors that actually exist albeit they are not directly observable or measured, and even their number is often unknown. While this might be seem a purely philosophical distinction, there are practical implications on the application of the method and especially on the choice between factor analysis and PCA as discussed in section 10.3. The contents of box 10.1, illustrating the origin of factor analysis, should give a better idea of the meaning attached to the extracted factors. Besides psychology, the discipline which originated the method, examples of 'latent structures' searched through factor analysis include concepts such as social classes, quality and lifestyles (compare with chapter 1).

It is clear that the interpretation or labeling of the extracted factors plays a key role in the successful application of factor analysis. Unfortunately this is another step which involves a degree of subjectivity. Interpretation is based on the value of the factor loadings for each factor. The factor loadings are the correlations between each of the original variables and the factor or – in other words – the importance of the common factor for each variable. So, they assume a value between –1 and 1, where high (positive or negative) values indicate that the link between the factor and the considered variable is very strong. One might wish to look at the statistical significance of the factor loadings, but this involves some complications, namely that the standard errors are usually so high that too few variables emerge as relevant in explaining a factor. The statistical relevance of a factor loading depends on its absolute value and the sample size, so that the same value for a factor loading assumes different relevance depending on the sample size. A study by Hair et al. (1998) suggests that with a sample of 100 units a reasonable threshold for detecting relevant factor loadings is 0.55 while with 200 units the value falls to 0.40 and with 50 units it rises to 0.75.

Clearly these thresholds have a relative value if one considers the indeterminacy problem, which implies that the factor loading estimates are not unique and any orthogonal rotation may be accepted. This can be seen either as a big limitation, as different rotations will lead to different factor interpretations, or as a useful mean to improve interpretability. It becomes possible to exploit factor rotation to obtain factors that – for example – reduce the number of factor loadings with high values and increase their values at the same time, thus making interpretation more straightforward.

The methods for estimating factor loadings described in box 10.2 return an estimate which is optimal according to the assumptions behind the estimation method. However, the optimal estimate is not necessarily the easiest to be interpreted. Hence, starting from such a solution it is possible to apply rotation techniques and obtain an alternative set of estimates which is easier to be interpreted. The rotated factor matrix is not the optimal one any more, at least not in absolute terms. However, it is optimal considering an additional criterion, which is the one which makes interpretation easier. The trade-off is between mathematical optimality and meaningfulness/interpretability of the factors. Again, the 'philosophical' grounds of the analysis become relevant. If it is assumed that the latent factor exists but cannot be directly measured, then meaningfulness of the results is a priority. If the objective is only data reduction, regardless of the factor meaning, there is no need for rotation (and possibly principal component analysis should be preferred).

In mathematical terms orthogonal rotation derives directly from the indeterminacy problem. Given any orthogonal matrix H, the factor analysis equation can be substituted by

$$X - \mu = \Gamma_R F_R + E$$

where the matrix of errors E (specific variability) is the same as in the original factor analysis equation, while $\Gamma_R = H\Gamma$ and $F_R = H'F$ are the rotated factor matrix and the rotated factor score vector, respectively. Rotation methods are defined according to the criterion for choosing the matrix H (box 10.4). Rotation methods can be classified in *orthogonal rotation methods* and *oblique rotation methods*. Orthogonal methods assume that the factor remain uncorrelated. In some circumstances the researches may want to relax such assumption to make interpretation even easier. Oblique rotation methods allow for correlated factors. There are major implications in using oblique rotation methods. Although they can lead to further simplification of the factor matrix, it becomes less straightforward to evaluate the contribution of each factor in explaining the original variability. This also means that the factor loadings are not directly comparable across factors unless one takes into account additional information (the factor structure and the reference structure). Oblique rotation methods should be only used when the researcher wants to assume correlation in the underlying factors.

BOX 10.4 *Rotation methods*

When interpretation of factors is not straightforward, it is possible to exploit rotation methods to obtain alternative estimates of the factor loadings that are easier to be interpreted. Rotation methods can be orthogonal (maintaining uncorrelation of the factors) and oblique (allowing for correlation).

ORTHOGONAL METHODS

Type of rotation	Description	Available in SPSS/SAS
Varimax	*Maximizes variance of loadings on each factor.* The matrix **H** is chosen to maximize the variability of the squared loadings in each *column* of the factor matrix. This leads to a reduced number of high factor loadings in each column, that is each factor is explained by few variables (but possibly the same variables can be relevant to explain two or more factors).	Both
Quartimax	*Maximizes variance of loadings on each variable.* Here the matrix **H** is the one maximizing the variability of the squared loadings in each *row* of the factor matrix, that is each variable is expected to explain a reduced number of factors. Again, this does not exclude that one factor needs to be interpreted by a large number of variables.	Both
Equamax	As a compromise between Varimax and Quartimax, the matrix **H** correspond to an equidistant solution between the two.	Both
Orthomax	As for Equamax, a compromise between Varimax and Quartimax is sought – here the researcher decides the relative weight of the Varimax component versus the Quartimax one.	SAS only
Parsimax	A specific case of orthomax rotation where the relative weight of the Varimax and Quartimax rotations depends on the number of variables and factors.	SAS only

OBLIQUE METHODS

Orthoblique	A Quartimax (orthogonal) rotation is applied to the rescaled factor loadings – so that the original non-scaled factors might be correlated.	SAS only
Procrustres	The researcher sets a target factor matrix and the H matrix is the one which best approximates the target solution. This rotation can be seen as a test of the hypothesized factor loadings.	SAS only
Promax	Starting from an orthogonal solution, further oblique rotation aims to drive to 0 those factor loadings that are smaller. This is achieved by raising to powers the orthogonal loadings.	Both

10.3 Principal component analysis

Factor analysis and principal component analysis show many similarities, but also some key differences (see box 10.5). They can be alternatively applied to the same problems and not infrequently they return very similar solutions, but their foundations are extremely different and it is important that researchers bear in mind these distinctions.

Principal component analysis shares with factor analysis the starting point – the original variables are correlated and it is possible to identify a reduced set of

---

BOX 10.5 *Factor analysis versus PCA at a glance*

While factor analysis and principal component analysis certainly share lots of common features, not least the fact that they are interchangeable for the objective of data reduction, there are several specificities that the researcher should take into account before choosing. A thorough comparison and reviews of contributions in comparing the techniques can be found in the works by Gorsuch (1990) and Velicer and Jackson (1982, 1990a, 1990b). This box lists the key differences.

Factor analysis	Principal component analysis
Number of factors pre-determined	Number of components evaluated ex post
Many potential solutions	Unique mathematical solution
Factor matrix is estimated	Component matrix is computed
Factor scores are estimated	Component scores are computed
More appropriate when searching for an underlying structure	More appropriate for data reduction (no prior underlying structure assumed)
Factors are not necessarily sorted	Factors are sorted according to the amount of explained variability
Only common variability is taken into account	Total variability is taken into account
Estimated factor scores may be correlated	Component scores are always uncorrelated
A distinction is made between common and specific variance	No distinction between specific and common variability
Preferred when there is substantial measurement error in variables	Preferred as a preliminary method to cluster analysis or to avoid multicollinearity in regression
Rotation is often desirable as there are many equivalent solutions	Rotation is less desirable, unless components are difficult to be interpreted and explained variance is spread evenly across components

---

uncorrelated linear combinations (the components) which can summarize most of the initial variability.

However, the statistical development is substantially different. One could refer to PCA as a mathematical rather than statistical technique, since its application is deterministic and there is one and only solution for each set of variables. This becomes clear when looking at the equation of PCA:

$$\mathbf{C} = \mathbf{XA'} = \begin{cases} \mathbf{c_1} = a_{11}\mathbf{x_1} + a_{12}\mathbf{x_2} + \ldots + a_{1p}\mathbf{x_p} \\ \mathbf{c_2} = a_{21}\mathbf{x_1} + a_{22}\mathbf{x_2} + \ldots + a_{2p}\mathbf{x_p} \\ \ldots \\ \mathbf{c_p} = a_{p1}\mathbf{x_1} + a_{p2}\mathbf{x_2} + \ldots + a_{pp}\mathbf{x_p} \end{cases}$$

where $\mathbf{C} = \{\mathbf{c_j}\}_{j=1,\ldots,p} = \{c_{ij}\}_{i=1,\ldots,n;j=1,\ldots,p}$ is an $n \times p$ matrix of *principal component scores*, with p columns (one for each principal component $\mathbf{c}_j$) and n rows (one for each statistical unit), $\mathbf{X}$ is the usual $n \times p$ data matrix and $\mathbf{A} = \{\mathbf{a_j}\}_{j=1,\ldots,p} = \{a_{ij}\}_{i=1,\ldots,p;j=1,\ldots,p}$ is the $p \times p$ matrix of component loadings. Two further constraints complete the definition of the PCA model, the first states that the component scores are uncorrelated across components:

$$Corr(\mathbf{C}) = \mathbf{I} \text{ or } Corr(c_i, c_j) = 0 \; \forall i \neq j \text{ and } Corr(c_i, c_j) = 1 \; \forall i = j$$

which corresponds to the condition $\mathbf{A}'\mathbf{A} = \mathbf{I}$ and the second requires that

$$\sum_{j=1}^{p} a_{ij}^2 = 1 \; \forall i = 1, \ldots, p$$

The second rule constrains the coefficients size across components, otherwise the variances of the principal components would depend on the coefficients sizes as well, while the aim is that they are only determined by the different weight of the original variables.

The objective of the PCA is to compute $\mathbf{A}$. It can be shown (see Krzanowski, 2000) that the following relationship holds for each component:

$$\mathbf{Var}(c_j) = \mathbf{a}'_j \mathbf{Var}(\mathbf{X})\mathbf{a}_j = \mathbf{a}'_j \Sigma \mathbf{a}_j = \lambda_j$$

which explains the relationship between the variability of the components and the data variability. Since the objective is to maximize the variability explained by the components, PCA consists in finding the $\mathbf{a}_j$ vectors that maximize the right-hand side of the above equation subject to the above specified constraints on $\mathbf{C}$ and $\mathbf{A}$, starting from the first component (the one with the largest variance), until the last (the one with the smallest variance). This allows an easier selection of a sub-set of components. The total variability of the p extracted components equals the total variability of the p original variables. In fact, the first step of PCA simply consists in passing from a set of p variables ($\mathbf{X}$) to an alternative set of p variables (the principal components in $\mathbf{C}$) that are linear combinations of the former and have the desirable characteristic of being uncorrelated between each other. One may then decide to select a sub-set of components.

For a complete mathematical treatment the reader is invited to refer to Dunteman (1989) or Krzawnoski (2000). However, it is important to know that the vectors $\mathbf{a}_j$ are the latent vectors or characteristic vectors or *eigenvectors* (see appendix) of the covariance matrix Σ, while the values λ_j are the corresponding latent roots or characteristic roots or *eigenvalues*. The eigenvectors and eigenvalues are unique and can be computed algebraically. The eigenvalues can be used as a criterion to decide the number of components to be retained for data summarization and subsequent analyses.

Once the matrix $\mathbf{A}$ has been computed, the component scores can be also (exactly) computed by simply applying $\mathbf{C} = \mathbf{X}\mathbf{A}'$. Sometimes it is convenient to standardize the component scores so that they all have a variance equal to 1. In this case the coefficient score a_{ij} is rescaled as follows:

$$\hat{a}_{ij} = a_{ij}/\sqrt{\lambda_j}$$

SPSS automatically provides standardized component scores; thus, if one wants to maintain the original component variability and the correct balance among components, the inverse operation is needed:

$$a_{ij} = \hat{a}_{ij}\sqrt{\lambda_j}$$

10.3.1 Standardized variables and correlation versus covariance matrix

So far the mathematical treatment of PCA has been based on the covariance matrix of the data. The alternative solution is based on the correlation matrix and yields different results. For an illustration of alternative situations see box 10.6.

The covariance matrix depends on the measurement units of the original variables. If these are different, those with a wider measurement scale will necessarily have a larger weight on the extracted component – which is undesirable. In such a situation the original variables are standardized by subtracting their mean and dividing the result by their standard deviation as shown in the appendix. Standardized variables all have zero means and variance equal to 1 and their covariance matrix corresponds to the correlation matrix of the original variables. Thus, standardizing the original variables corresponds to doing PCA on the correlation matrix. This means that the correlation matrix $\mathbf{R}_X$ is used instead of Σ in the principal component procedure, then computation proceeds as before with the eigenvalue and eigenvector extraction.

While the correlation matrix sorts out the problem of different measurement units, it necessarily induces another bias; now all variables have exactly the same weight on the principal components and the final PCA results will only depend on their correlations. The variability of each of the original variables plays no role in the extraction. This might be an undesirable circumstance as shown in the examples of box 10.6.

10.3.2 Choice of the number of components

By construction PCA starts by extracting all of the p principal components. However, unless the only objective is to eliminate correlations between the original variables, this does not reduce the dimension of the data matrix. Thus, the following step is to decide how many components should be retained without incurring an excessive loss of information as compared to the original variability. This issue is analogous to the choice of the number of factors in factor analysis, which explains why the methods described below are also useful for factor analysis. Note that these methods are rules of thumb rather than objective techniques, unless multivariate normality of the original variables can be safely assumed. In the latter, case, the distribution of the likelihood ratio test statistic is known and a more objective and transparent choice becomes possible.

As noted in the previous sections, the order of extraction in PCA follows the variability of the principal components. Hence, the objective is to set the appropriate value k, so that the first k components are retained and the remaining $p-k$ are discarded.

A standard PCA output returns, for each of the p components, an eigenvalue (see appendix) corresponding to the variance of the component. The sum of the eigenvalues equals the total variance of the original set of variables. When working on the correlation matrix, it follows that the sum of the eigenvalues equals p, since each of the p standardized variables has a variance equal to 1.

A first criterion is the *Kaiser's rule* on eigenvalues (Kaiser, 1960), which suggests retaining only those components with an eigenvalue larger than the average. This means that the extracted components explain a proportion of total variance which is larger than the quantity explained by the average variable. When PCA is run on the correlation

BOX 10.6 *Correlation or covariance matrix?*

An initial dilemma for the researcher is the choice between correlation and covariance matrix. A practical guideline could be:

1. When the measurement unit is the same for all variables, the covariance matrix should be used
2. When the measurement unit differs across variables and\or it is desirable that all variables have the same weight on the analysis, regardless of their variability, the correlation matrix should be used
3. When the measurement unit differs across variables and the researcher wants to take into account the original variability, a solution might be a rescaling of the original variables:

$$\tilde{x}_j = \frac{x_j}{\mu_j}$$

The resulting variables are now expressed as indices with respect to their mean values, but maintain the ranges of the original variables. The covariance matrix of $\tilde{x}_j$ could be used.

Example 1 – The aim of a research is to plot on two components a data-set which records information on five variables for a sample of New York citizens: age (in years), weight (in kilograms), height (in meters), distance from the nearest supermarket (in kilometers) and yearly income (in dollars). The age ranges between 18 and 105, the weight between 50 kilograms and 120 kilograms, the height between 1.4 meters and 2.1 meters, the distance in kilometers from 0 to 2, the income from 0 to 240,000 dollars. If one applies PCA on the covariance matrix, income will clearly have a very large influence on the results, followed by weight and age, while height and distance from supermarkets will be barely relevant. The correlation matrix should be preferred, bearing in mind that all variables will have equal weight. So, even if citizens are very similar for weight and extremely different for income, these two variables will equally influence the component extraction process. As an alternative, one may wish to divide each of the variable for the sample mean, so that the resulting scaled variables will have no measurement unit and will simply represent ratios, still maintaining the information on the original variability.

Example 2 – A questionnaire on customer satisfaction for cable TV has 20 items, all measured on a 1-7 Likert scale. The objective is to define one or two linear combinations of these variables which would synthetically describe each of the customers. In this case, the covariance matrix should be preferred to the correlation matrix. In fact, suppose that all customers have responded 1 or 2 to the question 'How much do you like the broadcast programs (1=not at all; 7=very much),' while the same customers have answered with a wide range of different values between 1 and 7 to the question 'How expensive are the subscription rates in your opinion (1=extremely expensive; 7=extremely cheap),'. If the researcher uses the correlation matrix, the synthetic indicator of customer satisfaction will tend to give the same weight to the differences between 1 and 2 in the first question and to the differences between 1 and 7 in the second one. This is undesirable, because 1 and 2 both reflect dissatisfaction, while 1 and 7 are very different in the second circumstance, as the former reflects extreme dissatisfaction, while the second reflects extreme satisfaction with the price. Instead, the covariance matrix will adequately represent the smaller variability of the first question.

matrix (or the original variables are standardized), this average equals to one, so that components with eigenvalues larger than 1 are retained. A more conservative threshold is suggested by Jolliffe (1972), where the average eigenvalue is multiplied by 0.7.

A second approach is to explore the *scree diagram*. The scree diagram is a bi-dimensional plot with the number of components on the x-axis and the eigenvalues on the y-axis, as in figure 10.1.

The principle is to detect the point at which the scree plot becomes relatively flat, so that the components on the right side of this *elbow point* are rather undifferentiated in terms of their contribution to total variability. In many situations the identification might become a highly subjective operation. A third rule of thumb relies on a prior definition of the amount of total variance (as a bracket) that the researchers want to explain with the extracted principal components. A common target is to retain those components which together explain between 70 and 80% of the total variance. Research for more sophisticated and less subjective ways of detecting the ideal number of components has progressed since these rules of thumb (see for instance the out-of-sample prediction method described in Krzanowski, 2000).

As for factor analysis it is possible to compute *communalities* for principal components, that is the proportion of total variance accounted for by the principal components. In the first instance, when all components are computed, communalities are all equal to 1, as the *p* principal components explain all of the variance of the original data-set. Once a decision on how many components are retained, it is possible to quantify how much of the variance of each original variable is summarized by the selected components.

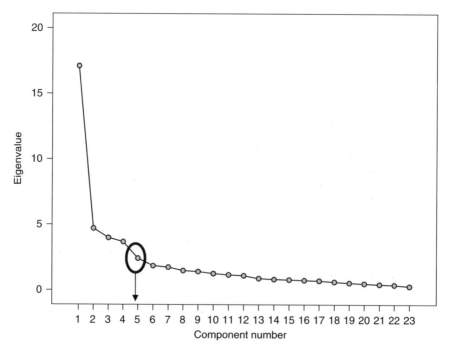

Figure 10.1 *Choice of the number of components according to the scree plot*

The relation which holds is the following:

$$Var(\mathbf{x}_i) = \sum_{j=1}^{k} a_{ij}^2 Var(\mathbf{c}_j) + \sum_{j=1}^{p-k} a_{ij}^2 Var(\mathbf{c}_j) = \sum_{j=1}^{k} a_{ij}^2 \lambda_j + \sum_{j=1}^{p-k} a_{ij}^2 \lambda_j$$

where the λ_i are the eigenvalues and a_{ij} are the elements of the eigenvectors as before. The first term on the right-hand side of the equation represents the amount of variance which is summarized by the retained components, while the second term is the amount of variance discarded by selecting only the first k components. The relative communality for each variable is then computed as:

$$v_i = \frac{\sum_{j=1}^{k} a_{ij}^2 \lambda_j}{Var(\mathbf{x}_i)}$$

10.3.3 Component rotation

As for factor analysis, components can be also orthogonally or obliquely rotated, but there are differences and drawbacks to be taken into account (Jolliffe, 1989). It is especially important to notice that rotation should concern all of the components and not only the selected ones. Otherwise there is no guarantee that those k selected components are still those which explain the larger amount of variability. Disparities in explained variability after rotation are generally smaller than before. Thus, a good practice is to apply rotation to all components, then select the sub-set of principal components to be retained.

Rotation is less problematic when explained variability is already evenly spread at the extraction of principal components. There are many rotation methods for principal components (Richman, 1986); those discussed for factor analysis can be also applied to PCA.

10.4 Theory into practice

To illustrate the application of factor analysis and principal component analysis in SPSS we refer to the Trust data-set and to the example introduced in section 10.1.

10.4.1 Factor analysis

The objective of the factor analysis could be the identification of 'latent dimensions of trust in sources of food safety information.' Rather than measuring specific trust toward a single source, the researcher is interesting in exploring whether, on average, respondents trust groups of sources that are similar in some respect. The following tasks can be easily implemented in SPSS by clicking on ANALYZE/ DATA REDUCTION/FACTOR, selecting the variables for the analysis (from q43a to q43w).

Tasks

1. Explore the correlation matrix for the original data matrix
2. Perform factor analysis and estimate factor loadings by maximum likelihood
3. Choose the number of factors
4. Examine the factor matrix
5. Rotate the factor matrix to improve readability
6. Estimate factor scores

1. Correlations. Those not familiar with the concept of correlation can refer to Section 8.1 of this book. In SPSS it is possible to explore the correlation matrix by clicking on DESCRIPTIVES in the initial dialog window and asking for the *correlation coefficients* (and possibly the *significance level*). The original variables are highly and significantly correlated. Looking at bivariate correlations, the highest correlation is observed between trust in information provided by animal welfare organization and environmental organization (0.73), but also trust in government is highly correlated to trust in political groups (0.71). There are no negatively correlated variables and the lowest correlation is about 0.1, but still significant. This anticipates a successful implementation of data reduction methods. As expected, the Bartlett's test of sphericity (also available through the DESCRIPTIVES dialog box) confirms that the correlation matrix is far from an identity matrix, as correlations outside the diagonal are high and significant.

2. Estimation of factor loadings. We exploit the maximum likelihood method, which is independent from the measurement scale. This can be chosen by clicking on EXTRACTION in the main dialog box and selecting *maximum likelihood* as a method. Note that alternative methods require a choice between covariance and correlation matrix; if the former is chosen, then two sets of results are shown for factor loadings and communalities and the first (raw) depends on the measurement unit of the original variables, while the second is re-scaled. Judgments should be based on re-scaled values.

3. Choosing the number of factors. The Kaiser's (eigenvalue) rule (which can be adopted in the EXTRACTION menu, where it is also possible to ask for the *scree plot*) results in the extraction of five factors. These represent about 56% of the original variability (as shown in the 'extraction sum of squared loadings' column of table 10.1). The scree diagram is relatively consistent with the Kaiser's rule, as it becomes pretty flat around the 6th factor. Sometime the adopted rule is to explain at least 65–70% of the original variability, which would require estimation of at least four additional factors. In this example we choose to estimate five factors. The table of *communalities* (a standard output of SPSS, see table 10.2) shows that the loss of information is especially relevant for trust in product labels (the value is 0.29), while for trust in political groups the extracted factor still accounts for 82% of the original variability (a communality of 0.82).

4. The unrotated factor matrix returned by SPSS is shown below in table 10.3, deleting all values below 0.40 (which may be chosen by clicking on OPTIONS) to simplify reading.

The first factor strongly depends positively on almost all of the original variables, thus it is not easy to be interpreted unless one regards it as 'general trust toward sources of food safety information.' The second factor is characterized by only two variables. High (low) values in this factor reflect high (low) values of trust in radio information, and conversely low (high) value of trust in the national food safety authority. This does not allow an easy interpretation. The two factor is mainly linked to two variables, reflecting trust in the main retailers, supermarket and shops. It should

Table 10.1 *SPSS output for factor analysis: Bartlett's test and variance explained*

KMO and Bartlett's test

Kaiser-Meyer-Olkin measure of sampling Adequacy.		0.893
Bartlett's test of Sphericity	Approx. Chi-square	4315.522
	df	253
	Sig.	0.000

Total variance explained

Factor	Initial eigenvalues			Extraction sums of squared loadings		
	Total	% of variance	Cumulative%	Total	% of variance	Cumulative %
1	8.402	36.531	36.531	7.890	34.304	34.304
2	2.138	9.294	45.824	1.492	6.489	40.792
3	1.824	7.931	53.755	1.523	6.622	47.415
4	1.522	6.619	60.375	1.208	5.253	52.668
5	1.082	4.702	65.077	0.794	3.453	56.120
6	0.832	3.619	68.696			
7	0.758	3.296	71.992			
8	0.733	3.186	75.178			
9	0.660	2.869	78.047			
10	0.626	2.722	80.768			
11	0.536	2.329	83.097			
12	0.499	2.169	85.266			
13	0.455	1.979	87.245			
14	0.423	1.841	89.086			
15	0.380	1.653	90.739			
16	0.367	1.597	92.336			
17	0.337	1.465	93.801			
18	0.278	1.211	95.011			
19	0.277	1.206	96.217			
20	0.256	1.111	97.328			
21	0.229	0.997	98.325			
22	0.195	0.849	99.175			
23	0.190	0.825	100.000			

Extraction Method: Maximum Likelihood.

be emphasized that the factor matrix doesn't say anything about the average values within the sample, it is simply an 'interpretation matrix.' We only know that this factor is positively correlated with trust in retailers. The fourth factor is correlated with trust in organic shops, animal welfare organizations and environmental organizations, suggesting trust in less conventional and 'greener' sources of information. The fifth factor is linked negatively to a single variable and high values in this factor reflect distrust in information from politics.

However, the fact that many variables show a high value in the first column and only a few of them are high in the remaining factors suggests that rotation might help.

5. Rotation. If we want a reduced number of large values to appear in each column, the preferred rotation is VARIMAX, which can be chosen by clicking on ROTATION and selecting VARIMAX in the main dialog box for factor analysis.

The rotated matrix in table 10.4 is clearly much easier to be interpreted. Now the first factor groups together positive correlation with all the mass media, the second factor emphasizes trust in public and academic institutions and consumer organizations, the third factor shows high values for all the food chain actors (including processors

Table 10.2 *Communalities*

	Initial	Extraction
Shopkeepers	0.501	0.568
Supermarkets	0.590	0.627
Organic shop	0.455	0.463
Specialty store	0.525	0.473
Farmers / breeders	0.447	0.412
Processors	0.447	0.414
Doctors / health authority	0.569	0.568
University scientists	0.486	0.448
National authority in charge of food safety	0.662	0.785
Government	0.683	0.720
Political groups	0.634	0.821
Environmental organizations	.0.610	0.711
Animal welfare organizations	0.603	0.725
Consumer organizations	0.452	0.448
European Union authority in charge of food safety	0.603	0.561
Television documentary	0.562	0.475
Television news / current affairs	0.655	0.611
Television adverts	0.474	0.458
Newspapers	0.548	0.564
Internet	0.471	0.450
Radio	0.642	0.704
Magazines	0.578	0.609
Product label	0.310	0.292

Extraction Method: Maximum Likelihood.

and organic shops). The fourth factor still includes 'greener' source of information and the fifth keeps together government, political groups and processors. Processors are relevant for two factors and this is a consequence of the VARIMAX method. Should this be a problem, we could try QUARTIMAX, which spreads values across columns.

6. Estimate factor scores. Once we are satisfied with our factor matrix, the next step is the estimation of the actual factor scores. This is a key step, as we will use these values instead of the original variables for subsequent analysis. If the explanatory power of the extracted factors is good enough, each factor's score will provide a synthetic information about the values of the original variables characterizing that factor. By choosing the appropriate options (*save scores* and *estimation method*) after clicking on the SCORES button in the main dialog box, we can exploit the regression estimation method, which – consistently with the maximum likelihood method for estimating factor loadings – assumes multivariate normality of the data. Five additional variables, the factors, now appear in the SPSS data-set. To see the average factor scores by country (as in table 10.5) we need to use the SPSS tabulation tools discussed in chapter 4.

Consider table 10.5, the estimated factor scores have a sample average of zero (and a variability of 1, as reflected by the last column). If we look at the first row, it would seem that Germans place more trust in mass media than respondents in Italy, France, UK and Netherlands. Table 10.6 below shows the average by country for the original variables. Consider table 10.5, the estimated factor scores have a sample average of zero (and a variability of 1, as reflected by the last column).

Table 10.3 *Unrotated Factor Matrix*

Factor Matrix(a)

	Factor				
	1	2	3	4	5
Shopkeepers			0.590		
Supermarkets	0.539		0.507		
Organic shop				0.522	
Specialty store	0.435				
Farmers / breeders	0.454				
Processors	0.557				
Doctors / health authority	0.603				
University scientists	0.592				
National authority in charge of food safety	0.696	−0.445			
Government	0.734				
Political groups	0.740				−0.429
Environmental organizations	0.617			0.421	
Animal welfare organizations	0.548			0.409	
Consumer organizations	0.547				
European Union authority in charge of food safety	0.682				
Television documentary	0.613				
Television news / current affairs	0.677				
Television adverts	0.609				
Newspapers	0.622				
Internet	0.569				
Radio	0.661	0.423			
Magazines	0.618				
Product label	0.441				

Extraction Method: Maximum Likelihood.
5 factors extracted. 5 iterations required.

Indeed, Germany has the highest values of trust for all of the original media variables but 'television adverts.' By looking at the factors instead than the original variables, we lose some information, for example the fact that the level of trust in food safety information through television adverts is high in the UK, while factor 1 showed a low average for media in general. If we are available to afford this cost for the sake of simplification, then factor analysis worked for us. Otherwise, we could decide to retain a larger number of factors to reduce the information loss.

Finally, factor analysis can be easily implemented in SAS by using the SAS/STAT procedure FACTOR, which has great flexibility in defining the analysis options and the statistical output.

10.4.2 Principal component analysis

Before implementing principal component analysis, it may be useful to have a look at box 10.7, which clarifies some differences in the terms used in SPSS compared to SAS. Furthermore, let us recall the main differences from factor analysis – first,

Table 10.4 *Rotated Factor Matrix (VARIMAX rotation)*

Rotated Factor Matrix[a]

	Factor				
	1	2	3	4	5
Shopkeepers			0.730		
Supermarkets			0.692		
Organic shop			0.566		
Specialty store			0.626		
Farmers / breeders			0.576		
Processors					0.422
Doctors / health authority		0.689			
University scientists		0.570			
National authority in charge of food safety		0.822			
Government		0.531			0.596
Political groups					0.771
Environmental organizations				0.743	
Animal welfare organizations				0.788	
Consumer organizations		0.489			
European Union authority in charge of food safety		0.617			
Television documentary	0.606				
Television news / current affairs	0.689				
Television adverts	0.480				
Newspapers	0.700				
Internet	0.615				
Radio	0.798				
Magazines	0.728				
Product label					

Extraction Method: Maximum Likelihood.
Rotation Method: Varimax with Kaiser Normalization.
[a] Rotation converged in 7 iterations.

Table 10.5 *Average scores by country*

	Country					
	UK	Italy	Germany	Netherlands	France	Total
	Mean	Mean	Mean	Mean	Mean	Mean
Trust in media	−0.11	−0.06	0.21	−0.03	−0.13	0.00
Trust in authorities	−0.27	0.08	0.15	0.16	−0.31	0.00
Trust in the food chain	0.12	−0.19	−0.03	−0.09	0.35	0.00
Trust in alternative source	−0.16	−0.08	0.27	−0.20	0.13	0.00
Trust in politics and processors	0.04	−0.14	0.14	0.28	−0.51	0.00

Table 10.6 *Average values for the original media variables*

	Country					
	UK	Italy	Germany	Netherlands	France	Total
	Mean	Mean	Mean	Mean	Mean	Mean
Television documentary	4.85	4.88	5.22	4.85	4.94	4.95
Television news / current affairs	5.00	5.25	5.55	5.26	4.87	5.19
Television adverts	4.17	3.58	3.29	3.13	3.00	3.42
Newspapers	4.43	5.05	5.27	4.93	4.89	4.92
Internet	4.56	4.11	4.74	4.82	4.37	4.53
Radio	4.73	4.64	5.32	4.84	4.94	4.90
Magazines	4.45	4.18	4.83	4.40	4.41	4.46

BOX 10.7 *The PCA dictionary for SAS and SPSS*

A source of confusion is sometime the variety of matrices and vectors in the outputs of statistical packages. Especially when looking at the generic *component loadings*, the differences in the definitions can be substantial. Here is a short guide to understand some differences between SAS (two SAS\STAT procedures, FACTOR and PRINCOMP) and SPSS.

Eigenvectors (SAS FACTOR, SAS PRINCOMP). These are the actual latent vectors A_j containing the elements a_{ij} where:

$$\sum_{i=1}^{P} a_{ij}^2 = 1 \text{ and } \sum_{i=1}^{P} a_{ij}^2 = 1$$

Raw component loadings/raw component matrix (SPSS). These loadings correspond to the values $\hat{a}_{ij} = a_{ij}\sqrt{\lambda_j}$. With these loadings

$$\sum_{i=1}^{P} \hat{a}_{ij}^2 = \lambda_j \text{ and } \sum_{j=1}^{P} \hat{a}_{ij}^2 = Var(X_i)$$

Factor pattern (SAS FACTOR) = scaled component matrix (SPSS). This is the matrix of the correlations between the original variables and the components and it is obtained by scaling the raw component loadings to the variance of the original variables, $\tilde{a}_{ij} = \hat{a}_{ij}/Var(x_i)$, so that

$$\sum_{j=1}^{P} \tilde{a}_{ij}^2 = 1$$

Standardized scoring coefficients (SAS FACTOR) = Component score coefficient matrix (SPSS). Standardized component scores are obtained by multiplying these standardized coefficients by the original variables (also standardized). They are obtained in a similar fashion to standardized regression coefficients:

$$\alpha_{ij} = a_{ij}\frac{\sqrt{Var(X_i)}}{\sqrt{\lambda_j}}$$

the number of components needs not to be chosen prior to the analysis, as this technique extracts a number of components equal to the number of original variables, postponing the choice. Second, while factor analysis is mainly targeted to search for an underlying (latent) structure, PCA is preferred for pure data reduction scopes.

The tasks undertaken are similar to those for factor analysis. In SPSS (and this is sometimes a source of confusion), the analysis is carried out using exactly the same dialog box as for factor analysis and simply specifying that the estimation method is 'principal component.' This also requires one to choose whether to use the correlation or covariance matrix for the analysis.

Tasks

1. Compute the component matrix
2. Choose the number of components
3. Examine the component matrix

Table 10.7   *Component Matrix*

Component Matrix[a]

	Component 1	2	3	4	5
Shopkeepers		0.656			
Supermarkets	0.593	0.512			
Organic shop	0.417	0.497	0.507		
Specialty store	0.490	0.526			
Farmers / breeders	0.513	0.500			
Processors	0.584				
Doctors / health authority	0.616		−0.411		
University scientists	0.614				
National authority in charge of food safety	0.671		−0.516		
Government	0.690				
Political groups	0.687				−0.520
Environmental organizations	0.598			0.526	
Animal welfare organizations	0.525			0.580	
Consumer organizations	0.568				
European Union authority in charge of food safety	0.685				
Television documentary	0.653				
Television news / current affairs	0.716				
Television adverts	0.645				
Newspapers	0.655				
Internet	0.607				
Radio	0.694				
Magazines	0.656		0.433		
Product label	0.500				

Extraction Method: Principal Component Analysis.
[a] 5 components extracted.

Table 10.8 *Rotated component matrix*

Rotated Component Matrix[a]

	Component				
	1	2	3	4	5
Shopkeepers			0.758		
Supermarkets			0.675		
Organic shop			0.686	0.434	
Specialty store			0.745		
Farmers / breeders			0.671		
Processors				0.655	
Doctors / health authority		0.768			
University scientists		0.694			
National authority in charge of food safety		0.812			
Government		0.525		0.618	
Political groups				0.749	
Environmental organizations					0.756
Animal welfare organizations					0.801
Consumer organizations		0.587			0.412
European Union authority in charge of food safety		0.676			
Television documentary	0.672				
Television news / current affairs	0.743				
Television adverts	0.509			0.511	
Newspapers	0.742				
Internet	0.666				
Radio	0.814				
Magazines	0.764				
Product label	0.482				

Extraction Method: Principal Component Analysis.
Rotation Method: Varimax with Kaiser Normalization.
[a]Rotation converged in 6 iterations.

4. Rotate the component matrix to improve readability
5. Compute component scores

1. **Computing the component matrix.** Note that this is a computation and not an estimation. The solution is unique, once we have decided whether to work on the correlation or covariance matrix. Since the variables are measured on the same scale, we opt for the covariance matrix.
2. **Number of components**. Applying the Kaiser's rule again, five components are selected. These explain about 66% of the original variability. It is interesting to recall that the maximum likelihood factor analysis with five factors only explained 56% of the original variability. Communalities show that university scientists are still the worst explained variable, but the value is now higher (0.34).
3. **The component matrix**. The unrotated component matrix is shown in table 10.7. As for factor analysis, the unrotated solution returns a first component which is correlated with almost all of the original variables, while the second component is positively related with trust in food chain actors, the third is negatively

Table 10.9 *Average values of component scores by country*

	Country					
	UK	Italy	Germany	Netherlands	France	Total
	Mean	Mean	Mean	Mean	Mean	Mean
Component 1	−0.12	−0.02	0.18	−0.05	−0.08	0.00
Component 2	−0.37	0.19	0.16	0.15	−0.35	0.00
Component 3	0.11	−0.30	−0.04	−0.07	0.52	0.00
Component 4	0.20	−0.06	0.07	0.21	−0.62	0.00
Component 5	−0.18	−0.28	0.36	−0.11	0.15	0.00

correlated with trust in authorities and positively to trust in magazines, the fourth is positively correlated with organic shops, animal welfare organizations and environmental organizations and the fifth is only a negative relation with political groups. As before, the VARIMAX rotation will improve the situation.

4. **Rotated solution.** The VARIMAX rotation returns a solution which is not far from the factor analysis one (see table 10.8), although the fourth component shows a relevant positive correlation with trust in television adverts as well and the fifth includes consumer organizations, together with the other 'alternative' source of information. In synthesis, while the interpretation is similar to factor analysis, the additional portion of variability explained by the first five component translates into a better characterization of the components.

Compute the component scores. This is straightforward for PCA; as scores are computed rather than estimated (and in SPSS it is irrelevant which estimation method is chosen). Table 10.9 summarizes the average value by country. Again, results are relatively similar to factor analysis.

Summing up

Factor analysis and PCA are statistical techniques for reducing the data dimension and summarizing sets of correlated variables in a reduced number of *factors (components)* which is still adequate to describe the phenomenon of interest.

Data reduction methods can be used as a preliminary step to other analyses (from simple tabulation to cluster analysis) or as self-contained methods to obtain indirect measures for concepts that are not directly measurable (e.g. quality, intelligence).

This chapter illustrates the technicalities behind the two methods, indicating rules for:

(a) choosing the factor estimation techniques;
(b) deciding the number of factors (components); and
(c) interpreting and rotating factors (components).

A discussion of the main differences and peculiarities between factor analysis and PCA is also provided to help choosing between the two.

Data reduction methods are easily implemented through commercial software, but the variety of options (and terminology) may lead to very different outputs, so that the analyst is recommended to become familiar with their implications.

EXERCISES

1. Open the Trust data-set
   a. Assume that questions q12a to q12k measure a single latent theoretical construct attitude toward purchasing chicken
   b. Using factor analysis, extract the single latent factor using image factoring as a method
   c. Check the scree diagram – is it consistent with the choice of retaining a single factor?
   d. Interpret the factor loadings – which items show a positive contribution to the factor and which ones a negative contribution?
   e. Save the factor scores and show the average by country. Which country shows the highest attitude? Do results change noticeably with different score estimation methods?
   f. How much of total variability does the factor reproduce?
   g. Exclude items with negative wording (q12c, q12i, q12k) and compute the alpha reliability index (HINT: use SCALE / RELIABILITY ANALYSIS).
   h. Run factor analysis again to look for a single factor measuring overall attitude. Do results improve?
2. Open the Trust data-set
   a. Consider the attributes of chicken safety (q21a to q21l) and assume they measure various dimension of safety
   b. Using factor analysis and the alpha factoring method on the correlation matrix, extract the number of factors that seems most appropriate and show the scree diagram
   c. Use the VARIMAX rotation and label the factors according to the factor loadings
   d. Consider the table of communalities – which variable is best explained? Which one is least explained by the retained factors?
   e. Draw a bar chart showing the level of the factor scores for each country
   f. How do result change if maximum likelihood is chosen instead of alpha factoring?
3. Open the EFS data-set
   a. Consider all food expenditure items between c11111 and c11941
   b. Summarize the 58 expenditure items into a reduced number of components able to explain at least 60% of the original variability. How many components are retained? How many should be retained to explain 75% of the variability?
   c. Try and interpret the component loadings for the first three components
   d. Run the analysis using alternatively the covariance and correlation matrix. Which one should be preferred? Which one produces the best results?
   e. Extract the component scores and check their correlation with anonymized income (incanon)

Further readings and web-links

❖ **Construct validity, negative wording of the questions and factor analysis** – The identification of a single factor from several questionnaire items aimed at measuring the same construct is generally considered as a validation of their internal consistency. However, sometimes this may not happen and two factors are identified simply because some of the items are positively worded and others are negatively worded. This may lead some respondents to answer inconsistently, which will ultimately result in the negatively worded questions to load into a different factor from the one collecting positively worded items. For a discussion of this issue see Spector et al. (1997), Schmitt and Stults (1985) or Schriesheim and Eisenbach (1995).

❖ **Rotation in principal component analysis** – To learn more on the pros and cons of rotation in PCA see chapter 11 of the book by Jolliffe (2002) or the articles by Jolliffe (1989 and 1995).

❖ **Over-extraction versus under-extraction in factor analysis and PCA** – When choosing the number of factors, one should be aware of the trade-off between exceeding the correct number of factors (over-extraction) and selecting an insufficient number of factors (under-extraction). Fava and Velicer discuss over-extraction and under-extraction in two separate articles (1992 and 1996 respectively). Further discussion on over-extraction can be found in Lawrence and Hancock (1999).

Hints for more advanced studies

☞ To what extent are items in a questionnaire measuring the targeted latent psychological construct? Discuss the contribution of factor analysis in evaluating the internal consistency of measurement of latent constructs

☞ Factor analysis and principal component analysis can also be explained with a geometric and graphical approach – find out how

☞ How important is the multivariate normality assumption to apply the maximum likelihood method in factor analysis?

Notes

1. SPSS also includes principal component analysis as an estimation method for factor analysis. However, for the reasons discussed in this chapter, we prefer to treat these methodologies as distinct.
2. As explained in the appendix, standardization is the process which transforms a variable into another one with zero mean and unity variance by subtracting its mean and dividing by the standard deviation.
3. Orthogonality of a matrix is briefly introduced in the appendix. An orthogonal rotation of a matrix is obtained by pre- or post-multiplying such matrix for an orthogonal matrix.
4. A comparison of the methods is provided by Lastovicka and Thamodaran (1991).

PART IV

Classification and Segmentation Techniques

This fourth part covers techniques that lead to the classification of statistical units in different sub-groups of the population, an operation which is especially relevant to the identification of market segments.

Discriminant analysis is the first of these techniques and is covered in **chapter 11.** Discriminant analysis starts from a pre-determined allocation into two or more groups and explains it on the basis of a set of predictors. Thus, rather than producing a classification, the objective is to explore its determinants. When more than two groups are considered, the technique is called multiple discriminant analysis (MDA). **Chapter 12** presents the class of techniques for cluster analysis. Cluster analysis forms groups (the clusters) which contain very similar observations, while those belonging to different groups are relatively different. Homogeneity within clusters and heterogeneity between clusters is computed on the basis of distance measures between observations and between clusters. In **chapter 13** the focus is on multidimensional scaling (MDS), a comprehensive set of methods which project on a low-dimensional map the differences in consumer perception and preferences measured on a set of products, brands or characteristics. This allows one to draw maps where all these elements are portrayed simultaneously and opens the way to the identification of market segments and spaces for new product development. While MDS mainly work with scale and ordinal variables, correspondence analysis, covered in **chapter 14,** is the technique which allows one to explore the association between categorical variables and display them jointly on a bivariate graph.

CHAPTER 11

Discriminant Analysis

T HIS CHAPTER explores a first classification technique. Discriminant analysis (DA) allows one to explain the allocation of observations into two or more groups on the basis of a set of predictors. This is especially useful in understanding the factors leading consumers to make different choices in order to develop marketing strategies which take into proper account the role of the predictors. DA looks at the discrimination between two groups, while multiple discriminant analysis (MDA) allows for classification into three or more groups. DA and MDA share similarities and overlap with several other statistical techniques, but the key differences are explained in this chapter.

Section 11.1 introduces DA by looking at its potential applications
Section 11.2 explains the two-groups technique through an applied example
Section 11.3 generalizes the method to the MDA case

THREE LEARNING OUTCOMES

This chapter enables the reader to:

➡ Appreciate the potential marketing applications of (M)DA
➡ Understand the functioning and the output of DA and MDA
➡ Know the difference between MDA and other statistical techniques

PRELIMINARY KNOWLEDGE: Familiarity with the basic principles of ANOVA (chapter 7), regression (chapter 8) and PCA (chapter 10) will facilitate understanding.

11.1 Discriminant analysis and its application to consumer and marketing data

What makes a customer loyal? What sort of shopper is expected to show up on a Saturday? Who will buy a wine labeled with a controlled denomination of origin (CDO)? Who likes the new flavor of a soft drink? What makes the difference between those shopping at Safeway and those shopping at Waitrose? These and many other questions can find an answer through *discriminant analysis*. As is shown in the following chapters (see especially cluster analysis in chapter 12 and discrete choice models in

---

BOX 11.1 *Discriminant analysis versus rest of the World (of multivariate statistics)*

Discriminant analysis shares similar objectives and problems with other multivariate statistical techniques. However, there are also some key differences. Here is a brief (and possibly incomplete) set of comparisons.

Analysis of Variance. This chapter shows how discriminant analysis exploits ANOVA and ANCOVA in stepwise multiple discriminant analysis and the relation with these techniques. It is also interesting to notice that DA can be seen as the mirror technique for factorial ANOVA. In factorial ANOVA a metric variable is measured for a combination of qualitative classification variables (treatments and explanatory factors). Here, the situation is reversed and the classification variable is explored for a combination of independent metric variables.

Principal Component Analysis. There are similarities between the discriminant functions and the principal components. The first discriminant function, as the first principal component, has the highest discriminating power. However, the techniques are very different. Multiple discriminant analysis considers the variability in the classification variable (the dependent one) and maximizes it conditional on the independent variables, while principal component analysis aims to summarize the variability of all the covariates by creating the principal components as linear combinations of the independent variables.

Discrete choice models. As it is clearer in chapter 16, discrete choice models are regression models where the dependent variable is categorical. It has been already mentioned in chapter 8 that *logistic regression* is to be preferred in the two-group case when one or more of the predictors are not continuous or normally distributed. Other generalizations of discriminant analysis to discrete choice model include the *multinomial logit* (more than two categories in the dependent variable) or the *ordered logit* (the dependent variable is ordered).

Cluster Analysis. This is another classification technique which is discussed in chapter 12, where cases are allocated to different groups according to a set of independent variables. However, the key difference from MDA is similar to the difference between MDA and PCA. While in MDA the researcher knows (from the sample) to which group each of the observation belongs, in cluster analysis the groups are unknown and are identified by the technique.

---

chapter 16), there are other techniques that answer to questions like the ones above and box 11.1 briefly illustrates the main similarities and differences.

DA is a statistical procedure which allows one to classify cases in the separate categories to which they belong, on the basis of a set of independent variables called *predictors* or *discriminant variables*. This implies that the target variable (the one determining allocation into groups) is a qualitative (nominal or ordinal) one, while the characteristics are measured by quantitative variables. To make a good use of the Trust data-set, let one first consider question q8d where respondents are asked whether 'in a typical week' they buy chicken at the butcher's shop. Thus, respondents may belong to one of two groups – those who purchase chicken at the butcher's shop and those who do not. One can try to 'discriminate' between these groups with a set of consumer characteristics. For example, the expenditure on chicken in a standard week (q5), age of the respondent (q51), whether respondents agree (on a seven point ranking scale) that butcher's sell safe chicken (q21d) and trust (again on a seven point ranking scale) toward supermarkets (q43b). Does a linear combination of these four characteristics allow one to discriminate between those who buy chicken at the butcher's and those who don't? In this case there are only two groups, so one of the final outputs of the analysis will be a single discriminating value. For each respondent we use the identified linear combination to compute a score. Respondents with a score above the discriminating value are predicted to belong to one group, those below to the other group. The linear combination value is usually termed as *discriminant score*. When this value is standardized to have zero mean and variance equal to 1, it is named *Z score*. The linear combination of the original variables is called *discriminant function* and allows one to predict to which category each case belongs on the basis of the independent variables. Finally, discriminant analysis provides information about which of the original predictors are most helpful to discriminate across groups.

Discriminant analysis may involve more than two groups in which case it is termed *multiple discriminant analysis (MDA)*. Consider question q6, where respondents were asked to identify the type of chicken they purchase 'in a typical week,' choosing among four categories, 'value' (good value for money), 'standard,' 'organic' and 'luxury' (that is associated with some high quality cues such as specific labels, albeit more expensive). Again, it could be useful to know whether one can use some independent variables to discriminate among consumers, for example age (q51), the stated relevance of taste (q24a), value for money (q24b) and animal welfare (q24k), plus an indicator of income (q60).[1]

In this case there is more than one discriminant functions. The exact number is equal to either $(g - 1)$, where g is the number of categories in the classification variable, or to k, the number of independent variables, whichever is the smaller. In this example there are four groups and five explanatory variables, so the number of discriminant functions is three (that is $g - 1$ which is smaller than $k = 5$).

As explained in the following section, the output of MDA shares many similarities with factor analysis or principal component analysis. For example, the first discriminant function is the most relevant for discriminating across groups, the second is the second most relevant, etc. In MDA as in PCA, the discriminant functions are also independent, which means that the resulting scores are non-correlated. Once the coefficients of the discriminant functions are estimated and standardized, they are interpreted in a similar fashion to the factor loadings. The larger the standardized coefficients are (in absolute terms), the more relevant is the respective variable to discriminating between groups.

As compared to the simpler case of two groups, the problem with MDA lies in the use of the discriminant scores. With two groups, it is straightforward, as a single value allows one to decide whether each unit should be assigned to one group or the other. With more than two groups, one needs first to compute the group means (usually termed as *centroids*) for each of the discriminant function to have a clearer view of the classification rule.

11.2 Running discriminant analysis

To understand the functioning of discriminant analysis, it is useful to proceed step by step. First, the two-group case is considered, and then a generalization to MDA is provided.

11.2.1 Two-groups discriminant analysis

Consider the example introduced in section 11.1. The purpose is discriminating between those who shop at the butcher's shop versus those who do not, using four independent variables or predictors (weekly expenditure on chicken, age, safety of butcher's chickens, trust in supermarkets). What we need is a *discriminant function*, viz.,

$$z = \alpha_0 + \alpha_1 x_1 + \alpha_2 x_2 + \alpha_3 x_3 + \alpha_4 x_4$$

where the x_i variables are the predictors and the coefficients α_i need to be estimated.

The above equation is the starting point for *Fisher's linear discriminant analysis*, which is the classic technique for DA. There are two key assumptions behind linear DA – (a) the predictors x are normally distributed and (b) the covariance matrices for the predictors within each of the groups are equal. Departure from condition (a) should suggest use of alternative methods (logistic regression, see chapter 16), while departure from condition (b) requires the use of different discriminant techniques, usually with *quadratic discriminant functions*. In most empirical cases, the use of linear DA should suffice. The first step is the estimation of the α coefficients, also termed as *discriminant coefficients* or *weights*.

11.2.2 Estimating the discriminant coefficients

Once the predictors have been defined (either by the researcher's decision or through a stepwise procedure), the next step is their estimation. It is not worth going into the technical details (see Klecka, 1980), but the rationale is very similar to factor analysis or PCA, as the coefficients are those which maximize the variability between groups. Thus, in multiple discriminant analysis, the first discriminating function is the one with the highest between-group variability, the second discriminating function is independent of the first and maximizes the remaining between-group variability and so on. Let us consider the two-groups example entering all the predictors. The coefficient estimates provided by SPSS are shown in table 11.1.

These coefficients depend on the measurement unit, thus it is preferable to look at standardized coefficients to compare the relative contribution of each predictor to the discriminant scores (table 11.2).

Thus, those who consider the butcher's shop as a safer chicken supplier have a higher discriminant score and this predictor has the larger positive impact, followed by age and expenditure on chicken. Instead, trust in supermarkets has a negative sign and reduces the discriminant score. It is intuitive from these values that higher discriminant scores increase the probability of being classified among those who shop at the butcher's shop, however one may look at the group centroids in table 11.3 (the discriminant score means for the groups) to have a clearer view.

As expected, the average value for those not shopping at the butcher's shop is negative, while for those who purchase chicken in the supermarket is positive. Since one can compute the discriminant score for each of the respondents, it is possible to

Table 11.1 *Discriminant coefficient estimates in SPSS*

Canonical Discriminant Function Coefficients

	Function
	1
In a typical week how much do you spend on fresh or frozen chicken (Euro)?	0.095
Age	0.025
Safer from the butcher	0.454
Supermarkets	−0.297
(Constant)	−2.515

Unstandardized coefficients.

Table 11.2 *Standardized discriminant coefficients*

Standardized Canonical Discriminant Function Coefficients

	Function
	1
In a typical week how much do you spend on fresh or frozen chicken (Euro)?	0.378
Age	0.394
Safer from the butcher	0.748
Supermarkets	−0.453

Table 11.3 *Group centroids*

Functions at Group Centroids

	Function
Butcher	1
No	−0.307
Yes	0.594

Unstandardized canonical discriminant functions evaluated at group means.

determine the *cut-off point* at that value which separates the two groups.[2] With equal group sizes, the average of the two centroids is the cut-off point, as the probability of being classified in either group is 50% and the distributions are normal. Instead, when group sizes differ as in the present example, it is necessary to compute a weighted average. The sample includes 277 respondents who answered 'no' and 143 who answered 'yes', thus the cut-off point is computed as $0.594 \cdot 143 - 0.307 \cdot 277 = -0.0002$. The value is close to zero, which needs to be the case when the scores are standardized.

Let us suppose one uses this discriminating rule, how precise would be the classification of the sample respondents?

Table 11.4 shows the number (and percentages) of correctly classified cases. 88.1% of those who answered 'no' are correctly classified, while 38.5% of those who answered

Table 11.4 *Classification results*

Classification Results[a]

		Butcher	Predicted Group Membership		Total
			No	Yes	
Original	Count	No	244	33	277
		Yes	88	55	143
		Ungrouped cases	1	1	2
	%	No	88.1	11.9	100.0
		Yes	61.5	38.5	100.0
		Ungrouped cases	50.0	50.0	100.0

[a]71.2% of original grouped cases correctly classified.

'yes' are correctly classified. In total 71.2% of the cases were correctly classified. This is a relatively satisfactory result. Given that 66% of respondents in the sample declared not to shop in butcher's shops, if one simply assumes that 'nobody buys chicken in butcher's shops,' 66% of cases would be still correctly classified. The use of discriminant analysis brought a small improvement in that respect. However, correct prediction of the sample cases is not the main aim of the analysis, what is useful is to understand the relative role of the predictors. The analysis shows that perceived safety of butcher's chicken is an important variable in discriminating between the two groups.

11.2.3 Diagnostics

As anticipated, one of the assumptions of DA is that the covariance matrices are equal across groups. This hypothesis is tested through the *Box's M test*. This test is very sensitive, as it is likely to reject the null hypothesis with large sample sizes and when some departure from normality is observed, thus one should use a very conservative significance level. Even when the null hypothesis is rejected, it is usually still safe to proceed with DA unless outliers affect the data.

Another statistic which appears in DA output is the *Wilks' Lambda* (or *U statistic*), which has values between zero and 1 and measures discrimination between groups. The Wilks' Lambda is related to analysis of variance, as it computes the ratio between the sum of squares within the groups and the total sum of square. The lower is the Lambda, the higher is the discriminating power of the examined function. Note (as this sometimes generates confusion) that the Wilks' Lambda can be computed both for the function as a whole or separately for each of the predictors in a discriminant function. Its *p*-value is drawn from an *F* distribution for one-way ANOVA (single predictors) and a chi-square distribution for multiple variables (overall function). The Wilks' Lambda and the *p*-value are provided for each of the predictor as in univariate ANOVA, and indicate whether there are significant differences in the predictor's means between the groups. An overall Wilks' Lambda (which is important when running multiple discriminant analysis) is provided for the discriminant function, accompanied by the chi-square *p*-value. In the latter case, a significant test means that there are significant differences in the group means for the discriminant function (the linear combination of the predictors). In MDA this allows one to discard those function who do not contribute to explaining differences between groups.

One of the first values returned by DA is the *eigenvalue* (or eigenvalues for MDA) of the discriminant function. This can be interpreted in a similar fashion to principal component analysis, although in the latter case there is a rule for selecting the most important principal components which does not apply here. In MDA, when more than one discriminant function is estimated, eigenvalues are exploited to compute how each function contributes to explain variability. The *canonical correlation* which SPSS shows next to the eigenvalue measures the intensity of the relationship between the groups and each single discriminant function (see section 9.3). Table 11.5 summarizes the diagnostics for the two-group discriminant analysis.

All the predictors are (individually) relevant to explain the discrimination between the two groups. The overall function is also significant. The assumption of equal covariances is rejected (which suggests to apply a quadratic discriminant function method).

Table 11.5 *Diagnostics for discriminant analysis*

	Statistic	p-value
Box's M statistic	37.30	0.000
Overall Wilks' Lambda	0.85	0.000
Wilks Lambda for		
Expenditure	0.98	0.002
Age	0.97	0.001
Safer for Butcher	0.91	0.000
Trust in Supermarket	0.98	0.002
Eigenvalue	0.18	
Canonical correlation	0.39	
% of correct predictions	71.2%	

11.3 Multiple discriminant analysis

While the bivariate case is quite useful to understand the aims and potential of DA, this technique is certainly more useful when there are more than two groups involved, as this second example where the objective is to identify the relevant predictors in explaining the choice of one type of chicken over the alternatives. Only a few generalizations are needed to understand the functioning of MDA.

Table 11.6 shows the mean of the predictors for each of the chicken types.

One may already notice that differences across the four groups are relatively higher for the income range than for other statistics. However, the first step should be to conduct a one-way ANOVA for each of the predictors.

As a matter of fact, table 11.7 shows that the income predictor has the lowest Wilks' Lambda (hence a high discriminating power), followed by the 'value for money' characteristic. These two predictors are significant at the 1% level, the preference for 'tasty food' is significant at the 5% level, while the other variables are above the threshold.

The Box's M statistic has a value of 65.2 and a p-value of 0.045. This is slightly below the 5% threshold, but as previously noted, the sensitivity of this test suggests that this p-value might be acceptable. As explained in section 11.1, with five predictors and four groups, we can estimate three discriminating functions. This does not mean that we need to retain all three functions. The first function is the one with most discriminating power by construction and the Chi-square statistic (see table 11.8) confirms that it is significant (at the 1% level). To evaluate the second function, the Chi-square test is run again, after elimination of the variability already explained by the first function. The second discriminant function is still significant if one chooses the 5% threshold. Instead, the third function has a non-significant Chi-square statistic, which means that it does not provide additional discriminating power.

The estimated coefficients are shown in table 11.9 for the first two functions, both in the unstandardized and standardized version.

As for factor analysis or PCA, it may be useful to label the discriminant functions according to the predictors that are most relevant in each of them. To this purpose, one might look at the standardized discriminant coefficients, but these reflect the presence of all other predictors in the function. A more appropriate evaluation is provided by the *structure matrix*, which contains the *structure coefficients* (or *discriminant loadings*), shown in table 11.10.

Table 11.6 *Descriptive statistics for MDA predictors*

Group Statistics

In a typical week, what type of fresh or frozen		Mean	Std. deviation	Valid N (list-wise)	
				Unweighted	Weighted
'Value' chicken	Value for money	5.585	1.3412	41	41.000
	Age	47.000	15.6668	41	41.000
	Tasty food	5.976	1.0365	41	41.000
	Animal welfare	4.512	1.7765	41	41.000
	Please indicate your gross annual household income range	0.707	1.0306	41	41.000
'Standard' chicken	Value for money	6.032	1.2005	157	157.000
	Age	42.847	15.4211	157	157.000
	Tasty food	6.376	0.8275	157	157.000
	Animal welfare	4.841	1.6506	157	157.000
	Please indicate your gross annual household income range	1.599	1.3628	157	157.000
'Organic' chicken	Value for money	5.343	1.3708	35	35.000
	Age	43.771	13.5323	35	35.000
	Tasty food	6.114	1.1317	35	35.000
	Animal welfare	5.371	1.4366	35	35.000
	Please indicate your gross annual household income range	2.200	1.3677	35	35.000
'Luxury' chicken	Value for money	5.981	1.1683	53	53.000
	Age	47.868	17.0272	53	53.000
	Tasty food	6.377	0.8140	53	53.000
	Animal welfare	4.906	1.8425	53	53.000
	Please indicate your gross annual household income range	1.566	1.4346	53	53.000
Total	Value for money	5.874	1.2558	286	286.000
	Age	44.486	15.6186	286	286.000
	Tasty food	6.287	0.9071	286	286.000
	Animal welfare	4.871	1.6882	286	286.000
	Please indicate your gross annual household income range	1.538	1.3855	286	286.000

Table 11.7 *Mean comparison tests for predictors*

Tests of Equality of Group Means

	Wilks' Lambda	F	df1	df2	Sig.
Value for money	0.960	3.878	3	282	0.010
Age	0.981	1.798	3	282	0.148
Tasty food	0.971	2.761	3	282	0.042
Animal welfare	0.982	1.679	3	282	0.172
Please indicate your gross annual household income range	0.919	8.272	3	282	0.000

Table 11.8 *Discriminating power diagnostics*

Wilks' Lambda

Test of function(s)	Wilks' Lambda	Chi-square	df	Sig.
1 through 3	0.851	45.098	15	0.000
2 through 3	0.938	17.904	8	0.022
3	0.986	3.818	3	0.282

Table 11.9 *Coefficients of the discriminant functions*

Discriminant functions' coefficients

	Unstandardized		Standardized	
	1	2	1	2
Value for money	−0.043	0.603	−0.053	0.746
Age	−0.009	−0.013	−0.148	−0.208
Tasty food	0.169	0.416	0.152	0.374
Animal welfare	0.186	−0.132	0.313	−0.222
Please indicate your gross annual household income range	0.652	−0.033	0.870	−0.044
(Constant)	−2.298	−4.868		

With some simplification, the first function discriminates using social characteristics (income, attention to animal welfare), the second one is related to the product's characteristics (value for money, taste) and the third one to the demographics (age).

As compared to two-group discriminant analysis, the use of scores and the computation of cut-off points for classification is not straightforward. The analysis of centroids (table 11.11) suggests that the first function especially discriminates between 'value' chicken and 'organic' chicken, since the centroids have opposite signs and high absolute values, while little distinction is provided between 'standard' and 'luxury' chicken, although they are both different from the other two types. Thus, one needs to use the second function as well and we may notice that consumers with preference for

Table 11.10 *Discriminant loadings*

Structure Matrix

	Function		
	1	2	3
Please indicate your gross annual household income range	0.929*	−0.021	0.078
Animal welfare	0.390*	−0.206	0.125
Value for money	−0.010	0.891*	0.168
Tasty food	0.241	0.660*	0.273
Age	−0.217	−0.204	0.944*

Pooled within-groups correlations between discriminating variables and standardized canonical discriminant functions
Variables ordered by absolute size of correlation within function.
*Largest absolute correlation between each variable and any discriminant function.

Table 11.11 *Function values at group centroids*

In a typical week, what type of fresh or frozen chicken do you buy for your household's home consumption?	Function	
	1	2
'Value' chicken	−0.673	−0.262
'Standard' chicken	0.058	0.156
'Organic' chicken	0.525	−0.470
'Luxury' chicken	0.003	0.052

Unstandardized canonical discriminant functions evaluated at group means.

standard chicken over luxury chicken tend to have a slightly larger discriminant score in function 2.

Having two functions only, similar indications can be drawn from a scatter-plot, as shown in figure 11.1.

The plot shows how function 1 is especially useful to discriminate between value chicken, organic chicken and the confounded group of luxury and standard chicken. To distinguish between these two types one may still decide to use the third discriminant function, which in facts shows relatively different centroids (−0.065 for standard chicken and 0.242 for luxury chicken). Thus, those confounded cases with a higher discriminant score in function 3 will be classified into the luxury category. We have already explored some of the diagnostic instruments (Box's M statistic and Wilks Lambda for predictors and for the discriminant functions); the remaining diagnostics relate to canonical correlations and correct predictions (table 11.12). Canonical correlations are not very high and the percentage of correct predictions is not satisfactory if one considers that 'standard chicken' represents 55% of cases (thus allocating all cases to the modal value would return a similar result). The problems in prediction are confirmed by looking at the correct prediction rates by type of chicken. The functions fail completely to classify luxury chicken and have low prediction rates for 'value' and 'organic' chicken.

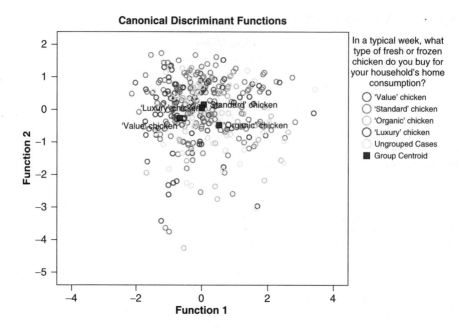

Figure 11.1   *Scatter-plot for two discriminant functions*

Table 11.12   *Canonical correlations and correct prediction rates*

	Function 1	Function 2	Function 3
Eigenvalue	0.10	0.05	0.01
Canonical correlation	0.30	0.22	0.12
% of correct predictions	56.30%		

		Predicted Group Membership				
		'Value' chicken	'Standard' chicken	'Organic' chicken	'Luxury' chicken	Total
Original	'Value' chicken	7.3	92.7	0.0	0.0	100.0
	'Standard' chicken	1.3	98.1	0.6	0.0	100.0
	'Organic' chicken	2.9	85.7	11.4	0.0	100.0
	'Luxury' chicken	1.9	96.2	1.9	0.0	100.0
	Ungrouped cases	0.0	94.4	5.6	0.0	100.0
	56.3% of original grouped cases correctly classified					

11.3.1   Validation

The examples above all started from the observation of the actual classification. Thus, prediction diagnostics were based on correctly classifying cases already used for the DA. If the model has to be generalized to a larger population, it is advisable to exploit a more reliable validation procedure.

One approach consists in the *leave-one-out classification procedure*. The discriminant analysis is run after dropping one case, and then this case is predicted using the model. This step is repeated for each of the observations, so that the analysis is run on a sample size practically equal to the original one, but predictions are made on out-of-sample observations. For our MDA, the correct prediction rate falls from 56.3 to 53.8%. With large samplea, one might also extract a random sub-sample which includes half of the observations and apply the model to predict the remaining half of the observations.

11.3.2 Stepwise discriminant analysis

As for linear regression (see chapter 8), it is possible to decide whether all predictors should appear in the equation regardless of their role in discriminating or a sub-set of predictors is chosen on the basis of their contribution to discriminating between groups. The latter approach is called the *stepwise method* (Klecka, 1980) and works sequentially as follows:[3]

1. A one-way ANOVA test is run on each of the predictors, where the target grouping variable determines the treatment levels. The ANOVA test provides a criterion value and test statistic (usually the *Wilks' Lambda*, but alternative values exist, see below). According to the criterion value it is possible to identify the predictor which is most relevant in discriminating between the groups (usually the one with the lowest Wilks' Lambda), while the *p*-value tells one whether the contribution of this variable is significant.
2. The predictor with the lowest Wilks' Lambda (or which meets the alternative optimal criterion) enters the discriminating function, provided the *p*-value is below the set threshold (for example 5%).
3. An ANCOVA test is run on the remaining predictors, where the covariates are the target grouping variables and the predictors that have already entered the model. The Wilks' Lambda is computed for each of the ANCOVA options.
4. Again, the criterion and the p-value determine which additional variable (if any) enters the discriminating function (and possibly whether some of the entered variables should leave the model)
5. The procedure goes back to step 3 and continues until none of the excluded variables has a *p*-value below the threshold and none of the entered variables has a *p*-value above the threshold (the *stopping rule* is met).

The stepwise results vary according to the *stopping rule*, as a higher threshold *p*-value leads to including more predictors. Furthermore, alternatives to the Wilks' lambda for evaluating the discriminating function as a whole exist, such as the *Unexplained variance, Smallest F ratio, Mahalanobis distance* or the *Rao's V*, all available in SPSS. We leave to the reader the application of stepwise discriminant analysis on the MDA example of this section. However, it is interesting to observe some of the main results. First, two predictors only are retained (value for money and income), which leads to the identification of two discriminant functions. Second, cross-validated predictions show a success rate of 52.6%, practically the same as for the four predictor case. Although this is not yet very satisfactory, it is interesting to notice that we achieve the same results using two independent variables only. Exploration on other potential predictors through the stepwise method may lead to improvement of results without necessarily increasing the number of independent variables.

Summing up

Discriminant analysis is a technique which allows one to test the relevance of a set of predictors in explaining why one observation belongs to a specific group of a categorical dependent variable. The generalization to the case where there are more than two groups is called multiple discriminant analysis. This technique opens the way to many useful applications in marketing, as it allows us to distinguish between those who make a specific purchasing decision and those who do not on the basis of some predictors that may be observed on potential customers outside the collected samples. Thus it becomes possible: (a) to anticipate/identify a potential consumer choice by looking at a set of observable variables; and (b) to influence consumer choice by changing some of the modifiable variables (for example product characteristics). In simple two-group DA, a single discriminant function is estimated. By looking at the coefficients and other statistics, it is possible to assess the discriminating power of the individual predictor. By using the estimated function it is possible to assign observations to either group according to the computed discriminant score and the cut-off value. The analysis is then validated by looking at the rate of correct predictions (possibly using out-of-sample observations). MDA simply generalizes DA to analyze target classifications with more than two categories. In this case more than two discriminating functions are estimated and classification is less straightforward. DA and MDA have much in common with discrete choice model, are related to ANOVA techniques and share some similarities with cluster analysis and PCA, although the use of these latter techniques is quite different.

EXERCISES

1. Open the EFS data-set
   a. Using discriminant analysis, explore how household income (incanon) and age of the household reference person (p396p) explain discrimination between those with a home computer and those without (a1661)
   b. How many discriminant functions are estimated?
   c. Are the group significantly different in terms of the predictor values?
   d. Which predictor is most relevant in explaining classification?
   e. Does the assumption of equality of group covariance matrices hold?
   f. What are the correct prediction rates?
   g. Does allocation of all units to the modal category generate a better or worse prediction rate?
2. Open the Trust data-set
   a. Consider the allocation of respondents according to their job status (q54)
   b. Using MDA, explore the role of the following predictors in explaining classification: age (q51), household size (q56), financial condition of the household (q61)
   c. According to ANOVA tests, which predictors are statistically different across groups? Which predictor has the highest discriminatory power?
   d. Which discriminant functions are significant according to the Chi-square test?

e. How much of the total variance does the first discriminant function explain?

f. Considering the structure matrix, try and label the discriminant functions

g. Compute the percentage of correctly predicted cases. Does allocation of all units to the modal category generate a better or worse prediction rate?

3. Open the EFS data-set

a. Consider the allocation of sampling unit according to the tenure type (a121)

b. Using MDA, explore the role of the following predictors in explaining classification: age (p396), household size (a049), income (incanon), total housing expenditure (p604)and number of children (a019)

c. Which predictors are irrelevant in explaining classification?

d. Repeat the analysis using the stepwise selection method and compare the results

Further readings and web-links

❖ **Unequal group covariance matrices and quadratic discriminant function methods** – For a clarification of the issue of different covariances and the use of the appropriate method, see Marks and Dunn (1974) and Dillon (1979).

❖ **Impact of sampling, small sample sizes (and small groups) on MDA** – See Sanchez (1974), or Morrison (1969). Frank et al. (1965) and Crask and Perreault (1977) review the problems in validating MDA for marketing research conducted on small samples.

❖ **Alternative methods for classification** – In recent years, neural networks and classification trees gained popularity as classification techniques. Neural networks are also known as non-linear DA. For an introduction to neural networks and their relationship with discriminant analysis see Garson (1998) or Abdi et al. (1998), while classification trees are explained in Breiman et al. (1984). For the application to direct marketing and a comparison between classification trees and neural networks see Linder et al. (2004). An extensive (but rather technical) discussion of neural networks and its relation with other classification techniques can be found in Ripley (1994). For a comparative assessment of alternative classification methods, see, for example, Kiang (2003). An Excel spreadsheet which illustrates the basic functioning of classification trees can be downloaded free at www.xlstat.com/en/support/tutorials/trees.htm.

Hints for more advanced studies

☞ What is the relationship between the canonical correlation computed in DA and canonical correlation analysis of chapter 9?

☞ What happens if the predictors in DA are strongly correlated (i.e. there is collinearity)?

☞ How is the Wilks' test built? What are the differences from the ANOVA F-test?

Notes

1. This raises the issue whether an ordered variable should be used as an independent variable in DA. There is little literature about the implications, but when the ordinal variable can be assumed as an approximation of a normally distributed metric variable (which should be the case for income), then DA is expected to provide reliable results. Even when some departure from the normal distribution is likely, if the ordered variable does not interact with other predictors, DA is still regarded as robust (Knoke, 1982). Otherwise, it is advisable to employ discrete choice models as those described in chapter 16.

2. SPSS and other software also provide the (Fisher's) *classification functions*, which should not be confused with the discriminant functions. A Fisher classification function is estimated for each group, with a separate set of coefficients. After computing the scores for each of the classification function, each case is classified to the group with the higher score.

3. SPSS only offers the stepwise option, while the SAS STEPDISC procedure allows to adopt also Forward and Backward selection rules (see chapter 16).

CHAPTER 12

Cluster Analysis

T HIS CHAPTER looks into the different methods that allow us to perform a cluster analysis, which is the grouping of observations into clusters where observations within the same cluster are similar and observations belonging to different clusters are quite dissimilar. This is a useful classification technique for market segmentation and consumer or product profiling. The many options, issues and tricks of an adequate cluster analysis are discussed in detail, together with examples and applications in SPSS and SAS.

Section 12.1   introduces cluster analysis and its use in marketing research with examples

Section 12.2   explores the features of different clustering methods according to the theory

Section 12.3   translates the theory into practice by running cluster analysis in SPSS and SAS

THREE LEARNING OUTCOMES

This chapter enables the reader to:

➡ Know when cluster analysis is useful for marketing and consumer research applications

➡ Become aware of the wide range of options which affect the output of cluster analysis

➡ Learn how to run cluster analysis in SPSS and SAS and interpret the output

PRELIMINARY KNOWLEDGE: Familiarity with the ANOVA concepts and the basic principles of PCA.

12.1 Cluster analysis and its application to consumer and marketing data

Anyone with some knowledge of marketing is familiar with the concept of market segmentation. Segmenting a market means dividing its potential consumers into separate sub-sets that are homogeneous within themselves, but relatively different between each other, with respect to a given set of characteristics. This operation opens the way to calibrating the marketing mix differently according to the target market.

The most studied case of market segmentation lies in price discrimination, which allows one to sell the same product at different prices to different consumers with a different willingness to pay for that product. A multinational company which sells a product in different countries is likely to adjust the price to the purchasing power in each specific country. Consider the following example. A producer of mineral water exploits a source which enables the production of 10,000 liters per day. The quality of the water, the brand value and the production costs allow it to sell each liter at a price of £2, which makes a turnover of £20,000. The company would like to increase profits by raising the price to £3, but marketing research suggest that in that way half of the customers would stop buying the water, reducing the turnover to £15,000 and leaving 5,000 liters unsold. If the company manages to sell 5,000 liters of mineral water at £3 and 5,000 liters at £2, they could sell the whole production and raise the turnover to £25,000 per day.

This marketing operation is made possible if one knows the differences between those available to buy at £3 and those who only buy at the £2 price. A segmentation study shows that the large majority of those who accept to buy at £3 shop in Waitrose, own a car, have children and buy about 12 liters per shopping trip, while most of those who would refuse to buy for a price higher than £2 are students, with no car, living in student flats and usually buy 2 or 3 liters per trip. The company decides to raise the price of the existing brand to £3 (price), improve the packaging by offering packs of six with a comfortable handle to lift the glass bottles (packaging), starts a promotion campaign targeted to families (promotion) and limits the distribution to Waitrose and a few other supermarkets in urban areas (place), which allows it to reduce distribution costs. At the same time, they introduce a new brand, sold at £2 (price), sold individually in plastic bottles and with very simple labels (packaging), advertised through TV ad in a popular music program for teenagers (promotion) and decide to distribute most of the production to supermarkets and newsagents next to universities. The content of the bottles is exactly the same, but segmentation allows the creation of two separate (sub-)markets.

If the company had the prior information about which customers are ready to buy water at £3 and those who would only accept buying at £2, then the most natural way to proceed would be to conduct a discriminant analysis as described in chapter 11. However, this information is not yet available and a survey could record something like the maximum price that respondents would be available to pay without turning to a substitute, which would be a continuous rather than a categorical variable. Furthermore, it could be possible that three segments exist, one with people that would buy at £2.50, but not at £3, who are not students and do not shop in Waitrose.

In this situation the way to go is *cluster analysis*, a technique which starts from measurements on a set of variables and allows classification of observations into homogeneous groups (clusters). The aim of cluster analysis is to classify the observations through a procedure which maximizes the homogeneity (that is it minimizes the variance) within the clusters and maximizes the heterogeneity (maximizes the variance) between the clusters.

Probably it is more appropriate to refer to cluster analysis as a class of techniques, since, as explained in this chapter, it includes a number of different definitions and procedures which may lead to sensibly different results. Market segmentation is only one of the many marketing actions that can be supported by cluster analysis. A second typical application is related to product characteristics and the identification of new product opportunities. By clustering similar brands or products according to their characteristics, one may identify the main competitors and potential market opportunities and available niches.

In the previous pages, cluster analysis has been explored as a classification technique. However, there is another use, which is data reduction. In chapter 10 we saw how factor analysis and principal component analysis allow us to reduce the number of variables. Cluster analysis allows us to reduce the number of observations by grouping them into homogeneous clusters. This is sometime useful when the number of observations is too large, or when the objective is to draw maps for profiling simultaneously consumers and products, to identify market opportunities and preferences (as in *preference* or *perceptual mappings* discussed in chapter 13). This chapter describes the essential steps in conducting a cluster analysis, with an overview of the main issues and developments.

12.2 Steps in conducting cluster analysis

In synthesis, five steps are required to run a cluster analysis:

1. Select a distance measure for individual observations
2. Choose a clustering algorithm
3. Define the distance between two clusters
4. Determine the number of clusters
5. Validate the analysis

12.2.1 Distance measures for individual observations

The key objective of cluster analysis is to group similar observations with respect to a set of characteristics. Thus, the starting point is the definition of *similarity*. If we had a single variable (say income), it would be easy to define two observations as similar when their income levels are similar. The smaller the difference in incomes, the more similar the observations are. However, if we have many characteristics (income, age, consumption habits, family composition, owning a car, education level, job...), it becomes more difficult to define similarity. To this purpose, it is first necessary to define a *distance measure*.

The most known measure of distance is the *Euclidean distance*, which is the concept we use in everyday life for the spatial co-ordinates. For a generic number of co-ordinates (variables) n, the generic equation for Euclidean distance between two observations i and j is the following:

$$D_{ij} = \sqrt{\sum_{k=1}^{n} (x_{ki} - x_{kj})^2}$$

where x_{ki} is the measurement of the k-th variable on the i-th observation.

This is only one of the potential definitions of distance. An alternative one is the city-block distance:

$$D_{ij} = \sum_{k=1}^{n} |x_{ki} - x_{kj}|$$

Thinking in term of bi-dimensional space, the *city-block* or *Manhattan* distance assumes that one can move between two points only horizontally or vertically, but not diagonally (as in a city block of Manhattan). Other types of distances provided by statistical packages include Chebychev, Minkowski, Mahalanobis (for details see Green and Rao, 1969). The choice of the distance measure should have little effect on the final output of the cluster analysis, unless data are affected by outliers (Punj and Stewart, 1983). Two potential issues need to be taken into account before proceeding with the computation of all potential distances (which are contained in the *proximity matrix*). The first one arises when the variables are measured in different measurement units. If age is measured in years and income in dollars, the straight application of a distance measure will necessarily attach a larger weight to income since a difference of 1,000 is very likely to occur, while it is almost impossible that a difference of 100 will occur for age measured in years. Thus, it is necessary to proceed with some standardization (see the appendix or chapter 6) by subtracting the mean and dividing by the standard error. This also comes at a loss, since all variables end up having the same variability (standardized variances are equal to 1), with the risk of ignoring the fact that differences were higher for some variables than for others. Before proposing a solution, let us consider the second issue. Some variables might be affected by double counting because of interactions. For example, if one measures the distance in terms of income, education level, job position, age, all these variables are likely to be correlated. It is undesirable to ignore these correlations because they will end up inflating the relevance of income as compared to the influence of other uncorrelated variables. This suggests a solution that also solves the issue of different measurement units. Running a principal component analysis allows us to purge the observed data from correlations and returns component scores that are standardized, hence unaffected by different measurement units (see chapter 10).[1] Finally, an alternative approach to distances is based on *correlation measures*, where correlations are not between the variables (as it is usually the case, see chapter 8), but between the observations. Each observation is characterized by a set of measurements (one for each variable) and bivariate correlations can be computed between two observations.

12.2.2 Clustering algorithms

Once all distances between all pair of observations have been computed and placed in the proximity matrix, the *clustering algorithm* can be applied to obtain the classification. Two basic types of algorithms exist, hierarchical and non-hierarchical.

As their name suggests, *hierarchical methods* follow a hierarchy in forming clusters. Two directions can be taken. To start the algorithm one may assume that each of the *n* observations constitutes a separate cluster. In the first step, the two clusters that are more similar according to the same distance rule are aggregated, so that in step 1 there are *n* − 1 clusters. In the second step, another cluster is formed (*n* − 2 clusters), by nesting the two clusters that are more similar, and so on. Thus, there is a merging in each step until all observations end up in a single cluster in the final step. As explained later in this chapter, it is then possible to choose the best partition, by defining the step at which the clustering algorithm should be stopped. This is an *agglomerative hierarchical method*, since the procedure proceeds by aggregation. An alternative hierarchical approach can work in the opposite direction. All observations are initially assumed to belong to a single cluster, and then the most dissimilar observation is extracted to form a separate cluster. In step 1 there will be two clusters, in the second step three clusters and so on, until the final step will produce as many clusters as the number of observations. Again, the optimal partition can be determined according to the same statistical rule. This type of approach is termed *divisive hierarchical method*. On the other hand, *non-hierarchical methods* do not follow a hierarchy and produce a single partition. Knowledge of the number of clusters (c) is required. In the first step, initial cluster centers (the *seeds*) are determined for each of the c cluster, either by the researcher or by the software, which may simply choose the first c observation. Then each iteration allocates observations to each of the c clusters, based on their distance from the cluster centers. Cluster centers are then computed again and observations may be re-allocated to the closest cluster in the following iteration. When no observations can be re-allocated or a stopping rule is met, the process stops and the final clustering into c clusters is provided.

While the flow of these algorithms is relatively simple, they clearly depend on the aggregation rule for pairs of clusters. Defining an aggregation rule for two observations is very simple, since one obviously wants to aggregate the two nearest observations, whatever the definition of distance chosen. The matter becomes slightly more complex for aggregating two clusters, or an observation with a cluster. Since a cluster already includes several observations, multiple definitions are possible for distances between clusters.

12.2.3 Distances between clusters

Algorithms vary according to the way the distance between two clusters is defined. The most common algorithm for hierarchical methods include: the single linkage method, the complete linkage method, the average linkage method, the Ward algorithm and the centroid method.[2]

The *single linkage method* defines the distance between two clusters as the *minimum* distance among all possible distances between observations belonging to the two clusters.

The *complete linkage method* nests two cluster using as a basis the *maximum* distance between observations belonging to separate clusters.

The *average linkage method* defines the distance between two clusters as the average of all distances between observations in the two clusters.

The *Ward method* is slightly more complex and computationally demanding. First, the sum of squared distances is computed *within* each of the clusters, considering all distances between observations within the same cluster. Then the algorithm proceeds choosing the aggregation between two clusters which generates the smallest increase in the total sum of squared distances. It is a computationally intensive method, because at each step all the sums of squared distances need to be computed, together with all potential increases in the total sum of squared distances for each possible aggregation of clusters.

The *centroid method* defines the distance between two clusters as the distance between the two centroids, which are the cluster averages for each of the variable, so that each cluster is defined by a single set of co-ordinates (the averages of the co-ordinates of all individual observations belonging to that cluster). Note the difference between the centroid method, which averages the co-ordinates of the observations belonging to an individual cluster and the average linkage method, which averages the distances between two separate clusters.

Table 12.1 *Hierarchical and non-hierarchical methods in cluster analysis*

Hierarchical methods	Non-hierarchical methods (K-means)
• No decision about the number of clusters • Problems when data contain a high level of error • Can be very slow, preferable with small data-sets • Initial decisions are more influential (one-step only) • At each steps they require computation of the full proximity matrix	• Faster, more reliable, works with large data-sets • Need to specify the number of clusters • Need to set the initial seeds • Only cluster distances to seeds need to be computed in each iteration

With non-hierarchical methods, the standard method is the *k-means algorithm*, where *k* stands for the number of clusters chosen to start the process. The *k*-means algorithm is an iterative procedure which works as follows:

1. An initial set of *k* 'seeds' (aggregation centers) is provided either by the first *k* elements of the data-set or by explicitly defining the initial seeds
2. Given a certain fixed threshold, all individual observations are assigned to the nearest cluster seed
3. A new seed is computed for each of the clusters
4. Go back to step 2 until no re-classification is necessary

In the *k*-means algorithm, the individual observations can be re-assigned in successive iterations. Other non-hierarchical methods are the *threshold methods*, where a prior threshold is fixed and units within that distance are allocated to the first seed, then a second seed is selected and the remaining units are allocated, etc. This method is called *sequential threshold*, while an alternative exists where more than one seed is considered simultaneously (*parallel threshold*). When reallocation is possible after each stage, the methods are termed *optimizing procedures*.

From the above discussion, it is clear that the main difference between hierarchical and non-hierarchical methods is the choice of the number of clusters. In hierarchical methods, there is no need for a prior choice and through some diagnostics it is possible to determine the optimal partition. Instead, with *k*-means clustering, the number of clusters needs to be pre-determined. This is not the only difference and possibly not even the most important, as the improvement in statistical packages now allows it to run quickly several *k*-means clustering for different cluster numbers, which opens the way to a comparison among the different resulting partitions. Table 12.1 summarizes the key characteristics of the two approaches.

On balance, the *k*-means approach seems to be preferable, but it requires arbitrary choices on the number of clusters and the initial seeds. We will get back to this issue after discussing the choice of the optimal number of clusters with hierarchical methods.

12.2.4 Number of clusters

The identification of the number of clusters (Everitt, 1979) is usually an important output of the research. In segmentation studies, it represents the number of potential separate segments. Thus, in most situations it is preferable to 'let the data speak' and identify the optimal partition through statistical tests. This is not easily done; sometimes data need to be tortured before they actually speak... Since the detection of the optimal number of clusters is subject to a high degree of uncertainty, if the research objectives allow us to choose rather than estimate it, non-hierarchical methods are the way to go. Suppose that a retailer wants to identify several shopping profiles in order to activate new and targeted retail outlets but the budget only allows three types of outlets to open. A partition into three clusters follows naturally, although it is not necessarily the optimal one. With a fixed number of clusters, the k-means non-hierarchical approach will do the job quickly and neatly.

On the other hand, take the case where clustering of shopping profiles is expected to detect a new market niche. For market segmentation purposes it is less advisable to constrain the analysis to a fixed number of clusters. A hierarchical procedure allows one to explore all potentially valid numbers of clusters and for each of them there are some statistical diagnostics to pinpoint the best partition. What is needed is a *stopping rule* for the hierarchical algorithm, which determines the number of clusters at which the algorithm should stop. Unfortunately these alternative statistical tests are not always unequivocal, still leaving some room to the researcher's experience and arbitrariness. One standard piece of advice is to balance the statistical rigidities with the informativeness of the final classification.

With hierarchical techniques, the optimal number of clusters can be chosen by looking at two plots. The first is the *dendrogram*, which allows us to follow the nesting process in a hierarchical method through the various steps. The individual observations are listed on the vertical axis and joined through lines, whose widths vary according to the distance at which the nesting occurs, as shown in figure 12.1.

In the first step, cases 231 and 275 are joined and the nesting distance is quite small (below 1). In the second step cases 117 and 136 are nested and in the following step the cluster made by 231 and 275 joins with observation 145. As one may expect, as the nesting process progresses the nesting distances increase. By deciding the maximum nesting distance, one can draw a vertical line to stop the process and identify the optimal number of clusters.

As for factor analysis and PCA, another graphical tool is the scree diagram (figure 12.2). By plotting the number of clusters against the nesting distance, it is (generally) possible to identify the 'elbow' where there is a jump suggesting to stop at that number of clusters.

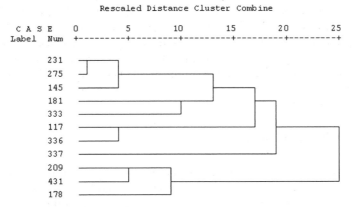

Figure 12.1 *Dendrogram*

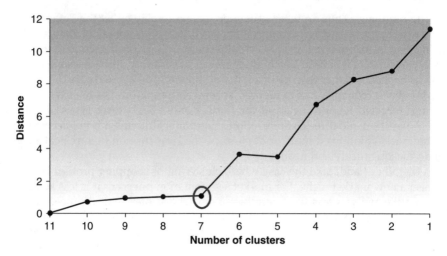

Figure 12.2 *Scree diagram*

The other option is to proceed through statistical tests similar to the analysis of variance, looking at which partition maximizes the sum of squares between clusters in relation to the sum of squares within clusters. However, it is not possible to simply exploit ANOVA F-tests, because hierarchical methods proceed in sequence and the probability distribution is not known. This aspect has been and is currently the subject of academic research. SPSS does not provide statistical diagnostics on the number of clusters, while SAS has a very complete toolbox. One statistical criterion is based on finding the minimum of the determinant of the within cluster sum of squares matrix W (Arnold's criterion as in Arnold, 1979). Other criteria include the *pseudo F statistic*, the *pseudo t^2 statistic* and the *cubic clustering criterion (CCC)*.[3] The ideal number of clusters should correspond to a local maximum for the pseudo F and the CCC, together with a small value of the pseudo t^2 which increases in the next step (preferably a local minimum). Unfortunately, these criteria are rarely consistent among them, so that the researcher should also rely on meaningfulness (interpretability) criteria. With SAS it is also possible to exploit *non-parametric methods* to determine the number of clusters, as the method devised by Wong and Lane (1983), named the *k-th nearest neighbor*.[4] The researcher needs to set a parameter (k) and for each k the method returns the optimal number of clusters. If this optimal number is the same for several values of k, then the determination of the number of clusters is relatively robust.[5] Finally, it is possible to compensate the rigidity of the results from hierarchical techniques with the flexibility of a non-hierarchical method. To this purpose, a good way to proceed is to choose the number of clusters according to a hierarchical approach and then apply the non-hierarchical method to obtain the composition of each group. This is the so-called *two-step procedure*, explored in greater detail in the next section.

12.2.5 Two-step procedures

Two-step procedures use a hierarchical method for a statistically-based definition of the number of clusters and a non-hierarchical method (the *k*-means method) for the actual clustering. The hierarchical method computes a partition for each possible number of clusters, hence allowing a statistical assessment to identify the best partitions.

However, a rigidity of hierarchical methods is the fact that once a unit is classified into a cluster it cannot be moved to other clusters in subsequent steps. Instead, as explained, the k-means method allows a reclassification of all units in each iteration. If some uncertainty about the number of clusters remains after running the hierarchical method, one may also run several k-means clustering procedures and use the statistical tests described in the previous section to choose the best one.

Recently, SPSS has introduced a two-step clustering procedure. The main purpose of this procedure is to reduce the computational burden of hierarchical methods. The number of clusters can also be automatically estimated. This method is different from the above and does not use non-hierarchical approaches. The observations are preliminarily aggregated into clusters using a hybrid hierarchical procedure named *cluster feature tree* (Zhang et al., 1996). This first step produces a number of *pre-clusters*, which is higher than the final number of cluster, but much smaller than the number of observations. In the second step, a hierarchical method is used to classify the pre-clusters, obtaining the final classification. During this second clustering step, it is possible to determine the number of clusters. The user can either fix the number of clusters or let the algorithm search for the best one according to *information criteria*[6] which are also based on goodness-of-fit measures.

12.2.6 Evaluating and validating cluster analyses

The objective of cluster analysis is to maximize homogeneity within clusters and heterogeneity between clusters. As seen, statistics based on this principle allow one to choose between partitions with different number of clusters. In general, it is possible to evaluate the *goodness-of-fit* of a cluster analysis by using the ratio between the sum of squared errors and the total sum of squared errors, in a fashion similar to the R^2 statistic used in regression (see chapter 8) and similar to the F test in analysis of variance. However, the standard F test cannot be applied, because the assumption of independence between groups is violated by the clustering algorithm. It follows that a second measure of goodness-of-fit is related to variability within cluster, the *root mean standard deviation* within clusters.

A further step is necessary to *validate* the cluster analysis, which means checking that the results are robust to the sampling error and to the choice of the clustering method. Validation can be based on the use of different samples to check whether the final output is similar, or the sample can be split into two groups when no other samples are available. When using non-hierarchical approaches, it is also advisable to check for the influence of the initial seeds by changing them and checking the final solutions. A good cluster solution should be robust to the method employed.

12.3 The application of cluster analysis in SAS and SPSS – empirical issues and solutions

To show the application of cluster analysis with SPSS and SAS, let us consider an example from the EFS data-set. Suppose the objective is to identify market segments on the basis of the following variables – household size, age of the oldest person in the household, household income and weekly expenditure in food, transport costs, clothing, housing and recreation. This set of variables has two problems. First, the variables are measured through different units (age in years, expenditure in pounds, etc.), secondly a strong correlation is likely to exist across variables, especially the expenditure values, income and the household size. The first problem can be overcome by standardization, while the second would require a prior principal component analysis. While we proceed with standardization to avoid meaningless results, we leave to the reader to run the cluster analysis on the extracted principal component scores and we ignore the correlation problem here, which means that results will over-emphasize difference between large and rich households and the smaller or poorer ones. In the rest of this example we use the *Euclidean distance* as the distance measure.

12.3.1 *SPSS clustering*

The SPSS ANALYZE/CLASSIFY menu (which we have already exploited for discriminant analysis) offers three options for clustering: hierarchical methods, the *k*-means method and two-step clustering. Let us start with a hierarchical method – the Ward algorithm.

Using the METHOD option button, we choose the clustering algorithm (Ward method), we opt for the Euclidean distance and we use standardized variables with zero mean and variance equal to 1 (choose the Z *scores* option – by variable – under the *standardize* item).

Through the STATISTICS option button we ask SPSS to provide an *agglomeration schedule*, which is necessary to evaluate the distance of each nesting step. By using the agglomeration schedule, it is also possible to draw the *scree diagram* (an option which is not directly provided by SPSS). Within the same window it is advisable to restrict the range of cluster numbers to be analyzed (*range of cluster solution*), since allowing for all potential partitions (which would be 500 given the sample size) would slow the process and produce unintelligible plots. We decide to explore solutions between two and ten clusters.

Within the PLOT option button, we can ask for the *dendrogram* to be plotted. SPSS can also produce an *icicle*, which shows which cases are nested at each step. A strong piece of advice – select 'no icicle' or limit the range of clusters being explored, as asking for a complete icicle may seriously slow down the clustering algorithm.

Finally, using the SAVE option button, it is possible to create an additional variable containing the cluster membership for each case. If a range of cluster numbers is selected, there will be more than one variable. At this stage, where no clear idea exists about the number of clusters, we omit to save the memberships.

Table 12.2 shows the last ten steps of the agglomeration process. The *agglomeration schedule* shows, for each step, the two clusters being merged and the nesting distance. For ease of interpretation we have added the number of clusters, which is not provided by SPSS but is equal to $n - s$, where s is the step number, and the difference in distance between two subsequent steps.

The ideal number of clusters should correspond to the step *before* a large increase in the nesting distance is observed. We notice that the increase in distance is quite uniform until step 494 (between 14.9 and 19), then we observe an increase of 26 units (at step 495) and another jump by 37 unit at step 497. If we consider the first jump, the targeted number of clusters should be six (that is stopping at step 494, before the first jump), otherwise the number of clusters is four. To see it graphically, let's check the scree diagram (plotting the number of clusters versus the distance).

Table 12.2 *Agglomeration process*

Stage	Number of clusters	Cluster Combined		Distance	Diff. Dist
		Cluster 1	Cluster 2		
490	10	8	12	544.4	
491	9	8	11	559.3	14.9
492	8	3	7	575.0	15.7
493	7	3	366	591.6	16.6
494	6	3	6	610.6	19.0
495	5	3	37	636.6	26.0
496	4	13	23	663.7	27.1
497	3	3	13	700.8	37.1
498	2	1	8	754.1	53.3
499	1	1	3	864.2	110.2

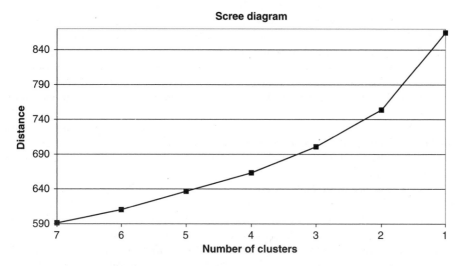

Figure 12.3 *Scree diagram for the EFS case study*

It is all but easy to see an elbow in the plot of figure 12.3, but it certainly seems that the distance increases linearly at least until four clusters are left. SPSS does not provide additional statistics, thus we stick to the four clusters solution.

Table 12.3 shows the cluster centroids for the four cluster solution using the Ward algorithm. To obtain the table with the cluster centroids, we simply create the cluster membership variable through the SAVE option and run the Ward algorithm again, this time fixing at four the number of clusters, then we use the TABLE command (see chapter 4).

According to this classification, the four clusters have similar sizes, ranging from 20.2% of the observations for cluster two to 29.4% for cluster four. Cluster one is made by elderly households with small household sizes, as the average age is 72 and the average household size is 1.4, the smallest of the lot. It also shows the lowest income (although this obviously also depends on the household size) and the lowest mean

Table 12.3 *Cluster centroids*

		Ward method				Total
		1	2	3	4	
Case number	N %	26.6%	20.2%	23.8%	29.4%	100.0%
Household size	Mean	1.4	3.2	1.9	3.1	2.4
Gross current income	Mean	238.0	1158.9	333.8	680.3	576.9
Age	Mean	72	44	40	48	52
EFS: Total food	Mean	28.8	64.4	29.2	60.6	45.4
EFS: Total clothing	Mean	8.8	64.3	9.2	19.0	23.1
EFS: Total housing	Mean	25.1	77.7	33.5	39.1	41.8
EFS: Total transport	Mean	17.7	147.8	24.6	57.1	57.2
EFS: Total recreation	Mean	29.6	146.2	39.4	63.0	65.3

expenditure for all items. On the other hand, cluster two has the largest household size and the highest expenditure for all items. Cluster three is the youngest one, both household size and expenditures are much lower than for cluster two. Finally, cluster four, the largest cluster, is not too dissimilar from cluster two in terms of household size, age and food expenditure, but income is almost half of the average for cluster two and all expenditures but food are sensibly smaller. If the objective is to target household with high incomes and expenditure we should certainly look after cluster two. If we are interested in food expenditure, cluster four is equally interesting.

Now let us consider non-hierarchical clustering. This requires an assumption on the number of clusters and we can exploit the results from the non-hierarchical analysis, choosing to fix the number of clusters at four. This is the typical procedure for a two-step clustering technique.

To run a k-means cluster analysis, choose ANALYZE/CLASSIFY/K-MEANS CLUS-TERS. After specifying the number of clusters, one should tick the box to *classify and iterate*, since this allows re-allocation of the units and improves the results. The initial seeds can be specified by creating an ad hoc SPSS file, with four values (rows) for each of the eight clustering variables (columns). In the first instance, we can let SPSS choose the initial seeds (four actual observations will be used, ensuring that they are as different as possible), then save the final centers as an SPSS file, which we can modify to test for different initial seeds. It is possible to specify the number of iterations and the convergence criterion by clicking on the ITERATE option button. Since the algorithm is quite fast, it is advisable to raise the number of iterations to 50 so that proper convergence is achieved. The algorithm can also be improved by requiring the cluster centers to be computed again whenever an observation is added (tick the box for *running means*). The OPTIONS button allows the provision of an ANOVA table. Finally, we can SAVE the cluster membership as for hierarchical methods. The final cluster centers and the number of observations in each cluster are reported in table 12.4.

The results already show a limitation of the k-means algorithm, sensible to outliers (see chapter 4). Since SPSS chose a peculiar initial seed for cluster two (an improbable amount for recreation expenditure, probably an outlier due to mis-recording or an exceptional expenditure), no other observation was close enough to be assigned to that cluster. Instead, we can use the output from the hierarchical method to modify the initial seeds.

Table 12.4 *Final cluster centers and number of cases per cluster (k-means)*

Final Cluster Centers

	Cluster			
	1	2	3	4
Household size	2.0	2.0	2.8	3.2
Gross current income of household	264.5	241.1	791.2	1698.1
Age of household reference person	56	75	46	45
EFS: Total food and non-alcoholic beverage	37.3	22.2	54.1	66.2
EFS: Total clothing and footwear	14.0	28.0	31.7	48.4
EFS: Total housing, water, electricity	34.7	100.3	47.3	64.5
EFS: Total transport costs	28.4	10.4	78.3	156.8
EFS: Total recreation	39.6	3013.1	74.4	125.9

Number of Cases in each Cluster

Cluster	1	292.000
	2	1.000
	3	155.000
	4	52.000
Valid		500.000
Missing		0.000

Table 12.5 *Final cluster centers after modification of the initial seeds*

		Cluster number of case				
		1	2	3	4	Total
Case number	N %	32.6%	10.2%	33.6%	23.6%	100.0%
Household size	Mean	1.7	3.1	2.5	2.9	2.4
Gross current income	Mean	163.5	1707.3	431.8	865.9	576.9
Age	Mean	60	45	50	46	52
EFS: Total food	Mean	31.3	65.5	45.1	56.8	45.4
EFS: Total clothing	Mean	12.3	48.4	19.1	32.7	23.1
EFS: Total housing	Mean	29.8	65.3	41.9	48.1	41.8
EFS: Total transport	Mean	24.6	156.8	37.4	87.5	57.2
EFS: Total recreation	Mean	30.3	126.8	67.9	83.4	65.3

Results are in table 12.5. The first cluster is now larger, but it still represents older and poorer households. The other clusters are not very different from the ones obtained with the Ward algorithm, indicating a certain robustness of the results.

Finally, we can apply the SPSS two-step procedure (ANALYZE/CLASSIFY/ TWO-STEP CLUSTER ANALYSIS). A potential advantage of this procedure is that it allows us to distinguish between categorical (nominal) variables and metric (continuous) variables, although in our example all variables can be regarded as metric. The metric variables are standardized, while it is possible to set (for example at ten) the *maximum number of clusters* (letting SPSS to identify the optimal number), the type of distance (we choose *Euclidean* for consistency with the other methods) and the criterion to be used for identifying the optimal partition (for example the BIC criterion). The OPTIONS

Table 12.6 *Distribution of cases in two-step clustering*

		\multicolumn{3}{c}{Cluster Distribution}		
		N	% of combined	% of total
Cluster	1	2	0.4%	0.4%
	2	5	1.0%	1.0%
	3	490	98.2%	98.2%
	4	2	0.4%	0.4%
	Combined	499	100.0%	100.0%
Total		499		100.0%

button allows to choose whether an *outlier treatment* is required (we omit this option) and some advanced options for the cluster feature tree (which we leave unchanged). The PLOTS option displays a range of charts on the final output, while through the OUTPUT button it is possible to obtain the descriptive statistics by cluster, the cluster frequencies and the cluster membership variable.

The results that we obtain with one click on the OK button are disappointing. Two clusters are found, one with a single observations and the other with the remaining 499 observations. It would seem that the problem is due to the abnormal recreation expenditure which affected the k-means procedure. If we allow for outlier treatment, results do not improve. If we run the two-step cluster analysis after deleting case 4953 (the one with the anomalous recreation expenditure), results still lead to a very large cluster (497 observations) and a second negligible cluster. Finally, if we set the number of clusters to four, the results in terms of cluster distribution are shown in table 12.6.

It seems that the two-step clustering is biased toward finding a macro-cluster. This might be due to the fact that the number of observations is relatively small, but the combination of the Ward algorithm with the k-means algorithm is certainly the most satisfactory.

12.3.2 SAS clustering

As compared to SPSS, SAS provides more diagnostics and the option of non-parametric clustering through three SAS/STAT procedures, the procedure CLUSTER and VARCLUS (for hierarchical and the k-th neighbor methods), the procedure FASTCLUS (for non-hierarchical methods) and the procedure MODECLUS (for non-parametric methods).[7]

The CLUSTER procedure is applied through the following command

```
PROC CLUSTER DATA = cluster OUTTREE = output METHOD = Ward STANDARD
    RSQUARE RMSSTD PSEUDO CCC
PRINT = 10;
VAR a049 p352 p396 p601 p603 p604 p607 p609;
RUN;
```

where the METHOD option allows one to choose the clustering algorithm, STANDARD implies standardization of the original variables (listed in the VAR command),

Table 12.7 Output of hierarchical clustering in SAS

The CLUSTER Procedure

Ward's: Minimum variance cluster analysis

Cluster history

NCL	- - Clusters joined - - -		FREQ	PMS STD	3 PRSQ	RSQ	ERSQ	CCC	PSF	PST2
10	CL12	CL24	38	1.0584	0.0222	0.595	0.540	11.20	80.0	13.2
9	CL25	CL15	124	0.6232	0.0271	0.568	0.525	7.92	80.6	48.2
8	CL17	CL23	26	1.1130	0.0276	0.540	0.508	5.90	82.6	18.9
7	CL1O	CL13	91	1.0257	0.0344	0.506	0.488	3.22	84.1	19.7
6	CL18	CL9	277	0.5763	0.0410	0.465	0.463	0.24	85.8	79.0
5	CL11	CL7	110	1.1237	0.0415	0.423	0.430	-0.99	90.8	19.1
4	CL5	CL14	196	1.0014	0.0616	0.362	0.386	-3.30	93.7	36.2
3	CL4	CL8	222	1.0829	0.0648	0.297	0.330	-4.10	105.0	31.3
2	CL3	OB366	223	1.1708	0.0905	0.206	0.251	-4.80	130.0	38.5
1	CL6	CL8	500	1.0000	0.2064	0.000	0.000	0.00		130.0

Table 12.8 *Tabulation of hierarchical clustering output in SAS: commands and results*

PROC TREE DATA=tree OUT=OUT N=7;
copy a049 p352 p396 p601 p603 p604 p607 p609;
run;

PROC TABULATE data=out out=table;
var a049 p352 p396 p601 p603 p604 p607 p609;
class cluster;
keyword mean;
TABLE n a049*mean p352*mean p396*mean p601*mean p603*mean p604*mean
p607*mean p609*mean,cluster;
run;

	Cluster						
	1	2	3	4	5	6	7
Case number	153	124	86	91	19	26	1
Household size	1.32	1.83	4.17	2.99	2.95	2.58	2.00
Gross current income of household	235.81	439.58	517.94	1200.71	1052.54	915.21	241.09
Age of household reference person	68.95	47.24	39.08	46.44	38.79	39.08	75.00
EFS: Total food & non-alcoholic beverage	28.23	40.01	57.64	68.70	56.04	44.19	22.22
EFS: Total clothing and footwear	8.18	10.32	15.94	59.26	33.84	60.40	28.00
EFS: Total housing, water, electricity	24.13	38.00	38.40	35.95	42.00	193.89	100.30
EFS: Total transport costs	16.28	43.42	39.92	89.64	377.01	75.94	10.40
EFS: Total recreation	27.79	52.95	52.63	120.37	96.62	57.76	3013.07

RSQUARE, RMSSTD, PSEUDO and CCC are instructions to obtain the statistical tests described in section 12.2 and PRINT = 10 specifies that output is required only for the last ten steps. The agglomeration schedule for the last ten clusters is provided in table 12.7, together with the relevant statistics.

The optimal number of clusters should correspond to local peaks for the cubic clustering criterion and the pseudo F statistic (or to the step prior to noticeable drops in these values), while the pseudo t^2 should be small and increase in the following step. There are no local peaks for the CCC or the Pseudo F statistic, while the Pseudo t^2 shows a sharp increase between seven and six clusters and between five and four clusters, suggesting partitions of seven or five clusters. As a second best for the CCC, a value larger than two is desirable (seven clusters), The diagnostics do not help much, although there are indications that suggest the use of seven clusters.

To obtain the dendrogram, one must use the TREE procedure, which also allows to get the cluster memberships at the desired level (seven clusters), with averages value for each cluster computed through the TABULATE procedure (see table 12.8).

The seventh cluster is still the outlier, while the remaining six show more disaggregated results as compared to the SPSS output. We still find a cluster with older and smaller households and low expenditure (cluster 1), one with larger sizes and expenditure (cluster 4) and one for younger and small households (cluster 2). Cluster 5 is not too dissimilar from cluster 4 apart from a sensibly higher expenditure in transport

Table 12.9 *Commands and output of k-means clustering in SAS*

PROC FASTCLUS MAXCLUSTERS=7 data=cluster out=kmeans maxiter=50 CONV=0.001;
var a049 p352 p396 p601 p603 p604 p607 p609;
run;

PROC TABULATE data=kmeans out=table2;
var a049 p352 p396 p601 p603 p604 p607 p609;
class cluster;
keyword mean;
TABLE n a049*mean p352*mean p396*mean p601*mean p603*mean p604*mean
p607*mean p609*mean,cluster;
run;

	Cluster						
	1	2	3	4	5	6	7
Case number	1	33	74	6	230	2	154
Household size	2.00	3.33	2.77	2.83	1.86	2.50	2.72
Gross current income of household	241.09	1787.92	1064.02	1077.13	213.61	3366.66	572.22
Age of household reference person	75.00	46.45	44.53	43.00	58.01	41.00	46.99
EFS: Total food & non-alcoholic beverage	22.22	70.94	58.06	56.93	35.96	72.85	47.44
EFS: Total clothing and footwear	28.00	47.97	36.53	99.16	13.15	103.09	22.06
EFS: Total housing, water, electricity	100.30	64.87	57.35	23.73	33.63	177.01	40.26
EFS: Total transport costs	10.40	128.22	86.14	572.88	24.55	292.04	54.09
EFS: Total recreation	3013.07	121.75	100.51	179.35	34.23	132.14	58.22

and housing and lower in other categories (commuters? Londoners?) and cluster 6 looks like the average cluster for most expenditure items. While we find some basic traits in common with the four cluster analysis obtained in SPSS, it is clear how the decision about the number of clusters can influence the results. To run the *k*-means algorithm we use the FASTCLUS procedure.[8]
Results in table 12.9 are once more different. This time only four clusters have non-negligible sizes and they resemble the output of the SPSS *k*-means procedure. By reducing the number of clusters to four, SPSS and SAS return (luckily) very similar results.

12.3.3 Discussion

The previous sections have probably left the reader with the feeling that cluster analysis is not too robust to the researcher's choices. While this problem can be in part imputed to the relatively small data-set and possibly to the fact that some of the clustering variables are strongly correlated (the reader may try clustering after PCA), it is probably more than a feeling. Still, all of the different outputs point out to a segment with older and poorer household and another with younger and larger households, with high expenditures. By intensifying the search and adjusting some of the properties, cluster analysis does help in identifying homogeneous groups. But the 'moral' learnt from this chapter should be that cluster analysis needs to be adequately validated and it may be risky to run a single cluster analysis and take the results as truly informative, especially in presence of outliers.

Summing up

While discriminant analysis starts from a given classification to explore the role of potentially discriminating variables, cluster analysis does not assume any prior classification. It can be used as a flexible classification technique, for example, for identifying existing market segments, or simply as a data reduction technique, for example, when the objective is to categorize a range of products into three classes according to a pre-determined set of characteristics. Cluster analysis requires many decisions to be taken by the researcher – the distance measure to define similarity between observations, the method to group similar observations and the number of groups (clusters) to be identified. These and other options can heavily influence the outcome, so it is necessary to validate cluster analysis by changing them and checking whether results are stable. The main distinction is between hierarchical methods, which work sequentially and provide a solution for all potential number of clusters, and non-hierarchical methods which exploit more flexible and efficient classification algorithms but require the prior choice of the number of clusters. The combination of these two approaches in combination with a range of statistical and graphical tools usually provides the best results. Both SPSS and SAS allow us to modify most of the available options, although the latter software is richer in providing statistical diagnostics to choose the number of clusters.

EXERCISES

1. Open the Trust data-set
   a. Segment the sample into four clusters according to the attitude variables (q12a to q12k), using the *k*-means algorithm
   b. Save the final seeds in a file
   c. Replicate the analysis using the final seeds as initial seeds
   d. Perform an ANOVA analysis to check whether the clusters are significantly different with respect to income (q60) and age (q51)
   e. Extract three principal components from the attitude variables, save the component scores and run the *k*-means algorithm on the component score
   f. Do the cluster averages differ from the previous ones with respect to attitudes, income and age?
2. Open the EFS data-set
   a. Segment the UK population with respect to household expenditure in various goods (use expenditure variables from p600 to p612)
   b. What is the optimal number of cluster according to the Ward algorithm?
   c. Run a *k*-means cluster analysis using the number of clusters chosen at point b
   d. Save the cluster-membership from the *k*-means clustering

    e.   Run cluster analysis again, using a hierarchical algorithm and choosing the same number of clusters as before

    f.   Save the cluster-membership from the hierarchical algorithm

    g.   Show the averages of p600 to p612 for both clustering methods. How different are they?

3.   Open the Trust data-set

    a.   Use SPSS two-step clustering to identify a number of clusters not larger than five on the basis of the risk variables (q27a to q27g)

    b.   Compute the average of the risk aversion variable q28 within each cluster

    c.   Do the cluster means of q28 show significant differences across clusters?

    d.   Run the analysis again choosing the number of clusters with the Ward algorithm and run the actual classification with k-means

    e.   Compare the results

Further readings and web-links

❖ **Outliers and cluster analysis** – The fact that cluster analysis is particularly sensitive to outliers makes it a good technique for outlier detection. K-means is especially sensitive to outliers (see Garcia-Escudero and Gordaliza, 1999) and the clustering fixed-point algorithm is seen as the most robust method, increasingly employed for outlier detection (see Hennig, 1998 and 2003). An interesting strategy to detect outliers through a combination of k-means clustering and discriminant analysis is explained in Neal and Wurst (2001).

❖ **Determining the number of clusters** – Tibshirani et al. (2001) have recently proposed a new method to determine the number of clusters, claiming that it outperforms other methods. It is based on the *gap statistic* which is based on the difference between the actual and expected variability within clusters. The method has been refined by Yan and Ye (2007).

❖ **Fuzzy cluster analysis** – In many situations there are observations which might end up in two different clusters. Traditional methods assign each observation to one and only cluster, but there is an alternative route, *fuzzy clustering*, where observations may be associated to more than one cluster and with a given degree of belonging. See Hoeppner et al. (1999). A freeware UNIX software for fuzzy clustering is available from the web-site fuzzy.cs.uni-magdeburg.de/fcluster/

Hints for more advanced studies

☞   Are there clustering algorithms that replicate regularly the same results when applied on the same data-set? Which ones?

☞   What are the consequences on the outcomes of cluster analysis of highly correlated data?

☞   Bayesian classification methods are becoming increasingly popular substitutes for cluster analysis. Find out more.

Notes

1. Note that one may opt for rescaling rather than full standardization, so that each component score has a variability proportional to the corresponding portion of explained variability. See box 10.6 and rescaling of principal component scores in paragraph 10.3.
2. A variety of other methods exist and the family of hierarchical methods continues to grow (see for example Everitt et al., 2001 or DeSarbo et al., 2004).
3. See Milligan and Cooper (1985) for an extensive review of methods for defining the number of clusters with hierarchical methods.
4. Actually SAS exploits a less computationally demanding adaptation of the true test.
5. Usually the value k is set to values around $2 \log_2 n$, where n is the number of observations.
6. Information criteria are the appropriate statistics for comparing models when they are estimated through hierarchical procedures, as in cluster analysis. They depend on how well the models fit the data and on the number of observations (degree of freedom). The most used information criteria are the Akaike Information Criterion (AIC), the Bayesian Information Criterion (BIC), also known as Schwartz Information Criterion.
7. See SAS/STAT Manual, SAS Inc.
8. For a complete list of options, including control on seeds (SEEDS and OUTSEEDS options) see the SAS/STAT Manual (chapter 27).

CHAPTER 13

Multidimensional Scaling

T HIS CHAPTER introduces the reader to the multi-faceted world of multidimensional scaling (MDS), a set of techniques allowing one to translate preference orderings and consumer perceptions toward products (objects) into graphical representations, usually in a two or three-dimension map where both consumers and product can be represented. A variety of MDS methods is exploited in consumer and marketing research, especially for purposes of new product development, brand repositioning and market segmentation according to consumer preference. The main outputs of MDS are perceptual and preference maps.

Section 13.1 gives an intuitive explanation of MDS and examples of its potential applications

Section 13.2 provides a discussion of the main principles, options and issues in MDS

Section 13.3 shows multidimensional unfolding in SPSS and introduces SAS procedures

THREE LEARNING OUTCOMES

This chapter enables the reader to:

➡ Appreciate the potential of multidimensional scaling in consumer research
➡ Become aware of the different options for choosing the appropriate MDS method
➡ Know the introductory steps to run MDS with commercial software

PRELIMINARY KNOWLEDGE: Familiarity with measurement scales (in chapter 1), distance measures (as defined in chapter 12) and the notions of principal component analysis (chapter 10).

13.1 Preferences, perceptions and multidimensional scaling

The most intuitive description of multidimensional scaling (MDS) in consumer research is that MDS is a statistical procedure to project consumer perceptions and preferences in a graph. In other words, MDS allows one to compute the co-ordinates of brands

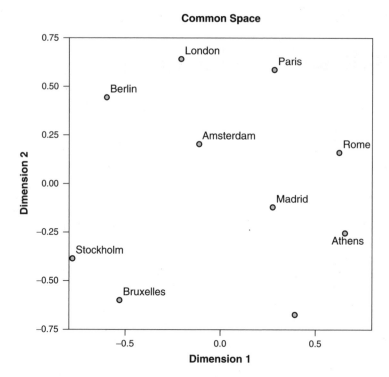

Figure 13.1   *Holiday destinations in two dimensions*

(or consumers) in a multidimensional space. While the name refers to multidimensions, we are unlikely to appreciate the interpretative power of this class of techniques unless we limit the analysis to two or three dimensions. Before getting into a more precise discussion of what is covered by MDS, it may be helpful to have a look at the final output. Figure 13.1 shows a bi-dimensional representation of tourist preferences for ten European cities. Each of the respondents was asked to rank the cities, without necessarily specifying why one city was preferred to another.

The idea is that by looking at similarities in ranking across an adequate number of respondents, it becomes possible to detect perceived similarities. Thus, London is more similar to Berlin than to Athens (as the points are more distant in the bi-dimensional graph). If the two dimensions can be labeled according to some criterion, as for principal component or factor analysis, then it becomes possible to understand the main perceived differences. For example, a researcher's prior knowledge about the characteristics of the cities could relate the first dimension to climate, while the second could be related to how trendy is that location for holidays.[1]

This very simple example only shows a part of the potential contribution of MDS and perceptual/preference mapping to consumer and marketing research.[2] The most successful application is probably related to sensory evaluation and new product development. For example, a company which is developing a low-salt soup is interested in evaluating the market situation. For this purpose an evaluation panel is asked to assess a set of existing soup brands according to several criteria concerning taste, smell, thickness, storage duration, perceived healthiness and price. Besides evaluating the market brands, consumers are asked to identify their ideal product in terms of the same characteristics, which may not coincide with one of the existing soups. The final

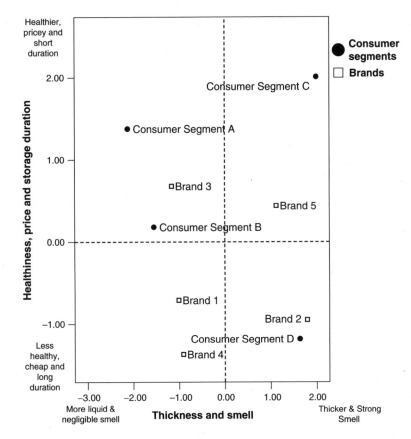

Figure 13.2 *Consumer segments and brand positioning*

output will be a perceptual map displaying both consumer preferences (in terms of their ideal products) and the current positioning of the existing brands. A concentration of consumers' ideal points identifies a segment (cluster analysis might also be used as a tool to segment respondents) and if no brands appear in the neighborhood of a segment, then there is room for the development of a new product in that area. If the perceptual dimensions have been clearly identified, this also allows one to choose the characteristics of the new products.

Figure 13.2 shows the output of the analysis. The two dimensions are the output of some reduction technique, typically principal component analysis or factor analysis for interval (metric) data (see chapter 10) and correspondence analysis for non-metric data (see chapter 14). Co-ordinates for brands are obtained by running PCA (or factor analysis) on sensory assessments (usually through a panel of experts unless objective measures exist), while consumer positions (as individuals or as segments) can be defined in two ways, either by using their 'ideal brand' characteristics (so that the co-ordinates are computed using the PCA equations estimated for actual brands) or by translating preference ranking for brands into co-ordinates, as explained in the next sections.

To appreciate the potential of MDS, consider the potential marketing conclusions of the example above. There are five existing brands and four consumer segments (identified by cluster analysis on their preference rankings or ideal brands). Consumer segment D is probably happy with brand two and consumer segment B is relatively close to brand three. Brands one and four are perceived to be relatively similar by consumers (thus they are competitors), but these products do not find any correspondence with any specific market segment. Those in consumer segment A probably choose brand three, although it is not that close to their preferences. Brand five probably survives because of consumer segment C, but the distance between its sensory properties and consumer preferences is quite large. If the company develops a product which is close to the preferences of segment C this is likely to be successful. If it is positioned on the left-hand side of the point identified for segment C, it is even possible to attract some consumers from consumer segment A. Note that both segments C and A have a preference for healthier products, although those in A prefer more liquid soups (and not too smelly), while consumers in C want a thick and smelly soup. The new product should definitely focus on perceived healthiness. In terms of thickness, if segment C is large enough to ensure a profitable share, then it is probably best to provide the soup they want. If a larger market share is needed, then probably a point between C and A should be chosen, provided it remains closer to those segments than brands three or five.

The same perceptual/preference map (the distinction will be clearer in the next section) could be exploited for brand repositioning. If brand five had this marketing research information, they could improve their performance by focusing on the perceived healthiness of their product, for example by reducing the salt content and through a targeted advertising campaign. This would move brand five closer to segment C.

If consumer perceptions are compared through MDS before and after an advertising campaign aimed at changing perceptions, it becomes possible to measure the success of the advertising effort. Finally, MDS could be exploited to simplify data interpretation and provide some prior insight before running psycho-attitudinal surveys. The rest of this chapter tries to shed some light on the underlying mechanism of MDS.

13.2 Running multidimensional scaling

MDS is not a statistical technique per se, it is rather a container for statistical techniques used to produce perceptual or preference maps. To understand the range of options and choices, it is probably useful to list a series of distinctions on the types of MDS data.

The key element in MDS is the *object* of the analysis, which can be a product, a brand or any other target of consumer behavior, like tourism destinations in the initial example. This object can be depicted as a set of characteristics, which can be either *objective dimensions* (for example salt content in grams) or measured through *subjective dimensions* as declared by respondents (*subjects*) in a survey. In the latter case, consumer evaluations can be based on preferences or perceptions. Measurement through *preferences* requires the subjects to rank several objects according to their overall evaluations (for example ordering of soup brands). Instead, to obtain *perceptions* (*perceptual or subjective dimensions*), the respondent must attach a subjective value to an object's feature (like an evaluation on a Likert scale of the thickness of each soup brand). When attribute perceptions are measured, it is also possible to ask respondents to state explicitly the combination of an object's features that correspond to their *ideal object* (translated into an ideal point in the spatial map). As is discussed later in this chapter, the ideal point can alternatively (and preferably) be derived from respondents' preferences through a statistical model. According to which type of measurement is used, the multidimensional spatial map is termed as *preference map* or *perceptual map*, respectively.

This brief introduction into the terminology of MDS shows the relevance of the measurement step. A description of alternative measurement scales is provided in chapter 1. Preferences can be measured through rank order scales, Q-sorting or other comparative scales. Perceptions require non-comparative scales like Likert, Stapel or Semantic Differential Scales. Finally, objective characteristics are measured through quantitative scales. These measurements can generate two broad types of variables for MDS, usually classified into metric or non-metric. Non-metric variables just reflect a ranking, so that it is not possible to assess whether the distance between the first and second object is larger or smaller than the distance between the second and the third. Instead, metric variables reflect respondent perception of the distances. Generally, preference rankings are classified as non-metric and perceptions and objective dimensions are metric. This distinction can be very important, as it leads to two different MDS approaches. The output of *non-metric MDS* aims to preserve the preference ranking supplied by the respondents, while *metric MDS* also takes into account the distances as measured by perceptions or objective quantities. This distinction is often overcome by the use of techniques which allow one to transform non-metric variables and treat them as they were metric, like the PRINQUAL procedure in SAS[3] or correspondence analysis (see chapter 14). Readers should now have some intuition of the MDS objectives, output and terminology. The actual MDS procedure can now be scrutinized following these steps:

1. Decide whether mapping is based on an aggregate evaluation of the objects or on the evaluation of a set of attributes (*decompositional* versus *compositional* methods)
2. Define the characteristics of the data collection step (number of objects, metric versus non-metric variables)
3. Translate the survey or objective measurements into a *similarity* or *preference* data matrix
4. Estimate the perceptual map
5. Decide on the number of dimensions to be considered
6. Label the dimensions and the ideal points
7. Validate the analysis

13.2.1 Types of MDS: decompositional versus compositional

The researchers must decide whether the spatial maps should reflect the subject evaluations and comparisons of the objects in their integrity (*decompositional* or *attribute-free approach*), or they should be asked to assess a set of attributes (*compositional* or *attribute-based approach*). In the former case, the advantages are that the respondent assessment is made easier and it is also possible to obtain a separate perceptual map for each subject or for homogeneous groups of subjects. The limit is that no specific indication emerges on the determinants of the relative position of the objects, thus limiting the potential for studies aimed at new product development or brand positioning. Also it is not possible to plot both the objects and the subjects in the same map. Finally, a limit of the attribute-free approach is the difficulty in labeling the dimensions, which can only be based on the researcher knowledge about the objects. Thus, the attribute-based approach is preferred when it is relevant to describe the dimensions and explain the positioning of objects and subjects in the perceptual map. On the other hand, the attribute-based approach suffers from some risks, as it is essential that the attributes are relevant to define the objects. This requires a careful consideration of all the relevant attributes, while avoiding including irrelevant ones. Furthermore, the assumption is that their combination is adequate to reflect the overall object evaluation. The choice between the compositional and decompositional approach determines the methodology to be used.

13.2.2 Number of objects and types of variables

The higher the number of objects, the more accurate is expected to be the output of MDS in statistical terms. However, the opposite could be said about the reliability of collected data, because it might be difficult for subjects to provide large number of comparisons. The number of objects required for the analysis increases with the number of dimensions being considered. In general, for two-dimensional MDS, it is advisable to have at least ten objects, and about fifteen for three-dimensional MDS, although it has been noted that as the number of objects increases, goodness-of-fit measures become less reliable (see Hair et al., 1998). The second decision is whether perceptions and/or preferences should be measured through metric or non-metric variables. This also leads to the choice of different MDS methods, because the starting matrix is different. With non-metric data (generally ordinal variables or paired comparison data), the initial data matrix does not consider the distance between the objects and only takes into account the ranking, while the matrix reflecting metric variables preserve the distances observed in the subject evaluations. Most MDS methods can also deal with mixed data-set with both metric and non-metric data.

13.2.3 Similarity and preference matrices

The data for applying MDS need to be in the form of *similarities* between objects or *preference* (ranking) of objects. If the approach is decompositional, then a matrix for each subject exists, which translates into a matrix comparing all objects. If the compositional approach is chosen, a matrix for each subject and attribute exists, and this translates into a matrix comparing all objects for each attribute. Sometimes data are directly collected in the form of *similarity data*, where the subject compares all potential pairs of objects (which can be rather complex with many objects) and ranks the pairs in terms of their similarity. For example, a subject might state to perceive brand A and C to be more similar between each other than brand A and B and so on. This generates a non-metric measurement of similarity (unless the respondent is asked to rate the degree of similarity on a scale). Another method for the direct collection of similarity data is the *Q-sorting* method described in section 1.2.1. Alternatively, the similarity (or dissimilarity) matrix can be computed from metric evaluations (rating) of the objects. If the approach is compositional, the similarity matrix is derived by summarizing (for example by averaging) the distances between the objects across the subjects, assuming that all subjects have the same weight. If measures are provided for each attribute, a synthetic measure of similarity between objects is computed for each subject (it is possible to use weights if available and appropriate), then the similarity matrix across the subjects is derived. SPSS and SAS compute the similarity matrix directly from observed data, depending on the characteristics of the collected data.

13.2.4 Estimating the perceptual map

It is not worth to get into the technicalities of MDS estimation methods, because the range of techniques is quite large and continuously evolving, although readers interested in the math can refer to Cox and Cox (2001), to the excellent and concise technical discussion in Krzanowski (2000, chapter 3) or to the review by Mead (1992).

Here it might be useful to describe the main steps involved in estimation and cite the algorithms provided by commercial software. In extreme synthesis, estimation starts from a proximity or similarity matrix and produces a set of n-dimensional co-ordinates, so that distances in this n-dimensional space reflect as closely as possible the distances recorded by the proximity matrix. The first distinction is between *metric scaling*, based on a proximity matrix derived from metric data, and *non-metric scaling*, which projects dissimilarities based on ranking (ordinal variables) preserving the order emerging from the subjects' preferences. Non-metric scaling should be applied also to metric distances, when the researcher suspects that collected data might be affected by relevant measurement errors, for example, in cases where respondents may encounter difficulties in stating their perceptions with precision, while ordering can be regarded as reliable.

13.2.5 Metric scaling

When the variables are all metric the most natural way to proceed is to exploit factor analysis or PCA (see chapter 10) to reduce the dimensions, then use the scores as co-ordinates. However, PCA and factor analysis start from a data matrix with the final objective to find the co-ordinates that allow the best representation of the original data-set in terms of variability. Instead, metric scaling techniques aim to construct co-ordinates so that the distance between two points is equal or as close as possible to the distance as measured in the proximity matrix. The *classical multidimensional scaling* technique (CMDS), also known as *principal co-ordinate analysis* is based on the work by Torgerson (1958) and implies a decompositional approach, hence a unique similarity matrix comparing all objects. Starting from the data matrix, the *proximity* or *similarity matrix* summarizing distances between objects across cases can be obtained using the *Euclidean distance* (see chapter 12) or other distance measures as those employed for cluster analysis. If d_{ij} is the distance between object i and object j on the original data-set, CMDS will provide an n-dimensional configuration of points, where the distance between these points is defined as d_{ij}^*. The objective of estimation is to minimize the following equation:

$$\sum_{i=2}^{p}\sum_{j=1}^{i-1}\left(d_{ij}-d_{ij}^*\right)^2$$

which means that the optimal configuration is the least distant from the original one in quadratic terms. The problem of the identification of n – the optimal number of dimensions – is discussed later in this chapter.

13.2.6 Non-metric scaling

Metric scaling allows for analytical solutions, which means that an equation exists to compute the optimal co-ordinates which only have to meet a *monotonicity* requirement.

The latter requirement means that the inter-point distances have a relationship with the initial (*input*) dissimilarities which is as monotonic as possible (it follows the same ranking). When ordinal variables are used, which is the case for non-metric scaling and preference data, there is no analytical solution and co-ordinates and perceptual maps can be obtained through computational algorithms. This explains the variety of procedures for non-metric scaling. The original method indicated by Shepard (1962a, 1962b) and Kruskal (1964a, 1964b) develops as follow:

a. given a number of dimensions n, the p objects are represented through an arbitrary initial set of co-ordinates;

b. a function S is defined to measure how distant is the current set of co-ordinates from the original ordering (monotonicity); and

c. using an iterative computer numerical algorithm, the values that minimize S are found. As discussed later in this chapter, the procedure can be extended to include a search for the optimal number of dimensions n. Since the introduction of the original non-metric scaling algorithm, many other computer algorithms have been developed. The best known is probably ALSCAL (Takane et al., 1977, available in both SAS and SPSS), but many others exist, like those included in the MDS procedure in SAS (for different types of metric and non-metric scaling) or INDSCAL.

13.2.7 Goodness-of-fit, and choice of the number of dimensions

Those who do not like statistics too much will probably appreciate the name of the key value for MDS, which Kruskal (1964a) called STRESS (STandardized REsiduals Sum of Squares). The STRESS value allows one to evaluate the goodness-of-fit of a MDS solution by comparing the derived distances $\hat{d}_{ij}$ from the original ones (d_{ij}) contained in the similarity matrix:

$$STRESS = \sqrt{\frac{\sum_{i=1}^{p-1} \sum_{j=i+1}^{p} (d_{ij} - \hat{d}_{ij})^2}{\sum_{i=1}^{p-1} \sum_{j=i+1}^{p} d_{ij}^2}}$$

The smaller the STRESS function, the closer are the derived distances to the original ones. The STRESS value decreases as the number of dimensions increases. Thus, for a proper evaluation of the number of dimensions, it is useful to plot a *scree diagram* of STRESS against the number of dimensions as shown for factor analysis, PCA or cluster analysis and look for an 'elbow' in the curve. Since the main objective of MDS is to simplify a multidimensional analysis, the preferred number of dimensions is usually two or three, which allows for graphical examination. In general, the search goes from one to five dimensions. The identification of the optimal number of dimensions can be also incorporated directly into the metric and non-metric iterative algorithm, where an additional step evaluates the STRESS function and stops when the addition of a further dimension does not reduce the STRESS value to a perceptible extent. With two dimensions, a STRESS value below 0.05 is generally considered to be satisfactory.

13.2.8 Labeling the dimensions and finding ideal points

As shown in the examples of section 13.1, when the dimensions can be interpreted so that the co-ordinates have a specific meaning, MDS maps for consumer research are much more informative. However, this is not always an easy task and researchers often rely on their own subjective evaluations or on the interpretation of experts (sometimes the respondents themselves). When the compositional approach is adopted (or attribute ratings are otherwise available), there are more objective methods which look into the relative weight of each attribute in defining each dimension (something similar to the factor loadings seen in chapter 10), making interpretation easier.

Another problem when dealing with preference data relates to the objective of positioning ideal points (for each subject) together with the actual brand evaluations (the objects). The *ideal point* is that set of co-ordinates which represents the stated optimal combination of attributes (under the compositional approach). If no precise statement is made by the subject, it is possible to locate the ideal point with a procedure ensuring that distances of the objects from a subject's ideal point reflect the preference ordering as much as possible. There are two approaches for the identification of the ideal point and the representation of subjects and objects on the same map – internal and external preference mapping.

Internal preference mapping (IPM) starts from the preference orderings of the subjects and computes the ideal point or ideal *vector* for each subject through *unfolding*. In principle, internal preference mapping can be based on preference data (ordering of objects) only, not on perceptual variables (measurement scales). Basically the ideal point for the subjects is derived by looking at the ordering of their preferences. For example, consider a situation where four brands (A, B, C and D) are evaluated. Consumer one states a preference ordering as D, B, C and A, where D is the most preferred brand and A the least preferred brand, while the sequence for Consumer two is C, B, D, A. The ideal point for Consumer one will be closer to D and further away from A, while for Consumer two the ideal point will probably be still further away from A, but closer to C than to D. The distance of the ideal point from the objects in the product space should reflect as much as possible the ordering of the consumer preferences. This rough example provides a basic intuition of the functioning of IPM. Note that the ideal product is not necessarily a precise point in the preference map, but could be represented as a line (or an arrow) going from the least preferred objects toward the most preferred ones, since the assumption is that the favored objects (products) will possess more of the desirable characteristics as reflected by the dimensions of the preference map. Thus, the underlying statistical method[4] starts by decomposing all preference orderings for a given set of objects (products), so that the same products can be represented in a lower dimensional space. Then, once the products are positioned on the preference map, it is possible to see where the subjects (consumers) are positioned. The principle is that while the dimensions reflect some product characteristics that are the same for all consumers, each consumer attaches a different weight to those dimensions, so consumers' ideal points are placed in different positions. If perceptual (usually sensory) evaluations are available for a set of attributes, these can be used to understand and label the map dimensions, but they are not exploited for the actual statistical analysis.

External preference mapping (EPM) follows a different philosophy and strictly requires the use of perceptual (metric) data, that is evaluations of the product characteristics on a measurement scale rather than their simple ordering, usually based on analytic or objective evaluations or expert evaluations (thus external to the set of consumers which provide their evaluation of the products). The input matrix contains the (quantitative) measurements of all attributes for each product. A data reduction technique, generally PCA, allows a set of co-ordinates to attach (the principal component scores) to each of the products. The principal components define the dimensions of the map and can be interpreted (labeled) as described in chapter 10, using the component loadings. Once the perceptual space has been defined on the basis of the attribute evaluations, an algorithm like those described in Cox and Cox (2001, chapter 8) or PREFMAP (see Carroll, 1972) allows one to elicit the position of subjects (consumers) in the map.

While the above discussion might be somewhat confusing, especially considering that both approaches could be applied to the same data-set, the point here is that they reflect different philosophies and gave rise to a wide literature. To appreciate the difference between IPM and EPM, consider a consumer who likes red full-bodied wine and white, sweet and sparkling wine very much. If IPM is used these two products share similar preferences and will be positioned next to each other. If EPM is used the product characteristics are very different, thus they will look distant on the perceptual map.

The choice between IPM and EPM is mainly related to the choice of prioritizing either the preferences of the subjects (in IPM) or the product characteristics (in EPM). When many dimensions are chosen, the two approaches produce similar results, but with reduced dimensions discrepancies are likely to emerge. Those supporting IPM claim that perceptual data are inadequate to reflect preferences, as it is not necessarily true that the combination of the product attributes, especially when they are chosen by the researcher, represent the overall product adequately. The idea is that the physical attributes as they are perceived by the consumer are processed into a number of 'perceived benefits' (see the means-end theory in Gutman, 1982), and these benefits are then translated into preferences through an abstraction cognitive process.[5] Thus, the relative weight of the physical attributes as compared to the abstraction process could drive the choice between IPM and EPM. For those goods where the cognitive evaluation is mainly based on the objective attributes, EPM seems to be preferable, whereas other goods where the connection between perceptions and preferences is not so natural (and affective processes play a major role) are better analyzed with IPM. In this perspective IPM and EPM can be seen as complementary.

13.3 Multidimensional scaling and unfolding using SPSS and SAS

For most techniques discussed in this book, the practice is generally easier than the theory. This might not be the case for MDS, given the multitude of alternative algorithms, especially when the objective is unfolding and the representation of subject and objects (or stimuli) within the same map. To illustrate MDS we cannot use our habitual data-sets – Trust and EFS – because MDS requires ad hoc data collection of preference or perceptual data. Thus, a third data-set is created specifically for this chapter, which we call MDS after a great effort in terms of imagination. While we briefly show examples of applications of MDS in SPSS and SAS, we recommend those readers interested in an effective application of MDS to refer to specialized bibliography (also on MDS algorithms and routines) e.g. the books by Kruskal and Wish (1978) and Borg and Groenen (2005).

13.3.1 The MDS data

As it should be clear from this chapter, multidimensional scaling is a multi-faceted procedure that can be applied to a variety of data collection approaches. In this example, the data-set is the following. Fifty individuals (the subjects or consumers) were asked to rank ten sports (the objects or products) according to their preference, while a panel of expert sport journalist provided an evaluation of the attributes of each sport (the product characteristics) in terms of strategy, suspense, physicality and dynamicity. The final data-set (MDS) has one row for each sport and one column for each consumer, plus four columns for the sport's attributes.

13.3.2 SPSS

Since version 14.0 SPSS has introduced an algorithm (PREFSCAL) for unfolding of preferences which allows to run internal preference mapping. To apply unfolding, the menu is ANALYZE/SCALE/MULTIDIMENSIONAL UNFOLDING. Within the initial menu one should define the *proximities* (the fifty variables whose value measure the ranking of the ten sports for each consumer). The *row* variable should be a nominal variable which provides the labels for each of the sports. For this example, we ignore the *weights* (a set of variables for providing relative weights for each consumers, the number of weighting variables being equal to the number of subjects) and the *sources* (the sources variable defines whether measures for the same set of objects are provided by different sources).

Clicking on the MODEL button allows to define:

a. what scaling model is adopted when multiple sources are available (in our case a single source leads to the identity model);

b. whether the proximity matrix contains similarity or dissimilarity measures (a ranking is a dissimilarity measure, the higher the value, the more dissimilar from the top choice);

c. the number of dimensions (unlikely to be above two, given the complexity of the process); and

d. whether the proximities should be transformed in order to estimate the co-ordinates and how (we specify that proximities are *ordinal* and do not request to *untie* tied observations, but with metric data one might opt for different types of transformation, for example a linear one).

In some cases one may wish to fix some of the product co-ordinates, which can be done by using the RESTRICTION button (it is not in our case). The OPTIONS set allows the user to choose the configuration of co-ordinates which starts the algorithm (see the SPSS manual for details about the alternative choices) and in general it is advisable to check whether the final solution remains stable when changing the initial configuration. For our example, we go for the *Classical Spearman* initial configuration. Within the options menu it is also possible to fix the convergence criteria and the maximum number of iterations. As discussed in section 13.2, the STRESS function is the usual convergence criterion; when the new configuration does not lead to perceptible changes in the STRESS value, then the algorithm stops. One may also wish to stop when the STRESS value is very small, even if the changes are not yet below the threshold. The Options toolbox also shows the possibility of changing two parameters

Table 13.1 *Diagnostics of the MDS analysis*

	Measures	
Iterations		992
Final function value		0.3835645
Function value	Stress part	0.0410912
Parts	Penalty part	3.5803705
Badness-of-fit	Normalized stress	0.0016885
	Kruskal's stress-I	0.0410908
	Kruskal's stress-I I	0.1905153
	Young's S-stress-I	0.0720164
	Young's S-stress-II	0.1781156
Goodness-of-fit	Dispersion accounted for	0.9983115
	Variance accounted for	0.9666225
	Recovered preference orders	0.8471837
	Spearman's Rho	0.8617494
	Kendall's Tau-b	0.7273984
Variation coefficients	Variation proximities	0.5043544
	Variation transformed Proximities	0.3322572
	Variation distances	0.5071630
Degeneracy indices	Sum-of-Squares of DeSarbo's	0.4694185
	Intermixedness indices	
	Shepard's rough nondegeneracy index	0.5609796

(*Strength* and *range*) of a *Penalty Term*. This can be very useful to avoid degenerative solutions (those where points can hardly be distinguished from each other). The penalty term adjusts the STRESS-I statistics (see Cox and Cox, 2001) on the basis of the variability of the transformed proximities. The influence of this correction on the STRESS criterion depends on the strength parameter (the weight of the penalty term increases as the strength becomes smaller). The penalty is not necessarily used for all iterations of the algorithm. If one sets the penalty range to zero, no correction is made to the STRESS-I criterion, while larger range values lead to solutions where the variability of the transformed proximities is higher. The penalty term seems to be pretty useful to improve the results of the analyses, but again it is better to check how relevant are the changes in results when the strength and range are modified.

The PLOTS button allows the user to select which graphs should be drawn. Obviously, the *Final Common Space* is very desirable. Finally, the OUTPUT menu lets you choose among a number of tables and outputs (for example besides the final co-ordinates one might want to see the transformed proximities). It is quite useful to save the estimated co-ordinates in a data-set (for example *coord.sav*), as this allows some further processing of the final output, as shown later in this discussion.

The PREFSCAL algorithm has converged after 992 iterations and the final solution has resulted in a STRESS-I value of 0.04, which is acceptable. Other measures of 'badness-of-fit' and 'goodness-of-fit' are provided in table 13.1 and confirm that the results are acceptable. The risk of unfolding is that the original variability is lost in the map so that points are too close to each other (which would mean a *degenerated solution*). To check that this is not the case an indicator is the change in variability observed when moving from the original proximities to the transformed proximities and distances. In this case the variation coefficient of the transformed proximities is 0.33 as compared to the 0.50 of the original ones, which means that most of the variability is retained after transformation. Furthermore, the distances show a variability which is more or less equal to the original one, indicating that the points in space should be scattered enough to reflect the initial distances. The *DeSarbo's Intermixedness index* and the *Shepard's RNI* also provide warning signals for degenerated solutions, where the former should be as close to zero as possible and the latter as close to one as possible. There are no strong signals for a degenerated solution, although one may wish to try different parameters for the penalty term to see whether these indicators improve.

SPSS output includes many other tables, like the co-ordinates for the subjects and the objects and the original and transformed proximities. The plots in figure 13.3 show the mapping of sports, consumers and the joint mapping, respectively.

According to our (fake) sample, basketball, baseball and cricket share similarities in subjects' perceptions and so do American football, motor sports and ice hockey. A third 'cluster' (and the similarity with the objectives of cluster analysis should be clear here) is provided by handball, waterpolo and volleyball, while football seems to be equidistant from all other sports. Consumers are also grouped in clusters according to their preferences and the joint representation allows showing which sports (products) are closer to the preferences of different segments, and also which sports need to be repositioned to attract more public, like the cluster with volleyball, waterpolo and handball. If one can attach a meaning to dimensions one and two, it becomes possible to understand what characteristics of the products should be changed (or whether is it possible to achieve a new market position through advertising). A rough method to obtain an interpretation of the co-ordinates consists in looking at the correlations betweens the co-ordinates of the sports and the object characteristics that can be measured objectively or through the evaluation of expert panelists. The algorithm has created an output file, *coord.sav* which contains the two co-ordinates for each sport and consumer. We can copy and paste the sport attributes' measurements of the *MDS sport* file into the *coord.sav file*, then look at the bivariate correlations (see chapter 8).

Although table 13.2 should be considered with care, given that there are only ten observations, it would seem that sports on the left side of the graph are likely to be more strategic, while those on the right are more dynamic. Considering the second dimension, as values move toward the top, the sports are expected to become more physical and strategic, while negative values seem to indicate a lack of suspense. Ideally those who want to bring people closer to volleyball, waterpolo or handball should try and move the points toward the top-left area, thus trying to persuade 'consumers' that these sports are more strategic (especially), dynamic and physical than currently thought.

This is only one application of many alternative options for multidimensional scaling, with an example focusing on internal preference mapping. SPSS also provides the ALSCAL and PROXSCAL algorithms.

13.3.3 SAS

The SAS procedure for multidimensional scaling (*proc MDS*) is a generalization of the ALSCAL algorithm. This procedure allows one to apply multidimensional scaling as a mapping technique for objects, but does not perform unfolding. To unfold preferences in SAS one can exploit the *TRANSREG* procedure, which has an option

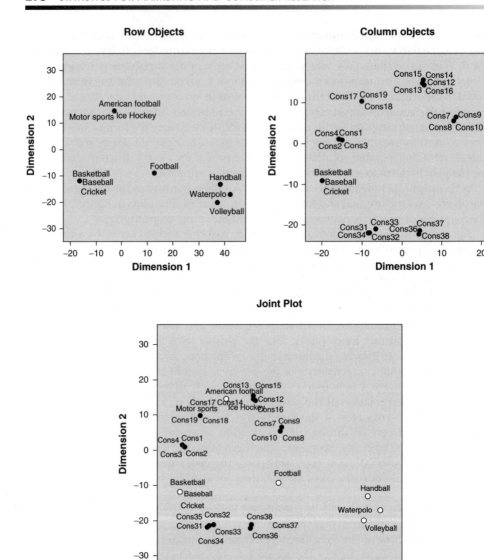

Figure 13.3 *MDS maps in SPSS*

Table 13.2 *Bivariate correlations between MDS dimensions and object characteristics*

	DIM_1	DIM_2	Strategy	Suspense	Physicity	Dinamicity
DIM_1	1.000	0.000	−0.839	−0.167	0.130	0.362
DIM_2	0.000	1.000	0.338	−0.180	0.330	0.168

(COO) which returns the co-ordinates of the ideal point or vector for internal preference mapping. Another procedure which is very useful for mapping purposes is in *proc PRINQUAL*, which basically performs principal component analysis on qualitative data (see chapter 14 and further readings at the end of this chapter). The combined use of PRINQUAL and TRANSREG to obtain preference maps is described with an example in the SAS-STAT Manual v8 (chapter 65).

Summing up

MDS is a collection of methods and algorithms to project on a low-dimensional map the differences in perceptions and preferences measured on a set of products, brands or their characteristics. MDS is especially useful in marketing research when the objective is to identify market opportunities or understand the positioning of several brands with respect to consumer perceptions and preferences. MDS starts by exploring the similarities (or dissimilarities) of objects (products) measured through consumer statements of their (ordinal) preferences or metric evaluations of the objects (perceptions). The various algorithms allow one to replicate the disparities in consumer perceptions through a reduced set of graphical co-ordinates (usually with a two or three-axis representations), so that the plots reflect as closely as possible the original differences. Besides representing the objects on a map, MDS unfolding techniques also allow one to estimate the 'ideal point' (or vector) for each subject (or for groups of subjects), so that consumer and products can be portrayed on the same spatial map, which allows one to show concentrations of products/consumers and available market niches. When the map dimensions can be interpreted according to a combination of product characteristics, repositioning on the map can be achieved by changing the product characteristics or through advertising. There are many options and difficulties in applying MDS techniques, which led to the development of many computational algorithms, whose output depends on the way preferences and perceptions are measured, on the transformation of preferences into distances and on the iterative procedure to obtain the final solution. SPSS and SAS have relatively flexible routines for estimating perceptual (preference) maps.

EXERCISES

1. Open the MDS data-set
   a. Using the ALSCAL program, assume that the data are metric (interval) and compute the distance matrix between pairs of sports for the 50 respondents with two-dimensional MDS
   b. Which sports are the most similar according to consumer perceptions? Which ones are the most different?
   c. Save the final co-ordinates and compute the bivariate correlations with the expert panel evaluations
   d. Try the three-dimensional configuration. Which solution works better according to the STRESS value and R^2?

    e.   Repeat the analysis, this time looking for similarities between respondents for the ten sports on two-dimensions

    f.   How many clusters of consumers appear from the graph?

    g.   Run a principal component analysis on the 50 consumers (hint – transpose the data-set), extract the first two components and run a cluster analysis. Do results look similar to the ones of MDS?

    h.   Save the final co-ordinates and compute the bivariate correlations with the expert panel evaluations

2.   Open the MDS data-set

    a.   Using the PROXSCAL program, assume that the data are ordinal and compute the distance matrix between pairs of sports for the 50 respondents with two-dimensional MDS

    b.   Which sports are the most similar according to consumer perceptions? Which ones are the most different?

    c.   What are the differences between this distance matrix and the one produced by ALSCAL (see exercise 1)?

    d.   Try and change the type of measure, the initial configuration (the latter from the OPTION menu) and the number of dimensions and find the best output according to the STRESS function

    e.   Repeat the analysis, this time looking for similarities between respondents for the ten sports on two-dimensions

3.   Open the MDS data-set

    a.   Run the MDS UNFOLDING program (as in this chapter's example) using the following assumptions (OPTIONS menu):

        i.   Identity scaling model

        ii.   Proximities are dissimilarities

        iii.   No transformation of proximities

        iv.   No intercept

        v.   'Correspondence' initial configuration

        vi.   Strength of penalty term: 0.3

        vii. Range of penalty term: 2.0

    b.   Look at the output and comment on the results, comparing them with the ones in the chapter's example

    c.   Change the following options:

       •   Ordinal transformation of proximities

       •   Include an intercept

       •   'Ross-Cliff' initial configuration

       •   Strength of penalty 0.1

       •   Range of penalty term 1.0

    d.   Compare the results with the previous ones. Try changing the options one by one to see which one has the largest impact on the results

    e.   Save the final co-ordinates in a separate file and compute the correlations with the expert panel evaluations

Further readings and web-links

❖   **The ideal-point model** – Consumers' ideal points projected in the same perceptual space as products allow one to identify market niches and spaces for new product development. To learn more about unfolding the ideal points, see chapters 14 and 15 in Borg and Groenen (2005). Vectors (lines) may be preferred to points and a method for estimating them is *preference regression*. For a discussion of alternative methods see e.g. Hauser and Koppelman (1979). Neslin (1981) reviews the alternative of asking consumer directly for the weights of each attribute.

❖ **Learn more about the scaling algorithms** – To understand options and the underlying techniques of the algorithms, see the technical guides from the SPSS student support web-site at: support.spss.com/Student/ Studentdefault.asp. A presentation of ALSCAL is given in his developer's web-site (Prof. Young), at forrest.psych.unc.edu/research/alscal.html. For the multidimensional unfolding technique (and the meaning of the penalty term), see Busing et al. (2005).

❖ **From attributes to overall product evaluation through Means-End chains** – The step between the specific attribute evaluations and the overall judgment on a product is quite big and sometimes hides inconsistencies. There is a whole modeling theory which tries to explain this step, based on Means-End chains. See for example Zeithaml (1988), Reynolds and Gutman (1988), Grunert and Bech-Larsen (2005), Reynolds and Olson (2001) or the seminal article by Gutman (1982).

Hints for more advanced studies

☞ How should a questionnaire for MDS be constructed? How many products should be taken into account simultaneously in an effective perceptual mapping study? On how many respondents?

☞ When is it necessary to transform proximities before applying MDS? What sort of transformations are applied? What are the implications?

☞ There are various measures for STRESS and goodness-of-fit of MDS. Find out the differences, pros and cons.

Notes

1. Data were made up and the interpretation is very subjective.
2. For a more comprehensive discussion and history of MDS for marketing research see also Carroll and Green (1997) and Cooper (1983).
3. See SAS/STAT Manual (SAS Institute Inc., 2004), chapter 53.
4. Usually the Singular Value Decomposition, see for example Carroll and Chang (1970) or Green and Carroll (1978).
5. Van Kleef (2006) provides an excellent (and well referenced) discussion of internal versus external preference mapping. See also Van Kleef et al. (2006).

CHAPTER 14

Correspondence Analysis

T HIS CHAPTER describes the functioning and application of correspondence analysis, a technique which allows one to look into the association of two or more categorical variables and display them jointly on a bivariate graph. As multidimensional scaling does for metric and ordinal variables, correspondence analysis is especially useful to analyze consumer preferences and characteristics in relation to products when the variables are not quantitative. Within the same graph it is possible to display brands and consumer preferences according to a set of characteristics measured on nominal scale.

Section 14.1   introduces the basic rationale of correspondence analysis
              through examples
Section 14.2   reviews the technical steps to run a correspondence analysis
              and its options
Section 14.3   illustrates the application of simple correspondence analysis
              in SPSS

THREE LEARNING OUTCOMES

This chapter enables the reader to:

➡ Understand how correspondence analysis works with categorical variables
➡ Run correspondence analyses to obtain joint bivariate plots of consumers
  and brands
➡ Interpret the output of correspondence analyses in SPSS

PRELIMINARY KNOWLEDGE: Review section 9.1 for contingency tables and association measures. PCA (chapter 10) and MDS (chapter 13) are essential to the understanding. A review of log-linear analysis (chapter 9) may also help.

14.1   Principles and applications of correspondence analysis

For those who survived chapter 10 on factor analysis and PCA and digested the previous chapter on MDS, getting through this chapter should be relatively easy. They might even wonder why correspondence analysis is presented here as a stand-alone technique and not – as in most marketing research textbooks – as a specific technique within MDS. While an extra chapter in this book does not increase the publisher's profits (it may rather have the opposite effect), the idea is that a short but independent presentation of correspondence analysis may serve as a useful linkage between those methods mainly aimed at data reduction (as PCA) and the graphical techniques under the heading of MDS.

The peculiarity of correspondence analysis is essentially the nature of the data being processed. Factor and PCAs are only applied to metric (interval or ratio) quantitative variables. Traditional MDS deals with non-metric preference and perceptual data, when those are on an ordinal scale. Correspondence analysis closes the circle, by allowing data reduction (and graphical representation of dissimilarities) with non-metric nominal (categorical) variables.

The issue with categorical (non-ordinal) variables is how to measure distance between two objects. Correspondence analysis exploits contingency tables and association measures as described in sections 4.2.2 and 9.1.

Consider the following example based on the Trust data-set. The objective is to check whether consumers with different jobs show preferences for some specific type of chicken. The job position is measured by variable q55, while the type of chicken purchased in a typical week is recorded in q6.

Table 14.1 shows the frequencies for each cell crossing the occupation with the typically purchased chicken. It is a *correspondence table*, as it displays a measurement of correspondence between the row and column variables, in this case the frequencies (as in contingency tables), but other measures are possible. As discussed in section 9.1, if the two characters are independent, then the number in the cells of the table should simply depend on the row and column totals. It is then possible:

1. to measure the distance between the expected frequency in each cell and the actual (observed) frequency; and
2. to compute a statistic (the chi-square statistic) which allows one to test whether the difference between the expected and actual value is statistically significant.

As discussed in section 9.1, the elements composing the chi-square statistic are standardized metric values, one for each of the cells, and they become larger as the association between two specific characters increases. These elements can be interpreted as a metric measure of distance, and the resulting matrix is similar to a covariance matrix, which opens the way to applying a method similar to PCA for reducing the number of dimensions. Since the principal component scores provide standardized values that can be used as co-ordinates, one may apply the same technique first by rows (synthesizing occupation as a function of types of chicken), then by column (thus synthesizing types of chicken as a function of occupations). Considering the first two components, one obtains a bivariate plot (figure 14.1) which shows both the occupation and the type of chicken in the same space.

The association between occupation and the type of preferred chicken is shown in the bi-dimensional plot. Executives are closer to 'luxury' chicken, farmers seem to prefer 'standard' chicken, unemployed respondents are closer to 'value' chicken.

This is an example of two-way correspondence analysis, but the technique can be applied to multiple variables. A natural application is the representation of preferences for combinations of product attributes. For example, it would be possible to represent on the same graph consumer preferences for different brands and characteristics of a specific product, like car brands together with color, power, size, etc. This would allow one to explore brand choice in relation to characteristics, opening the way to product modifications and innovations to meet consumer preferences.

Correspondence analysis is particularly useful when the variables have many categories. The application to metric (continuous) data is not ruled out, but data need to be categorized, first (see banding in box 4.1).

Table 14.1 *Job position and chicken choice: a correspondence table*

Correspondence Table

If employed, what is your occupation?	In a typical week, what type of fresh or frozen chicken do you buy for your household's home consumption?				Active margin
	'Value' chicken	'Standard' chicken	'Organic' chicken	'Luxury' chicken	
I am not employed	17	50	10	17	94
Non-manual employee	11	74	14	28	127
Manual employee	6	19	4	8	37
Executive	0	7	6	14	27
Self-employed professional	1	18	7	3	29
Farmer/agricultural worker	1	1	1	0	3
Employer/entrepreneur	0	4	2	3	9
Other	11	31	1	1	44
Active margin	47	204	45	74	370

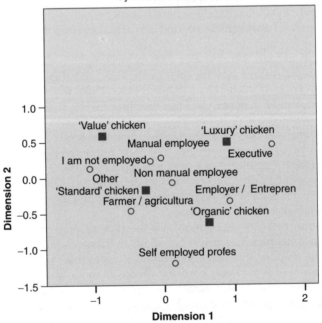

Figure 14.1 *Bivariate correspondence plot*

14.2 Theory and techniques of correspondence analysis

According to the terminology used in chapter 13 for MDS, correspondence analysis is a compositional technique, which starts from a set of product attributes to portrait the overall preference for a brand.[1] For a recent (and excellent) technical review of correspondence analysis with the key technical details, we suggest you to refer to Beh (2004). As mentioned in the previous section, this technique is very similar to PCA and can be employed for data reduction purposes or to plot perceptual maps. Because of the way it is constructed, correspondence analysis can be applied either on purposes the rows and columns of the data matrix. For example, if rows represent brands and columns are different attributes, by applying the method by rows one obtains the co-ordinates for the brands, and then the attributes can also be represented in the same graph after obtaining their co-ordinates through the application on the columns.

Correspondence analysis consists of the following steps:

1. Represent the data in a contingency table (see sections 4.2.2 and 9.1)
2. Translate the frequencies of the contingency table into a matrix of metric (continuous) distances through a set of chi-square association measures, on the *row* and *column* profiles
3. Extract the dimensions (in a similar fashion to PCA)
4. Evaluate the explanatory power of the selected number of dimensions
5. Plot row and column objects in the same co-ordinate space

Consider the representation of two categorical variables, x (with k categories) and y (with l categories) in a contingency table, where absolute frequencies have already been transformed into relative frequencies as shown in table 14.2.

The last column of the table (the row totals) represent the *row profile* (and in correspondence analysis its values are the *row masses*) and the last row (the column totals) are the *column masses*. The basic idea of correspondence analysis is that the categories of the x variable can be seen as different co-ordinates for the points identified by the y variable and the other way round. Thus, it is possible to represent the categories of x and y as points in space, imposing (as in multidimensional scaling) that they respect some distance measure. This measure can be obtained by exploiting the associations between two categories as defined in section 9.1, that is through a *chi-square* association measure.

Correspondence analysis for representing data in both ways, (which means having both the categories of x and those of y on the same graph), starts by considering the row profile (the categories of x) and plotting them in a bi-dimensional[2] graph, using the categories of y for defining the distances. This allows us to compare nominal categories

Table 14.2 *Relative frequency table for correspondence analysis*

	y_1	y_2	$\cdots$	y_j	$\cdots$	y_l	
x_1	f_{11}	f_{12}		f_{1j}		f_{1l}	f_{10}
x_2	f_{21}	f_{22}		f_{2j}		f_{2l}	f_{20}
$\cdots$							
x_i	f_{i1}			f_{ij}		f_{il}	f_{i0}
$\cdots$							
x_k	f_{k1}	f_{j2}		f_{kj}		f_{kl}	f_{kl}
	f_{01}	f_{02}		f_{0j}		f_{0l}	1

within the same variable – those categories of x which show similar levels of association with a given of y can be considered as 'closer' than those with very different levels of association with the same category of y. Then the same procedure is carried out transposing the table, which means that the categories of y can be represented using the categories of x to define the distances.

When the co-ordinates are defined simultaneously for the categories of x and y, the chi-square value can be computed for each cell as follows (compare section 9.1) – first, the expected table frequencies are computed with

$$f_{ij}^* = \frac{n_{i0} \cdot n_{0j}}{n_{00}} = \frac{f_{i0} \cdot f_{0j}}{f_{00}} = f_{i0} \cdot f_{0j}$$

where – as usual – n_{ij} and f_{ij} are the absolute and relative frequencies respectively, n_{i0} and n_{0j} (or f_{i0} and f_{0j}) are the marginal totals for row i and column j (the *row masses* and *column masses*), respectively) and n_{00} is the sample size (hence the 'total' relative frequency f_{00} equals 1). The chi-square value is now computed for each cell (i, j):

$$\chi_{ij}^2 = \frac{(f_{ij} - f_{ij}^*)^2}{f_{ij}^*}$$

The matrix χ^2 (containing all of the χ_{ij}^2 values) measures all of the associations between the categories of the first variable and those of the second one. Note that a generalization to the multivariate case, known as multiple correspondence analysis (MCA) is possible by *stacking* the matrix, which consists in composing a large matrix made by blocks where each block represents the contingency matrix for two variables, so that all possible associations are taken into consideration. This stacked matrix is referred to as the *Burt Table*.

When we extract the square root of the elemental chi-square values, preceded by the appropriate sign (the sign of the difference at the numerator before calculating its squared value), the resulting matrix contains measures of *similarity*, as large positive values correspond to strongly associated categories and large negative values identify those categories where the association is strong but negative, indicating dissimilarity.

We have obtained a matrix which contains metric and continuous similarity data which we call **D**. This opens the way to the use of PCA to translate such a matrix into co-ordinates for each of the categories, first those of x, then those of y. Before PCA can be applied some normalization is required, so that our input matrix becomes similar to a correlation matrix. The use of the square root of the row masses or column masses for normalizing the values in **D** represents the key difference from PCA.

The rest of the discussion on the output of correspondence analysis draws on the results of the PCA. As for PCA we have *eigenvalues*, one for each dimension, which can be used to compute the proportion of dissimilarity maintained by that dimension.

Another key term in interpreting the output of correspondence analysis is *inertia*. To introduce this concept it is important to notice that the effectiveness of correspondence analysis depends on the degree of association between x and y. If x and y are independent, while it is still possible to run the procedure, its results will be not very meaningful. *Total inertia* is the sum of the eigenvalues and is a measure of the overall association between x and y. It corresponds to the Chi-square value divided by the number of observations. One would expect an inertia above 0.20 for adequate representations. An inertia value can be computed for each of the dimensions. The inertia for each dimension represents the contribution of that dimension to the association (Chi-square) between the two variables. For example, an inertia of 75% (or 0.75) for the first dimension means that it represents 75% of the Chi-square.

14.3 Running correspondence analysis

To illustrate the use of correspondence analysis with SPSS (while the SAS procedure is summarized in box 14.1 at the end of the chapter), let us consider the following two categorical variables from the EFS data-set – the 'Economic position of the household reference person' (a093) and the type of tenure (a121). As expected, these two variables show a very strong association (the Pearson chi-square value is 274, which means significant association at the 99.9% confidence level). To explore graphically this associ-ation, select ANALYZE/DATA REDUCTION/CORRESPONDENCE ANALYSIS and indicate that a093 should define the rows and a121 the columns of the contingency table. It is also necessary to 'define the range' of categories we want to consider, that is one to six for the economic positions and one to eight for the tenure type. For some of these categories, it is possible to include them only for graphical representation, with no influence on the actual estimation of the scores. This is done by specifying that those categories are 'supplemental' using the dialog box below the list of variables. For example, we can specify the tenure type 'free rental' to be supplemental, since we presume that it depends from non-relevant determinants and it is not likely to be related to socio-economic condition. We also set as supplemental the row category 'Work related government training program,' which has no data and might influence the analysis.

By clicking on the MODEL option, we can specify how to run correspondence analysis. First, one should choose how many dimensions should be considered. The maximum number of dimensions for the analysis is equal to the number of rows minus one or the number of columns minus one, whichever the smaller. In our example, the maximum number of dimensions would be five, which reduces to four due to missing values in one row category. As shown later in this section, one may then choose to graphically represent only a sub-set of the extracted dimensions (usually two or three), to make interpretation easier. Then one can specify whether to use the chi-square distance (as presented in section 14.2) or, alternatively, an Euclidean measure which uses the square root of the sum of squared differences between pairs of rows and pairs of columns. The latter distance also requires us to choose a method for centering the data (see the SPSS manual for details). For this example we run standard correspondence analysis (with the chi-square distance), so that it is not necessary to choose a standardization method.

The *normalization method* actually defines how correspondence analysis is run, since total inertia is spread over the scores for the categories of x and y in a sort of weighting. Basically it is a choice on whether to privilege comparisons between the categories for x (the row variable) or those for y (the column ones). This choice influences the way the distances are summarized by the first dimensions.

For example, if one chooses the 'row principal' normalization, it means that the Euclidean distances in the final bivariate plot of x and y will be as close as possible to the chi-square distances between the *rows*, that are the categories of x. In this case, we are primarily interested in portraying the distances within x. The opposite is valid for the 'column principal' method. Otherwise one could aim to a solution where the distances on the graph resemble as much as possible distances between both the categories of x and y by spreading the total inertia symmetrically, as in the 'symmetrical method.' With the 'principal option,' the inertia is first spread over the scores for x, then over those for y. This does not allow for a meaningful comparison between the categories of x and y, so SPSS does not produce the bi-plot. It is also possible to choose a 'weighted' approach, by defining a value between -1 and 1, where -1 is the column principal, zero is symmetrical and 1 is the row principal. By choosing intermediate values one decides whether to focus more or less on the row categorical differences as compared to the column categorical differences. While the final plot will show categories from both the variables, strictly speaking the distance between categories of x and categories of y cannot be interpreted as a measure of association between these categories, although when categories of x and y appear in the same area of the bi-plot some 'correspondence' can be gathered.

For our example, let us opt for the *row principal* method, since what is more meaningful is to look at how differences in socio-economic conditions impact on the tenure type, rather than identifying distances between tenure types. On the STATISTICS menu, it is possible to ask for specific outputs, like the correspondence (contingency) table and the separate row and column profiles. Correspondence analysis is a non-parametric analysis and this should prevent it from computing statistics like those provided under the option 'confidence statistics.' However, SPSS gives the possibility of computing standard deviations and correlations under the assumption of multinomial distribution of the cell frequencies, which is acceptable when the data are obtained as a random sample from a normally distributed population. Another output that can be quite interesting is provided through 'permutations of the correspondence table,' which allows ordering the categories of x and y using the scores obtained through the correspondence analysis. For example, we may think that the tenure types and the socio-economic conditions follow some ordering, but we cannot define it with sufficient precision to consider these variables as ordinal. Thus, we can use the scores in the first dimension (or the first two) to order the categories and produce a permutated correspondence table. Under the STATISTICS options, we choose to see a permutation of the correspondence table using the first dimension only.

The last set of options comes from the PLOTS button. We are certainly interested in the bi-plot representing both tenure types and socio-economic conditions, but it is also possible to have separate graphs in SPSS. Note also that SPSS allows us to display a reduced number of dimensions, so that we may choose to perform correspondence analysis on a multidimensional level, then choose to draw only the first two dimensions.

The first output being produced by SPSS is the correspondence table (top part of table 14.3), while the summary results are reported as shown in the bottom part of the table 14.3.

The *singular value* is nothing else than the square root of the eigenvalue and inertia is actually the eigenvalue. From table 14.3 we note that: the total inertia is 0.53 (usually a value above 0.2 is good) and this is confirmed by the significance of the chi-square statistic, which is 231. The later value is usually similar to the Pearson chi-square statistic computed on the contingency tables (differences may arise by defining some variables as supplemental, etc.). Furthermore, by looking at the relative inertia, we observe that the first dimension accounts for 85% of total inertia and the first two dimensions for 93%.

Note this essential distinction from PCA. The first two dimensions explain 93% of the total inertia and not of the original variability as in PCA. The original variability is already mediated by the use of the Chi-square distances, so that the first component might be very good to explaining the inertia, but very bad to explain the original variability when the total inertia is very low.

In our case, a bivariate plot is more than acceptable to display differences and associations between categories, since both the total and relative inertia show satisfactory values. Under the assumptions of data normality and random sample extraction, table 14.3 also shows standard deviations and correlations for the singular values.

Table 14.3 Correspondence analysis output in SPSS

Correspondence Table

Economic position of Household Reference Person	Tenure type								
	Local Authority rented unfurnished	Housing association	Other rented unfurnished	Rented furnished	Owned with mortgage	Owned by rental purchase	Owned outright	Rent free[a]	Active Margin
Self-employed	2	5	0	1	22	0	5	0	35
Full-time employee	13	6	15	9	165	1	28	1	237
Pt employee	6	2	4	3	8	0	11	1	34
Unemployed	3	2	1	0	2	0	0	0	8
Work related govt train prog[b]	0	0	0	0	0	0	0	0	
Ret unoc over min ni age	19	14	2	1	4	0	86	2	126
Active Margin	43	29	22	14	201	1	130		440

[a]Supplementary column.
[b]Supplementary row.

Summary

Dimension	Singular value	Inertia	Chi-square	Sig.	Proportion of inertia		Confidence singular value			
					Accounted for	Cumulative	Standard deviation	Correlation		
								2	3	4
1	0.669	0.447			0.850	0.850	0.031	0.094	-0.032	-0.022
2	0.209	0.044			0.083	0.933	0.055		0.011	0.081
3	0.173	0.030			0.057	0.990	0.055			-0.042
4	0.072	0.005			0.010	1.000	0.053			
Total		0.526	231.402	0.000[a]	1.000	1.000				

[a]24 degrees of freedom.

Table 14.4 Row scores

Overview Row Points[b]

| Economic position of household reference person | Mass | Score in dimension | | | | Inertia | Contribution | | | | | | | | | |
| | | 1 | 2 | 3 | 4 | | Of point to inertia of dimension | | | | Of Dimension to Inertia of Point | | | | |
							1	2	3	4	1	2	3	4	Total
Self-employed	0.080	0.296	0.025	0.433	-0.164	0.024	0.016	0.001	0.496	0.407	0.290	0.002	0.620	0.089	1.000
Full-time employee	0.539	0.527	0.049	-0.039	0.026	0.152	0.334	0.030	0.027	0.071	0.984	0.008	0.005	0.002	1.000
Pt employee	0.077	-0.239	-0.409	0.352	-0.143	0.028	0.010	0.295	0.318	0.300	0.156	0.453	0.336	0.055	1.000
Unemployed	0.018	-0.154	-1.223	0.509	0.241	0.033	0.001	0.622	0.157	0.202	0.013	0.814	0.141	0.032	1.000
Work related govt train prog[a]	0.000	.	.	.	.	.	0.000	0.000	0.000	0.000	.	.	.	.	.
Ret unoc over min ni age	0.286	0.999	0.089	0.015	0.019	0.288	0.639	0.052	0.002	0.020	0.992	0.008	0.000	0.000	1.000
Active Total	1.000					0.526	1.000	1.000	1.000	1.000					

[a]Supplementary point.
[b]Row Principal normalization.

Table 14.5 Column scores

Overview Column Points[b]

Tenure type	Mass	Score in dimension				Inertia	Contribution								
		1	2	3	4		Of point to inertia of dimension				Of dimension to inertia of point				
							1	2	3	4	1	2	3	4	Total
Local authority rented unfurnished	0.098	−0.699	−1.993	0.051	1.106	0.039	0.048	0.388	0.000	0.120	0.548	0.436	0.000	0.016	1.000
Housing association	0.066	−0.781	−1.263	2.821	−1.273	0.039	0.040	0.105	0.524	0.107	0.462	0.118	0.405	0.014	1.000
Other rented unfurnished	0.050	0.487	−2.023	−2.190	0.891	0.022	0.012	0.205	0.240	0.040	0.245	0.413	0.333	0.010	1.000
Rented furnished	0.032	0.531	−1.098	−2.270	−4.585	0.014	0.009	0.038	0.164	0.669	0.284	0.119	0.349	0.248	1.000
Owned with mortgage	0.457	0.971	0.371	0.233	0.133	0.196	0.431	0.063	0.025	0.008	0.982	0.014	0.004	0.000	1.000
Owned by rental purchase	0.002	1.179	1.120	−1.287	5.002	0.002	0.003	0.003	0.004	0.057	0.725	0.064	0.058	0.153	1.000
Owned outright	0.295	−1.244	0.819	−0.382	0.018	0.214	0.457	0.198	0.043	0.000	0.954	0.040	0.006	0.000	1.000
Rent free[a]	0.009	−0.957	−1.039	−2.996	−3.705	0.007	0.000	0.000	0.000	0.000	0.512	0.059	0.338	0.090	1.000
Active total	1.000					0.526	1.000	1.000	1.000	1.000					

[a] Supplementary point.
[b] Row Principal normalization.

The next step is the computation of scores (co-ordinates) and possibly the interpretation of the dimensions, in a similar fashion to PCA. Consider rows first (table 14.4).

The *Mass* column simply reports the relative weight of each category on the sample. Scores are computed for each category but the supplemental one, provided there are no missing data. The *Inertia* column shows how total inertia has been distributed across rows. It may be seen as something similar to communalities and the relative importance of 'full-time employees' and 'retired or under age' is higher because they are more important categories in the original correspondence table. These two categories (especially the latter) strongly contribute to explaining the first dimension, while the second dimension is characterized by unemployed and part-time employees.

The same exercise is carried out on columns (table 14.5), remembering that in our case, where we chose to use the 'row principal' method, the procedure did not normalize by column.

Here we notice how the first dimension is also related to the 'owned by mortgage' and 'owned outright' categories.

Finally, we have the bivariate plot with the x and y categories represented on two dimensions as in figure 14.2.

In extreme synthesis, the plot shows that those retired or employed (whether full-time or self-employed) are closer to owned tenures, especially those owned with mortgage. Unemployed and part-time employees, instead are closer to rented houses and housing associations or other forms of rentals.

SPSS allows us to run MCA (three or more variables) by using the menu ANALYZE/DATA REDUCTION/OPTIMAL SCALING and choosing multiple nominal variables. We leave the reader (with the aid of the SPSS manual, see for example SPSS, 2006) to try more complex correspondence analyses, but figure 14.3 shows the

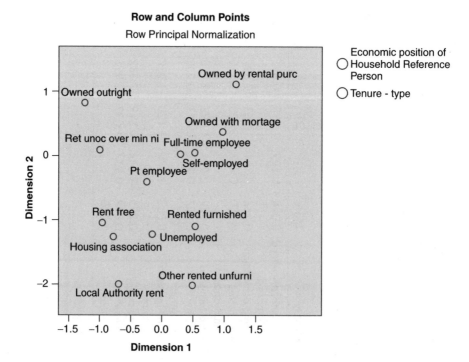

Figure 14.2 *Bivariate plot from correspondence analysis*

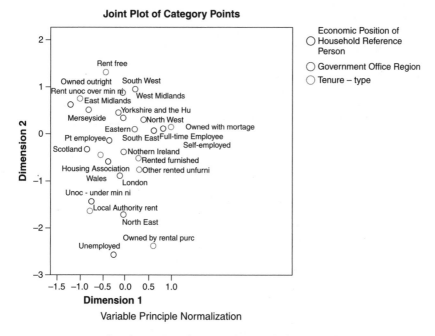

Figure 14.3 *Bivariate plot of correspondence analysis with three variables*

BOX 14.1 *SAS correspondence analysis*

The relevant procedure for running correspondence analysis in SAS is *proc CORRESP*, which performs both simple and multiple correspondence analysis, the latter by specifying the option MCA. The type of normalization encountered in the discussion of the SPSS analysis is defined by the SAS option PROFILE (where PROFILE can assume the value ROW, COLUMN or BOTH).

---

bi-dimensional plot for an analysis which extends the previous one by considering a third variable, the 'government office region' (gor).

Summing up

In chapter 13 it is shown how multidimensional scaling could work with metric and ordinal variables to display consumers (subjects) and products or brands (objects) in the same graph according to preferences and product characteristics. When those characteristics cannot be measured through ordinal or metric variables (as the color of a product or the type of music), correspondence analysis provides the solution. This is why this technique is often discussed as one of the multidimensional scaling approaches. Simple correspondence analysis works with two variables and exploits the fact that the categories within the first variable can be compared by looking at their individual association with the categories of the second variable. Thus, by using a contingency table (see section 9.1) and computing Chi-square distances it is possible to build a metric similarity matrix, which can be processed with a technique closely related to principal component analysis (see chapter 10).

Multiple correspondence analysis (with three or more variables) is a generalization which simply requires building a stacked (block) matrix, the Burt matrix, prior to running the analysis. Both techniques can be implemented relatively easily in SPSS and SAS.

EXERCISES

1. Open the Trust data-set
    a. Run a simple correspondence analysis between q54 (job status) and q61 (financial situation of the household)
    b. Show the bi-plot for the symmetric, row principal and column principal models – which one provides better result?
    c. Which dimension is most important in representing points for unemployed and fully employed individuals?
    d. Is the total inertia acceptable? Which category is least explained by the model?
2. Open the EFS data-set
    a. Using simple correspondence analysis plot tenure type (a121) and household composition (a062)
    b. Show the bi-plot
    c. If the Euclidean distance is used instead of the Chi-square one, do results change?
    d. Check the total inertia – is its value acceptable?
    e. Which category is best described in the first dimension? Which one is worst described?
    f. Try and look for a third variable and draw the tri-dimensional plot
3. Open the EFS data-set
    a. Your objective is to reduce into two components the following categorical variables – a062 (household composition) a121 (tenure type) gor (region) a093 (economic position) a094 (NS-SEC8 class)
    b. Use multiple correspondence analysis (DATA REDUCTION/OPTIMAL SCALING) to extract two components and save the scores
    c. Plot tenure type with region and economic position
    d. Use gender (a071) as supplemental variables. How do results change?
    e. Draw a table with the average scores by gender

Further readings and web-links

❖ **Correspondence analysis and log-linear analysis** – Compared to log-linear analysis, correspondence analysis is more effective when working with many categories. For a discussion of the relationship between correspondence analysis and log-linear analysis see Van der Heijden et al. (1989) and the discussion which follows the article (see Gower in particular).

❖ **Multiple correspondence analysis (MCA)** – generalizes simple correspondence analysis to the case of three or more variables. MCA starts with a different matrix (the *design matrix*), with observations as rows and a column for each of the categories of the various variables. A specific cell is 1 if the corresponding observation belongs to the category in that column and zero otherwise. More on MCA can be found in Greenacre and Blasius (2006) and Tenenhaus and Young (1985).

❖ **Qualitative principal component analysis** – When the objective is to reduce the dimension of the data-set, there are other techniques similar to correspondence analysis which can deal with nominal categorical variables. As for MDS they apply a transformation of the original variables. Alternating this transformation (via optimal scaling) and quantitative PCA does the job. See for example Young (1981) and the SAS PRINQUAL procedure (SAS, 2004, chapter 53). See also the SPSS CATPCA algorithm (categorical principal component analysis) at http://support.spss.com/Student/Documentation/Algorithms/14.0/catpca.pdf

Hints for more advanced studies

☞ Explore how correspondence analysis can be seen as a special case of canonical correlation analysis

☞ What happens if the chi-square distance matrix contains values that are non-significant (that is they reflect low associations)? And if some of the cells in the contingency table have zero or very low frequencies?

☞ How can the use of supplementary points in MDA help dealing with the presence of outliers?

Notes

1. A good introduction to correspondence analysis for marketing research can be found in Hoffman and Franke (1986).
2. As for PCA and MDS, more dimensions are possible, although not necessarily desirable when the objective is making interpretation easier.

PART V

Further Methods in Multivariate Analysis

This fifth part illustrate methods which extend the linear regression model to deal with more complex relationships, namely structural equation models and discrete choice models.

Chapter 15 illustrates structural equation models (SEM), which combine elements of factor analysis and linear regression to model multiple relationships simultaneously. SEM allow for the estimation and inclusion of latent (unobservable) variables and is a key method to test theoretically-driven models and compare alternative theories.

Chapter 16 introduces a variety of discrete choice models, where the dependent variable in a regression model can be non-metric with binary, ordinal or categorical nominal values. These models are useful to model consumer behaviors using variables which measure consumers' preferences as they state them. Among methods for eliciting preferences and choices, conjoint analysis allows one define appropriate data collection designs and compare different combinations of attributes.

Chapter 17, besides leaving room for some final personal considerations on the use and teaching of statistics, briefly introduces two approaches to data analysis which are becoming more and more relevant in marketing applications; one is mainly computational (data mining), the other requires a philosophical switch (Bayesian statistics).

CHAPTER 15

Structural Equation Models

T HIS CHAPTER provides the basic understanding of a complex method, structural equation modeling, which encompasses various techniques already encountered in this book. SEM involves the use of factor analysis to measure latent constructs through manifest indicators and the simultaneous estimation of various regression equations. SEM is a key method to test theoretically-driven models and compare alternative theories and can be used to explain a variety of consumer behaviors. An example of theoretical specification for purchasing intention using the Theory of Planned Behavior is provided.

Section 15.1   highlights the main features of SEM with examples
Section 15.2   reviews all the key concepts in SEM and illustrates how the
                 method works
Section 15.3   provides a practical example of estimation and testing of SEM
                 using the software AMOS

THREE LEARNING OUTCOMES

This chapter enables the reader to:

➥ Appreciate the functioning and potential application of SEM
➥ Understand the links with other techniques and the confirmatory approach
   of SEM
➥ Move the first steps into estimation of SEM using SPSS AMOS

PRELIMINARY KNOWLEDGE: This is all but an easy chapter. For a full understanding, it is essential to review the concept of latent variables in chapter 1, the functioning of factor analysis in chapter 10, and correlation and linear regression in chapter 8.

15.1   From exploration to confirmation: structural equation models

SEM is a powerful method to estimate multiple and simultaneous relationships involving several dependent and explanatory variables, and allows for the inclusion of latent variables which cannot be directly measured but can be expressed as a function of other measurable variables.

This short definition of SEM does not do justice to the complexity of the techniques involved in developing a structural equation model, but points out the peculiarity of SEM as compared to the statistical methods encountered so far in this book. In linear regression (see chapter 8), a single dependent variable is related to one or more independent (explanatory) variables, under the assumptions that these explanatory variables are fixed, independent from each other and exogenous (which means that they are determined outside the relationship). In SEM we can relax all of these assumptions as:

- Several dependent variables can be considered at the same time
- Explanatory variables can be assumed to be measured with a random error
- Endogenous variables can be used to explain dependent variables
- Correlation between explanatory variables is allowed for

And yet this is not all. Another key feature in SEM is the possibility of including in the model, as endogenous or exogenous variables, some latent (unobservable) variables. These are not directly measured, but can be approximated by a set of observable variables (see chapter 1). In other words, it is possible to incorporate factor analysis (see chapter 10) into a regression model, where latent factors appear in the structural equation model as explanatory and/or dependent variables.

At a first glance, the above characteristics of SEM make it look like the 'encompassing' (comprehensive) model for regression, factor analysis (and possibly canonical correlation analysis, given that the relationship between sets of dependent and independent variables are estimated). While this is partially true, one should not get over excited at the perspective of substituting all those specific methods with SEM. Because of its complexity, SEM is subject to many issues which are left to the researcher's arbitrary decisions. In other words, it is difficult to use SEM to 'let the data speak,' because this tool is so flexible that it may lead to hundreds of potential alternative models.

This is the reason why SEM is classified as a *confirmatory* rather than *exploratory technique* and is sometime referred to as *confirmatory factor analysis*. This distinction was introduced in chapter 10, but now there are all the ingredients to understand it properly. Exploratory techniques, like practically all multivariate methods examined so far, take the data, specify some parametric structure and use an estimation procedure to quantify the relationship among variables. Factor analysis starts by assuming a linear structure and defines the factors as unobservable variables that can be expressed as linear combinations of the observed variables. This allows one to estimate the factor loadings and the factor scores, together with some of the diagnostics (mainly the eigenvalues) which allow one to decide on the number of factors. In regression analysis a linear relationship is assumed between the dependent and independent variables and, after some further assumptions on the variables and the regression residuals, the data and the estimation procedure return the parameter estimates. With hypothesis testing and the stepwise method it is also possible to select the most informative among the potential explanatory variables. These are exploratory analyses, because they start from a single general form for the relationship and exploit the estimation technique to obtain the specific relationship.

Confirmatory analyses follow the opposite philosophy. The specific relationship among the variables needs to be specified prior to the analysis, based on some pre-existing theory. The technique is used to fit this pre-specified model to the data and diagnostics are used to assess whether the model is good or not. SEM is a confirmatory analysis, as it requires the model and all of the relationships between the variables to be chosen prior to estimation. This does not mean that a confirmatory approach precludes exploration for the best models, only that the strategy is different. With confirmatory analysis, it is still possible to compare alternative models through a *competing model strategy*, using some statistical criterion to choose the winning model. Some guidance is needed to read a chapter which might be confusing to many readers due to the number of terms and notions to be dealt with. Probably the best way to appreciate the functioning of SEM through this chapter requires first a quick read of section 15.2, which explains the theory, then a closer look at the application in section 15.3 before going back to section 15.2 to gain a better understanding of the approach. If this is not enough, you may choose to proceed iteratively.

15.2 Structural equation modeling: key concepts and estimation

Before illustrating the functioning of SEM through an appropriate example,[1] it is useful to clarify some of the founding principles. First, a little more on *latent constructs* should be said.[2] We are used to thinking in terms of variables, which represent concepts whose association with their measurement is straightforward. However, some concepts are not so easily measured. Think of attitudes, passion, trust, risk aversion... There is no unique measure, because they have many dimensions, and researchers usually exploit several items in a questionnaire to obtain some indirect measurement. For example, one could measure risk aversion by asking respondents to rate how dangerous they consider driving at 100 mph, diving from ten meters, swimming in the ocean, jumping off a plane with a parachute, etc. These are *indicators* (or *manifest variables*) which help in measuring the construct, although with some measurement error, as introduced in section 1.2.

Constructs are the foundation of structural equation modeling, because they represent the real actors of a *causal relationship*. Let us define what is causality in a relationship – when two constructs are significantly associated (correlated) to one another but one occurs before the other, then the construct occurring earlier is said to cause the 'outcome' construct, provided there are no other reasons for such result. So, one can expect a high risk aversion to precede the decision of avoiding risky actions, so that risk aversion 'causes' behavior. Note that the idea of causality has always sparked hot debates in the scientific community (still ongoing); in particular there is justified scepticism about the power of statistics in identifying causal relationships (see the recommended further readings at the end of this chapter). As a matter of fact, the causation direction from the explanatory variable to the dependent one is simply hypothesized by the researcher and statistical models merely assess whether the assumed relationships fit the data appropriately.

If we are able to measure the 'true' risk aversion, then we might succeed in evaluating its role in determining some specific behavior, like purchasing chicken when newspapers report stories about bird flu. It would be a nonsense to relate the fear of flying to chicken consumption, but it is not a nonsense to use the fear of flying and ten other risk aversion measures to quantify the construct of risk aversion, then relate the latter to chicken consumption. Thus, we move from variables to constructs because the latter can be seen as the outcome of several measurable variables. If one is still attached to the term 'variable,' an option is to distinguish between *manifest variables*, that can be directly measured and serve as indicators, and *latent variables*, that are not observable and are the actual components of the causal relationship. However, just like variables

BOX 15.1 *SEM in consumer and marketing research*

Scarcity and willingness to buy. Consumers are available to pay higher prices and make efforts to obtain products that are scarce, so that scarcity can become a marketing variable. For example, brands like Nike, Swatch and Apple have designed limited-availability products to draw consumer attention. Wu and Hsing (2006) look through a structural equation model at how perceived scarcity influences willingness to buy through a set of mediating variables, like perceived quality, perceived monetary sacrifice and perceived symbolic benefits. They conclude that scarcity enhances value both directly (through enhancing perceived quality and symbolic benefits), and indirectly (through the price-quality and price-symbolic benefits associations).

E-shopping. What are the determinants of on-line shopping? Using SEM, Mahmood et al. (2004) see the interaction of four variables as the driver for buying on line: education, technological know-how, economic condition and trust. After collecting data in 26 countries, using a set of proxy variables for these latent constructs and accounting for both direct and indirect effects, they conclude that wealth is the most relevant determinant.

Customer relationship. Dillon et al. (1997) evaluate the use of SEM in relation to customer satisfaction for credit cards. They define two latent construct, customer service and card member rewards, as measured by some manifest (questionnaire) variables, like being courteous, accurate in answering questions, correcting errors (for customer service) or ease of obtaining rewards. These latent constructs determine another latent construct (customer satisfaction) which in turn generates the card level spending.

Human brands. Why people become so attached to those famous personalities who are the subject of marketing communication? Thomson (2006) looks at the relative importance of the factors which make a basketball or movie star successful in generating attachment in consumers. Three latent determinants are considered, referring to the extent that the personality is perceived to fulfill a consumer's (a) autonomy (or self-determination), (b) relatedness (sense of closeness with others) and (c) competence (the tendency to seek for achievement). A consumer who feels autonomous, related and competent is more likely to develop attachment (a parallel can be made with kids, parents making them feel autonomous, related and competent are more likely to encourage a loving feeling). These three latent constructs produce the attachment strength, which is also latent but it can be measured indirectly through questionnaire items exploring the effects of 'separation' from the personality. Furthermore, attachment to human brands is found to produce high levels of satisfaction, trust and commitment.

in regression analysis, one can distinguish between *exogenous constructs* and *endogenous constructs* in a causal relationship. Instead, exogenous constructs only act as explanatory factors and do not depend on any other construct, and endogenous constructs play the role of the dependent variables and they appear on the left-hand side in at least one of the SEM causal relationships.

To appreciate the potential of SEM, one can look at the variety of applications in consumer and marketing research, as listed in box 15.1.

15.2.1 SEM related techniques

SEM is the comprehensive method which includes as special cases *confirmatory factor analysis, path analysis* and *multivariate regression* or *simultaneous equation systems*.

Confirmatory factor analysis is a factor analysis where the number of factors and the loadings of the original variables are assumed to follow some prior theory. Thus, the

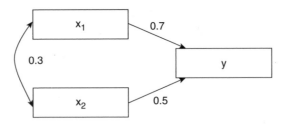

Figure 15.1 *An example of path diagram*

researcher runs the factor analysis on the basis of these assumptions on the number of factors and the loadings (constraining to zero the loadings for those variables that are not expected to load on a specific factor), then evaluates the result with some goodness-of-fit diagnostic as those discussed later in this chapter for SEM.

Path analysis is a generalization of the regression model to deal with the (discussed) causality concept. Path analysis is based on the *path diagram*, which is also the core of the SEM approach, as it is explained later in this chapter. When one refers to path analysis, the assumptions is that all variables are directly measured, which marks the distinction with SEM, where latent constructs are included instead. The path diagram represents the relationship between the variables through arrows and boxes. Boxes are the variables and the straight arrows leave the boxes containing predictors and point toward those containing the dependent variables. It is also possible that two variables are correlated without implying causation, in which case the arrows are curved. For example, figure 15.1 shows a model where x_1 and x_2 cause y and there is a non-zero correlation (there is collinearity) between x_1 and x_2.

The number next to the arrows indicates the bivariate correlation between the variables linked by each of the arrows. The utility of such a pictorial representation is that the overall effect of x_1 on y can be decomposed in the direct effect (which is 0.7, the direct correlation) and the indirect effect through x_2 (which is 0.15, that is, 0.3×0.5), leading to a total correlation of 0.85. Similarly, the total correlation of x_2 and y is 0.71.

The third technique which is embedded in SEM is *multivariate regression analysis* or *simultaneous equation systems*, which consists of a system of regression equation, where the dependent variable in one equation can appear on the right-hand side of other equations. Again, all variables in multivariate regression are manifest. Some of the key assumptions of standard regression analysis are violated in these models, because the endogenous variables which appear on the right-hand side are correlated with the residuals, leading to inconsistent least squares estimates. Thus, alternative estimation methods – like those employed for SEM – are needed. Note that the parallelism between simultaneous equation systems and SEM lies in the fact that the equations of a system can be defined as *structural equations* when their specification depends on some theoretical basis, although the *structural form* can generally be simplified into a *reduced form* by solving the system analytically. However, when one refers to SEM rather than simultaneous equation systems, it is necessary to bear in mind two key differences – the former method allows for the inclusion of latent constructs (unobserved factors) and uses a confirmatory approach, while the latter is based on observed (manifest) variables only and is often exploited for exploratory rather than confirmatory analysis.

15.2.2 *Structural equation models*

A structural equation model is composed by a *measurement model*, which links the latent constructs to the manifest indicators, and a *structural model* which summarizes the relationships linking the endogenous and exogenous constructs.

The *measurement model* corresponds to confirmatory factor analysis, so that it can be tested to check whether the measurement of the latent variables using the manifest indicators is acceptable.

A structural equation model can be represented through a path diagram, using the following rules:

a. manifest variables are shown in square or rectangular boxes;
b. latent variables (and measurement errors) are shown through ovals or circles;
c. causality relationships are indicated through straight arrows; and
d. correlation without causality is shown through a curved arrow.

The path diagram is drawn according to some theory. Before estimating the unknown quantities of the model (the latent constructs and the parameters defining the correlations and the causal relationships), one should be aware of a potential difficulty in estimating SEMs linked to the *identification* of the model (see box 15.2 for details). In summary, a model is identified when there is enough information (relationships as compared to the number of endogenous variables, manifest variables as compared to latent constructs) to estimate the model in an unique way. When the amount of information is exactly what is needed for unique estimation of the parameters, the model is said to be *just-identified*. In many cases there are a number of relationships which exceed what is needed for just-identification. In this situation, called *over-identification*, it is possible to exploit these relationships to test the validity of a theory. To check for identification and over-identification, one may look at the *degrees of freedom*, the 'free information' after all of the necessary information has been used. If there are more than zero degrees of freedom, the model is over-identified. However, this is only a necessary condition, not a sufficient one. For identification of the measurement model, a useful rule of thumb is to have at least three manifest indicators for each latent variable. Identification problems can emerge during estimation (see Hair et al., 1998) when standard errors look too large, when indicators are too highly correlated between each other or some of the estimates are unacceptable (like negative variance). Note that while SEM allows for multi-collinearity[3] (the curved arrows), when the correlation is too high (above 0.9) this could lead to identification problems.

Estimation can be achieved through *maximum likelihood estimation*, provided that the manifest indicators follow a multivariate normal distribution, which implies that they are normally distributed for any value of the other indicators. The latent constructs are also assumed to be normally distributed. While it is not necessary to deal with the complex details of estimation, it is important to emphasize a key difference with other multivariate techniques. SEM does not use the individual observations (cases) for the estimation of the parameters, but only exploits them to estimate the *covariance matrix*, which provides the basis for the actual parameter estimation process. This does not reduce the importance of having an adequate sample size. An identification problem may emerge when many of the elements of the covariance matrix are close to zero, unless the sample size is large enough. A simple rule of thumb (Stevens, 2002) requires at least 15 cases per measured variable or indicator.

The abovementioned degrees of freedom refer to the elements in the covariance matrix. When there are more elements in the covariance matrix than parameters to be estimated, then the model is over-identified and it is possible to test its theoretical foundation.

We have repeatedly mentioned tests of the theory, but what are they?

First, parameter estimates should be reasonable, both in terms of the founding theory and statistical acceptability (negative error variances are unacceptable). Beyond estimates of the parameters, the main output of SEM consists in a set (usually large) of *goodness-of-fit tests*. Which fit measures to use depends on whether one is testing a single theory (model) or producing a comparison across competing theories. For a detailed review of goodness-of-fit tests the reader should refer to the appendix 11B in Hair et al. (1998) or to Cheung and Rensvold (2002) for a comparison, but for a quick summary the main indices provided by SPSS/AMOS may be useful.

The *chi-square statistic* tests whether the observed covariance matrix is equal to the estimated one (which is what one hopes). The number of degrees of freedom indicates whether the model is just-identified (zero degrees of freedom) or over-identified (more than zero degrees of freedom). If the p-value of the chi-square test is larger than 0.05 (0.01), then the observed covariance matrix is not different from the estimated one at a 95% (99%) level of confidence. Non-rejection of the null hypothesis suggests that the theory is acceptable, although this does not rule out better models. As shown in section 15.3, the output from the tested model is compared to two boundaries, the *independence model* (which assumes no correlation between the endogenous and exogenous variables) and the *saturated model* (no constraints at all, perfect fit with the data, just like in log-linear analysis of chapter 9). The tested model lies between these two extremes.

The *minimum sample discrepancy* (CMIN) simply tests whether the model perfectly fits the data (very unlikely and not really useful as a test). When this measure is divided by the degree of freedom (CMIN/DF), one obtains the above-mentioned Chi-square test. The *root mean square residual* (RMR) refers to the residuals between the estimated and sample covariance matrices. It can be used to compare alternative models, where a smaller RMR indicates better fit. Another index is the *goodness-of-fit index* (GFI), which should be above 0.90 for acceptable theories (an *adjusted* version, AGFI, is also shown with similar interpretation). Other indices which are expected to be as close as possible to one (and generally not below 0.90) are the *normed fit index* (NFI), the *relative fit index* (RFI), the *incremental fit index* (IFI), the *Tucker-Lewis coefficient* (TLI) and the *comparative fit index* (CFI). The *non-centrality parameter* (NCP) and the *root mean square error of approximation* (RMSEA) consider both the discrepancy criterion as the CMIN and some parsimony criteria accounting for degrees of freedom. The RMSEA should be less than 0.05 for a good model. The hypothesis that RMSEA <0.05 is tested through the *test of close fit* (PCLOSE). Other measures for comparing alternative models are the AIC and BIC information criteria and other similar information indices. Finally, the *Hoelter's critical N* shows the largest sample size necessary to accept the model and it is a useful complement to the chi-square test, which tends to reject the model when the sample size is large. Better models require larger sample sizes to be rejected and generally one would expect a critical N of at least 200 for a good model.

15.3 Theory at work: SEM and the Theory of Planned Behavior

SPSS has produced a separate package, AMOS, for structural equation models, while other specialist packages allow estimation of SEMs (see box 15.3).[4] In this section we

BOX 15.2 *Identification*

Another issue that SEM and simultaneous equation systems have in common is the problem of *identification*, which refers to the fact that a single reduced form might be associated with different structural formulations. This means that the solution found might not be unique or optimal. To understand the problem of identification, consider a two-equations regression system in *structural form*, based on some theoretical structure which relates three manifest variables, y_1, y_2 and x_1:

$$\begin{cases} y_1 = \alpha + \beta x_1 + \varepsilon_1 \\ y_2 = \gamma + x_1 + \delta y_1 + \varepsilon_2 \end{cases} \quad (1)$$

Where ε_1 and ε_2 are error terms. If we solve the two system analytically, after some substitution we obtain the following system (the *reduced* form)

$$\begin{cases} y_1 = \alpha + \beta x_1 + \varepsilon_1 \\ y_2 = (\gamma + \delta \alpha) + (1 + \delta \beta) x_1 + (\delta \varepsilon_1 + \varepsilon_2) \end{cases} \quad (2)$$

The first equation is a standard linear regression. The second equation could be transformed into a regression equation by defining $\phi = \gamma + \delta \alpha$ and $\eta = 1 + \delta \beta$. Let us assume for simplicity that the error term of the second equation $\upsilon = \delta \varepsilon_1 + \varepsilon_2$ is well-behaved according to the regression assumption, we could then estimate

$$y_2 = \phi + \eta x_1 + \upsilon \quad (3)$$

But it would be impossible to derive from ϕ and η unique values for α, β, γ and δ, the *structural coefficients* that are needed to understand our structural (theoretical) model.

We could find one solution by setting γ equal to some value (for example 0), use the estimates of α and β from the first equation and obtain δ from the estimates of ϕ and η. This is one of the potential solutions, but we could have obtained infinite other solutions by simply assigning a different value to γ. In this case the model is *not identified*. If we had a third relationship:

$$y_1 = \gamma x_2 + \varepsilon_3 \quad (4)$$

Then the problem would be solved, since we could estimate γ using equation (4) and obtain the remaining parameters from (2). In this case there is a unique solution and the model is said to be *just-identified*. Note that in practice the estimate of the coefficients is simultaneous (and the precise error structure is taken into account), but the simplification helps showing the feasibility of estimation when the model is identified.

A third situation which is quite relevant to SEM is *overidentification*, which occurs when there is even more information than needed for proper identification. For example, suppose one has a fourth equation

$$y_1 = \beta x_1 + \gamma x_2 + \varepsilon_4 \quad (5)$$

Now one could estimate β and γ from this equation as well. If the structural system reflects a true theory, one should get the same results whether using either (2) and (4) or (2) and (5). An overidentified model allows to test the theory.

For complex systems with many equations, it is extremely difficult to understand whether the model is not identified, just-identified or over-identified prior to the analysis. Furthermore, identification problems may also arise in the measurement model (not enough indicators for estimating the latent variables), as discussed in the references suggested at the end of this chapter.

BOX 15.3   *SEM using other commercial software*

The widespread use of SEM in many disciplines is mainly due to its implementation with LISREL (http://www.ssicentral.com/lisrel/index.html), the first computer program which dates back to 1973 after the pioneering work of the statistician Karl Jöreskog (1967 and 1969), which has evolved together with the method. Jöreskog himself, together with Dag Sörbom developed the LISREL program.

SAS also provides a procedure for estimating structural equation models, *proc CALIS*. Other packages designed for structural equation models are EQS (http://www.mvsoft.com) and Mplus (http://www.statmodel.com). A comparison among EQS, LISREL and AMOS is provided in Byrne (2001a).

---

provide an overview of the use of AMOS on the Trust data-set. It should be clear by now that SEM starts where a theory exists. For our example we first need a short explanation of the theory on which the Trust questionnaire is based, the Theory of Planned Behavior. Basically, theory relates behavior (or better the intention to behave) to three determinants, attitudes, social influence and control on the behavior. More details are provided in box 15.4.

BOX 15.4   *The Theory of Planned Behavior*

The Trust data-set is framed within the Theory of Planned Behavior (TPB) introduced briefly in chapter 1, which has proved to be a successful analysis tool for a range of behaviors, often associated with risky or health-related actions such as smoking, risky driving, physical activities and exercise, or contraception. For an extensive explanation of TPB, bibliography and links see www.people.umass.edu/aizen/index.html.

The TPB framework defines human action as a combination of three dimensions–behavioral beliefs, normative beliefs, and control beliefs. Behavioral beliefs (beliefs about the outcome of the action), produce either a positive or a negative *attitude* (A) toward the behavior; normative beliefs refer to *subjective norms* (SN) or perceived social forces; control beliefs lead to *perceived behavioral control* (PBC). All these produce *intentions to behave* (ITB) (Ajzen, 2002). When possible, intentions to behave can be linked to actual behavior.

Attitude to behavior is noted as being principally different from the broader concept of attitude toward an object. For example, one may like chicken (an attitude to chicken), yet have a negative attitude toward eating chicken because of a specific dietary requirement. Subjective norm is a concept based on how one 'should' act in response to the views or thoughts of others. Subjective norm influences may include friends, family members, colleagues, doctors, religious organizations etc. Perceived behavioral control can be described as 'the measure of confidence that one can act,' for example one may perceive an obstacle in buying chicken because of laziness which prevents a walk to the supermarket.

These three global components of the Theory of Planned Behavior are latent variables, interpreted as the global outcome of the underlying beliefs.

- The global attitude to behavior is determined by a set of behavioral beliefs (BB)
- The global subjective norm is determined by a set of normative beliefs (NB)
- The global perceived behavioral control is determined by a set of control beliefs (CB)

- Each behavioral belief is the product of the *strength* of the belief (*b*) and the *evaluation of its outcome* (*o*)

BOX 15.4 *Cont'd*

- Each normative belief is the product of the *strength* of the belief (s) and the *motivation to comply* (m)
- Each control belief is the product of the *strength* of the belief (c) and the *power* of control beliefs (p)

Formally, this can be written as:

$$A \propto \sum_{i=1}^{n} BB_i = \sum_{i=1}^{n} b_i o_i$$

$$SN \propto \sum_{j=1}^{g} RB_j = \sum_{j=1}^{g} s_j m_j$$

$$PBC \propto \sum_{k=1}^{q} CB_k = \sum_{k=1}^{q} c_k p_k$$

Thus, the data are prepared in this way:

1. All b_i, o_i, s_i, m_i, c_i, and p_i are measured directly through the questionnaire items
2. Each BB_i, RB_i, CB_i is computed mathematically using the equations above

Now there is all we need for a typical *path diagram* for TPB, as shown below:

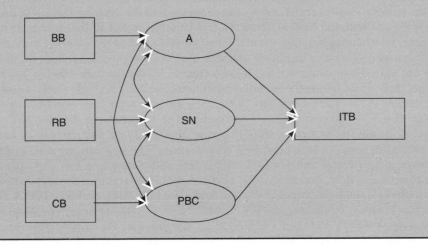

The Trust data-set is based to some extent on the Theory of Planned Behavior. Intentions to purchase chicken (*ITP*) are measured by question q7. A set of eleven behavioral belief strengths is measured by questions q12a to q12k and the relative outcome evaluations are measured by questions from q24a to q24k. By multiplying all pairs, we obtain the eleven behavioral beliefs (*bba* to *bbk* in the Trust data-set) which act as the manifest indicators for the latent construct attitude (*A*). For the sake of simplicity, although this is quite far from the ideal TPB recommendations, for the subjective norm (*SN*) and perceived behavioral control (*PBC*) we depart from the

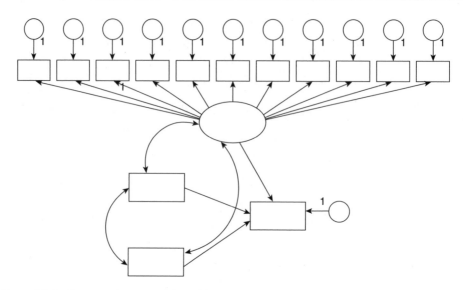

Figure 15.2 *Drawing the structural model*

broader expectancy-value formulation and assume that they can be measured directly (although with error) through individual measures, *PBC* and *SN* in the Trust data-set.

Now that we have selected the manifest variables, we can start drawing the path diagram by using the AMOS graphic program in SPSS AMOS. Using the graphic interface, we first build the theoretical model by selecting the desired icon (circle, square, straight arrow, curved arrow) and dragging it to the drawing surface. A useful icon is the one with the caption 'Draw a latent variable or add an indicator to a latent variable,' which helps building the first latent variable and all the manifest indicators. Another useful icon is the 'duplicate' one which allows to copy objects. Since the model does not fit in the drawing surface, it is possible to 'resize the path diagram to fit on a page.' According to our theoretical assumptions, we have a single latent construct (the attitude *A*), measured through 11 manifest indicators for behavioral beliefs (*bba* to *bbk*). The other two determinants are assumed to be measured directly, but with error. All determinants have a straight arrow directed to the intention to purchase and are linked between each other by a curved arrow. The structure we need to draw is shown in figure 15.2 and is based on the TPB figure shown in box 15.4 after the data adjustments described above. The next step requires naming of the known boxes, which can be done by associating the graph to a data file (click on FILE/DATA FILE and open Trust.sav by clicking on FILE NAME). Now, variable names can be associated by choosing VIEW/VARIABLE ON DATASET and dragging the names on the desired point of the drawing surface. The only objects that remain unnamed are the circles, since we have no measured variables for them. We first name the largest oval which represents the latent attitude by right-clicking on it, selecting OBJECTING PROPERTIES and writing *A* as the variable name. All the remaining latent objects are errors and we can easily name them by clicking on PLUG-INS on the toolbar and selecting NAME UNOBSERVED VARIABLES.

The resulting model is depicted in figure 15.3.

The model is almost ready for estimation. When missing data exist, a workaround is to click on the VIEW/ANALYSIS PROPERTIES MENU and specify that means and intercept need to be estimated. AMOS performs maximum likelihood estimation in the presence of missing data, although with a small proportion of missing data it might be preferable to apply list wise deletion on the SPSS file, prior to the AMOS analysis.

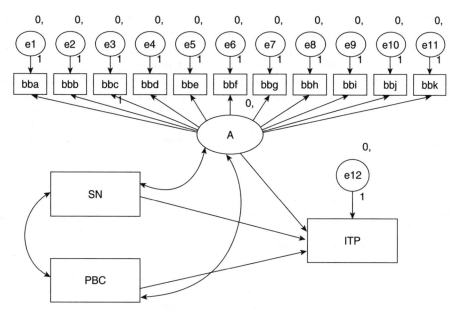

Figure 15.3 *Structural equation model*

The complete output can be seen by clicking on 'view text,' which in AMOS has the desirable feature allowing for interactive hints for interpretation (just click on any part of the output to see its meaning). Table 15.1 shows the estimation output.

The model has 27 variables including errors, 14 of them are observed, 13 are unobserved and 12 of them are endogenous (the only exogenous latent variable is the error of ITP). The covariance matrix has 119 elements and the model requires

Table 15.1 *AMOS estimation output*

Variable counts (Group number 1)

Number of variables in your model:	27
Number of observed variables:	14
Number of unobserved variables:	13
Number of exogenous variables:	15
Number of endogenous variables:	12

Default model (Default model)
Notes for model (Default model)
Computation of degrees of freedom (Default model)

Number of distinct sample moments: 119
Number of distinct parameters to be estimated: 45
Degrees of freedom (119 – 45):   74

Result (Default model)

Minimum was achieved
Chi-square = 619.428
Degrees of freedom = 74
Probability level = 0.000

estimation of 45 parameters, which means that the model (with 74 degrees of freedom) is over-identified. The chi-square statistic is not encouraging, as it is very large and significant. This means that the observed and estimated matrices are quite different. However, this is an extremely frequent result (in fact, few models in the TPB literature show non-significant chi-squares). The chi-square statistic has been criticized because it tends to be inflated when there are many degrees of freedom (or large sample sizes) and is very sensitive to the assumption of multivariate normality, so that one should first look at the overall model fit. As discussed in section 15.2, one may look at other diagnostics. Before looking at the goodness-of-fit, AMOS returns the parameter estimates shown in table 15.2.

Table 15.2 *Maximum likelihood estimates of the coefficients*

Maximum Likelihood Estimates

Regression Weights: (Group number 1 – Default model)

			Estimate	S.E.	C.R.	P
bba	< – – –	A	1.000			
bbb	< – – –	A	0.985	0.074	13.298	***
bbc	< – – –	A	−0.097	0.067	−1.461	0.144
bbd	< – – –	A	0.836	0.067	12.496	***
bbe	< – – –	A	1.045	0.078	13.485	***
bbf	< – – –	A	0.853	0.061	13.872	***
bbg	< – – –	A	0.887	0.074	11.953	***
bbh	< – – –	A	0.965	0.079	12.296	***
bbi	< – – –	A	−0.297	0.064	−4.628	***
bbj	< – – –	A	0.674	0.077	8.807	***
bbk	< – – –	A	−0.028	0.084	−0.327	0.744
q7	< – – –	A	0.073	0.012	6.307	***
q7	< – – –	sn	0.008	0.008	1.051	0.293
q7	< – – –	pbc	0.016	0.010	1.581	0.114

***P-value below 0.001

Standardized Regression Weights: (Group number 1 – Default model)

			Estimate
bba	< – – –	A	0.735
bbb	< – – –	A	0.656
bbc	< – – –	A	−0.072
bbd	< – – –	A	0.616
bbe	< – – –	A	0.661
bbf	< – – –	A	0.682
bbg	< – – –	A	0.594
bbh	< – – –	A	0.633
bbi	< – – –	A	−0.228
bbj	< – – –	A	0.447
bbk	< – – –	A	−0.017
q7	< – – –	A	0.324
q7	< – – –	sn	0.048
q7	< – – –	pbc	0.071

Table 15.3   *Estimates of model correlations*

Covariances: (Group number 1 – Default model)

			Estimate	S.E.	C.R.	P
sn	< – – >	pbc	3.712	4.219	0.880	0.379
pbc	< – – >	A	−2.880	3.548	−0.812	0.417
sn	< – – >	A	12.017	4.461	2.694	0.007

Correlations: (Group number 1 – Default model)

			Estimate
sn	< – – >	pbc	0.040
pbc	< – – >	A	−0.040
sn	< – – >	A	0.134

Estimates are relatively satisfactory in terms of what one expects from the theory. Note that some of the regression coefficients might be constrained to ensure identification of the measurement model parameters; here it is the case for *bba*. The *coefficient ratio C.R.* corresponds to the *t*-statistic in regression and the relative *p*-values (p) show that most of the manifest indicators for attitude have a significant weight at the 5% significance level. The signs all correspond to what was expected, since the only negative signs relate to the behavioral belief associated with difficulty of preparation (*bbc*), welfare concern (*bbk*) and (the only significant) agreement with the statement that chicken lacks flavor (*bbi*).[5] Instead, when one considers the impact of the global determinants on intentions to purchase chicken in the following week (*ITP*), only attitude (*A*) is significant at the 5% level, whereas *PBC* and *SN* have a larger *p*-value. To compare the different loadings one should refer to the standardized coefficients, which indicate that the manifest indicator which is most correlated with attitude is *bba* (related to chicken taste), followed by *bbf* (the perception of chicken as a food which works well with many ingredients and helps variety in meals). The coefficient of *A* on *ITP* reflects a (causal) correlation of about 0.32.

Next one can look at the 'curved arrows,' which represent the covariances and correlations in the model (table 15.3). Only Subjective Norm and Attitude seem to be interlinked significantly, for a correlation of about 0.13. The fact that the model returns the expected coefficient is a good start not to reject it, but it is necessary to evaluate the fit before reaching a conclusion. Table 15.4 shows the goodness-of-fit statistics computed by AMOS.

Our 'default model' does not perform well in terms of the CMIN, the ratio between the chi-square and the degrees of freedom (which is actually distributed as a Chi-square statistic). In general, a model with a CMIN/DF value above five tends to be rejected and good models show values below three.

Other goodness-of-fit indicators confirm that the model should be probably modified to be a good theory for explaining the data. In fact, as is seen in section 15.2, all of the indicators in the middle-lower section of table 15. 4 should be above 0.9 and all of them are below 0.7. Table 15.5 looks at the non-centrality parameters and other goodness-of-fit statistics accompanied by the 90% confidence limits. What needs to be evaluated is the RMSEA which should be below 0.05 and non-significant. We have another symptom that the model is not too good.

Table 15.6 contains information criteria to compare this specific SEM with other alternative models (the lower the better) and the Hoelter indicator, which should show a size of above 200 for the 0.01 significance level. We are very far from that threshold.

Table 15.4 *AMOS goodness-of-fit statistics*

Model Fit Summary

CMIN

Model	NPAR	CMIN	DF	P	CMIN/DF
Default model	45	619.428	74	0.000	8.371
Saturated model	119	0.000	0		
Independence model	14	1736.822	105	0.000	16.541

Baseline Comparisons

Model	NFI	RFI	IFI	TLI	CFI
	Delta1	rho1	Delta2	rho2	
Default model	0.643	0.494	0.672	0.526	0.666
Saturated model	1.000		1.000		1.000
Independence model	0.000	0.000	0.000	0.000	0.000

Parsimony-Adjusted Measures

Model	PRATIO	PNFI	PCFI
Default model	0.705	0.453	0.469
Saturated model	0.000	0.000	0.000
Independence model	1.000	0.000	0.000

Table 15.5 *Non-centrality parameters*

NCP

Model	NCP	LO 90	HI 90
Default model	545.428	469.732	628.594
Saturated model	0.000	0.000	0.000
Independence model	1631.822	1500.462	1770.572

FMIN

Model	FMIN	F0	LO 90	HI 90
Default model	1.241	1.093	0.941	1.260
Saturated model	0.000	0.000	0.000	0.000
Independence model	3.481	3.270	3.007	3.548

RMSEA

Model	RMSEA	LO 90	HI 90	PCLOSE
Default model	0.122	0.113	0.130	0.000
Independence model	0.176	0.169	0.184	0.000

Our theoretical model is rejected by the data, although there are positive indications from the coefficients. What should start from here is a *competing model* strategy. For example, one might try and remove all the non-significant components of the model or add some other explanatory variables. One problem might be related to the measurement of the latent construct for Attitude, because the presence of items

Table 15.6 *AMOS information criteria and Hoelter indicators*

	AIC	
Model	AIC	BCC
Default model	709.428	712.218
Saturated model	238.000	245.376
Independence model	1764.822	1765.690

		ECVI		
Model	ECVI	LO 90	HI 90	MECVI
Default model	1.422	1.270	1.588	1.427
Saturated model	0.477	0.477	0.477	0.492
Independence model	3.537	3.273	3.815	3.538

	HOELTER	
Model	HOELTER .05	HOELTER .01
Default model	77	85
Independence model	38	41

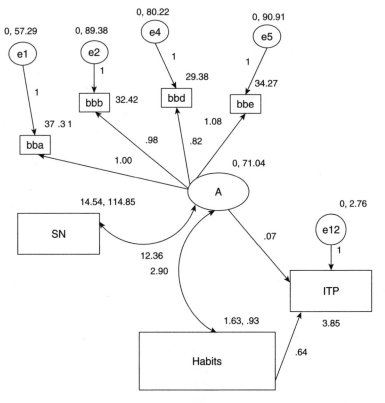

Figure 15.4 *The final structural model with estimates*

with negative wording might lead to the identification of more than one latent factor (see further readings in chapter 10). Another element which could improve the TPB theory is the inclusion of some variable explaining habit, which could be correlated with Attitude and influence ITP. For example, variable q2b measures the frequency of chicken purchases and we label it as habit.

Table 15.7 Estimates for the final model

			Raw estimate	S.E.	C.R.	P	Std. estimate
Intercept		bba	37.307	0.509	73.344	***	
Intercept		bbb	32.417	0.567	57.216	***	
Intercept		bbd	29.377	0.513	57.300	***	
Intercept		bbe	34.267	0.593	57.774	***	
Intercept		q7	3.848	0.169	22.763	***	
bba	< – – –	A	1.000				0.744
bbb	< – – –	A	0.975	0.080	12.131	***	0.656
bbd	< – – –	A	0.820	0.072	11.426	***	0.611
bbe	< – – –	A	1.083	0.086	12.647	***	0.692
q7	< – – –	A	0.066	0.012	5.490	***	0.291
q7	< – – –	q2b	0.637	0.091	7.022	***	0.319
Q2b	< – – >	A	2.901	0.448	6.479	***	0.358
sn	< – – >	A	12.357	4.456	2.773	0.006	0.137

***P-value below 0.001

Table 15.8 *Goodness-of-fit results for the final model*

Model	NPAR	CMIN	DF	P	CMIN/DF	NFI-Delta1
Final model	22	20.99	13	0.07	1.62	0.97

	RFI-rho1	IFI-Delta2	TLI-rho2	CFI	PRATIO	PNFI
Final model	0.93	0.99	0.97	0.99	0.46	0.45

	PCFI	NCP	LO 90	HI 90	FMIN	FO
Final model	0.46	7.99	0.00	24.62	0.04	0.02

	LO 90	HI 90	RMSEA	LO 90	HI 90	PCLOSE
Final model	0.00	0.05	0.04	0.00	0.06	0.80

	AIC	BCC	ECVI	LO 90	HI 90	MECVI
Final model	64.99	65.71	0.13	0.11	0.16	0.13
Saturated model	70.00	71.14	0.14	0.14	0.14	0.14
Independence model	652.53	652.75	1.31	1.15	1.48	1.31

	HOELTER	HOELTER				
	0.05	0.01				
Final model	532	659				
Independence model	33	38				

Figure 15.4 shows the final output of maximum likelihood estimation for a modified model. Next to each box one can see the average and standard error for that variable, while the values next to the arrows are the raw model parameters. Table 15.7 summarizes the raw and standardized model estimates and shows that now all the parameters are significant at the 5% level. The final confirmation that the model is now acceptable comes from table 15.8, which reports the goodness-of-fit statistics.

Summing up

SEM is one of the most widely applied techniques in consumer research, as it allows us to model simultaneously many relationships and include latent constructs in the analysis as dependent or explanatory variables. This latter feature enables the multivariate modeling of key consumer behavior determinants which cannot be measured directly, like attitudes, social pressure and lifestyles. SEM is a confirmatory technique as the specification of the starting model is driven by some pre-existing theory and the data are used to assess whether the chosen specification can be confirmed. This evaluation is made possible by the use of goodness-of-fit statistics, which can be exploited either to decide whether to reject or maintain the starting specification or, alternatively, to explore the fitness of alternative specifications. SEM is based on confirmatory factor analysis, which relates manifest indicators to latent constructs and on path analysis, which is a generalization of regression analysis emphasizing the role of assumed causal relationships. The starting point for the application of SEM is the drawing of a path diagram, which determines which variables are endogenous and which ones are exogenous, the direction of causal relationship and distinguishes between latent and manifest variables. Maximum likelihood estimation allows us to derive the regression coefficients, the correlations and the goodness-of-fit statistics.

EXERCISES

1. Using AMOS, estimate the following path model using EFS data, where beer consumed at home (c21311) is the dependent variable and purchase of sausages (c11251) and hot take-away meal eaten at home (cb1127) are the explanatory variables:

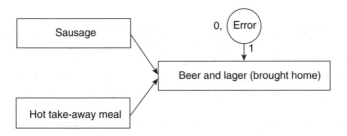

a. Compare parameter estimates with those of standard regression models
b. What is the correlation between the explanatory variables?

    c. Check model identification. How many degrees of freedom are available?

    d. Fix correlation to zero to create an extra degree of freedom and re-estimate the model

    e. Doing the goodness-of-fit diagnostics suggest that the model fits the data well?

2. Open the Trust data-set and run the following confirmatory factor analysis using AMOS, where it is assumed that variables q27a to q27g are manifest variables which measure the same latent construct, risk aversion:

    a. Estimate the above model (HINT: do not forget to fix the variance of the latent factors at 1 for identification)

    b. Can the model be regarded as satisfactory according to goodness-of-fit indicators?

    c. Maintain only three manifest indicators related to daily behaviors, driving, eating chicken and eating beef – do results improve? (HINT: set the regression weight 'eating chicken' to be one in order to have an over-identified model)

    d. Try changing the fixed regression weights (HINT: fix 'eating beef' or 'driving' instead of 'eating chicken,' with appropriate values as they emerged from the initial analysis)

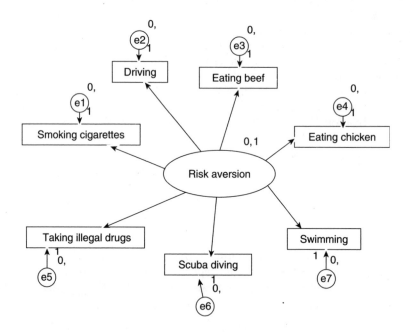

    e. Compute the scores of the latent factor (HINT: see AMOS help on 'how to estimate a latent variable score'). How do they correlate with directly measured risk aversion (q28)?

3. Consider the following theory – Chicken consumption (q4kilos) is the outcome of price (PRICE), income (INCLEVEL) and un-measurable attitudes as measured by *bba, bbb, bbd* and *bbe*.

    a. Draw the structural equation model, considering correlations across variables

    b. Using constraints and deleting non-significant elements, find an appropriate model

c. Compare the model with an alternative theory which explains chicken consumption (CONS) as the outcome price (PRICE) and habits, intended as a latent construct measured by q2b, q2c and q2d. Which theory works best? Can either of the two theories be retained as valid?

Further readings and web-links

❖ **AMOS** – A free trial version of AMOS which works with all features for two weeks can be downloaded from www.spss.com. Helpful instructions on using Amos can be found in the book by Byrne (2001b) and on the ITS University of Texas Austin web-site at www.utexas.edu/its/rc/tutorials/stat/amos/

❖ **More on identification** – For more details about the identification problem with SEM see Jöreskog (1969), Rigdon (1995), Davis (1993), the edited book by Bollen and Long (1993) and especially the chapter by Jöreskog and the book by Bollen (1989). See also the simple discussion in Prof. Ed Rigdon's web-page on SEM (www2.gsu.edu/~mkteer/identifi.html)

❖ **SEM journals and web-sites** – SEM is one of those methods which has a its own journal. See *Structural Equation Modeling* and the journal web-site at www.erlbaum.com. There are also various web pages on SEM. Beside the abovementioned page by Prof. Ed Rigdon (www2.gsu.edu/~mkteer), also visit the extensive Statnotes web-page on SEM by Prof. David Garson (*Statnotes: Topics in Multivariate Analysis*, www2.chass.ncsu.edu/garson/pa765/structur.htm)

❖ **Causal modeling and SEM** – A must is the well written paper by Bentler (1980), which takes into account the criticism in the amusing article by Guttman (1977). A more recent discussion, with statements from all the big players in consumer psychology research, can be found in Netemeyer, Bentley et al. (2001), or the book by Pearl (2000). See also Field's handouts on www.statisticshell.com/sem.pdf. Then make up your mind.

Hints for more advanced studies

☞ Maximum likelihood is just one of the methods for estimating a structural equation model? What are the others? What are the properties of different methods?

☞ Besides the confirmatory strategy on individual SEM, a competing model strategy can be adopted to identify the best model, whether among models from different theories or among nested models. Find out more.

☞ In other areas (for example in econometrics) SEM is not yet so popular. This may be due to the fact that SEM might be seen as a combination of existing techniques, like factor analysis and simultaneous equation models. However, there are some differences. Find out.

Notes

1. There is a growing literature on SEM. A good introduction is the book by Maruyama (1998) or chapter 11 of the book by Hair et al. (1998). More comprehensive textbooks include – among many others – Kaplan (2000) and Kline (2005).
2. See Bollen (2002) for a good review of the use of latent variables in social sciences.
3. See chapter 8 and box 8.3.
4. See Hoyle (1995) for a good set of directions on how to interpret the output from AMOS and other software.
5. As mentioned in chapter 10, the presence of items with negative wording to measure the same latent construct may induce a bias and reduce the capability of the measurement model to identify a single factor.

CHAPTER 16

Discrete Choice Models

T HIS CHAPTER generalizes linear regression to the case where the dependent variable is non-metric with binary, ordinal or categorical values. In consumer research, discrete choice models are closely related to the analysis of stated preferences and are helpful to explain consumer choices as a consequence of varying product attributes. A specific case where discrete choice models can be applied to elicit consumer preferences and attribute evaluations is conjoint analysis, a method for defining appropriate data collection designs and compare different combinations of attributes, which is especially useful in new product development.

Section 16.1   explains the extension of regression models to the case of discrete choice

Section 16.2   explains the underlying statistical principles and lists the main types of models

Section 16.3   illustrates the application of discrete choice models in SPSS

Section 16.4   provides an overview of conjoint analysis with links to discrete choice models

THREE LEARNING OUTCOMES

This chapter enables the reader to:

➡ Understand why and when the use of discrete choice models is preferred to regression

➡ Know the main types of discrete choice models and how to estimate them in SPSS

➡ Appreciate the potential uses of conjoint analysis and its links with choice models

PRELIMINARY KNOWLEDGE: For a proper understanding of this chapter review the regression model (chapter 8), the introduction to the binomial distribution, the logarithm and the exponential functions in the appendix.

16.1   From linear regression to discrete choice models

Choice modeling is central to consumer research because it is the natural methodological option for studies targeted at the elicitation of consumer preferences. In other words, choice models are closely related to an effective method for measuring consumer behaviors, attitudes and preferences, based on *stated preference theory*. In a stated preference survey, consumers state their choices among a potential set of alternatives (for example different brands, different product characteristics, different stores), which can include both real and hypothetical market alternatives. Starting from what consumers declare to prefer in the survey, one can use models to go back to the determinants of those choices. The alternative to stated preference is *revealed preference*, where consumers are not asked directly what they prefer or choose, but their actual choices and determinants are observed indirectly, for example considering what they purchase in different situations.

To understand the relevance of stated preference in consumer research, think of the following question – suppose that tomorrow you go to your usual supermarket and find out that the price of your favorite washing powder brand has doubled. Would you buy less washing powder, like a smaller pack? Or would you move to a different brand? Would you go back home without buying washing powder at all?

In this case, the researcher is interested in consumer reaction to price changes, but it would be rather difficult to define a model which explains your choice using revealed preference, that is what could be observed by waiting for you at the checkout till. For example, if you decided not to buy washing powder at all, how would it be possible to infer this choice simply from a look at the products in your shopping trolley? Or if you bought an alternative brand with exactly the same size and price as before the price increase, would a revealed preference model capture that consumer decision? With revealed preference, it might still be possible to model the above behaviors, after collection of information on the frequency, quantity and brands of washing powder purchases, but it would be a very expensive data collection effort. Now, consider the stated preference alternative, based on a survey where the consumer is asked to choose between a set of alternative choices, which differ by brand, pack size and price. Provided that the survey is designed in an appropriate way (which is not necessarily easy), the collected data open the way to a more effective model.

Generally, consumer models do not target individual behaviors, they rather focus on the average one. With revealed preference one could think of a regression model, where the purchased quantity is the dependent variable on the left-hand side and price and other explanatory variables are on the right-hand side.

With stated preference models, a *discrete choice* variable is on the left-hand side of the equation – for example, the choice whether to purchase washing powder or not or the choice among a set of alternative brands. In the former case, the dependent variable is binary (purchasing vs. not purchasing). In the latter case, the dependent variable is a qualitative one with a modality for each of the brands listed in the survey. In both cases regression analysis is not appropriate, for several reasons. Suppose the binary dependent variable y is coded to be zero for non-purchases and one for purchases, while x is a continuous metric variable. If this variable is regressed on the explanatory variable as in a standard regression model, we would have the following model:

$$y = \alpha + \beta x + \varepsilon \text{ with}$$

$$y = \begin{cases} 0 \text{ for non purchases} \\ 1 \text{ for purchases} \end{cases}$$

However, there are a few problems. If one estimates the model parameters by least squares (see chapter 8) and tries to predict the values of y using the value of x, predictions would include many other values than zero and one, including values below zero and values above one. Second, one might decide to code the binary variable as one and two, or zero and ten. Obviously this would lead to very different estimates for the α and β coefficients, which makes the interpretation of the regression parameters difficult. Finally, the above model does not meet the assumptions of the regression model (see chapter 8), since multivariate normality of the dependent variable for any value of the explanatory variables is broken.

With regression-like models where the dependent variable is a non-metric one, *discrete choice models* are the modeling route. This class of models includes a variety of specific models depending on the nature of the dependent and explanatory variables. Some of these are related to other techniques already explored in this textbook, for example discriminant analysis (see chapter 11) also deals with binary and categorical dependent variables. The main types of discrete choice models are listed in section 16.2, which also tries to shed some light on the terminology and uses for different models. As discussed in section 16.4, choice models are especially useful in designing new products and evaluating product characteristics through *conjoint analysis*. The above discussion has shown how the use and effectiveness of choice models depend on appropriate data collection and an accurate survey design. The questionnaire (or rather the experimental) design for stated preference and choice models can be quite complex and its discussion goes beyond the scope of this book, but the reader is referred to Train (2003) or Louviere et al. (2000) for an accurate treatment. Also, it might be interesting to mention the recent evolution of a data collection approach which is directly related to choice models and goes under the name of *choice experiments* and *experimental auctions*. This strategy builds a bridge between social science and experimental science by collecting data on human behavior in an experimental setting, controlling for relevant influential factors in order to isolate the experimental condition (see the handbook by Kagel and Roth, 1995).

16.2 Discrete choice models

Consider the usual regression model

$$y_i = \alpha + \beta x_i + \varepsilon$$

Discrete choice models generalize the regression model for the situations where y is a non-metric variable, for example a binary (0-1) variable, an ordinal variable (like a questionnaire item assuming the values completely disagree, disagree, neither, agree, completely agree), or a categorical variable (for example a variable recording the preferred holiday destination). The variable on the right-hand side is generally assumed to be metric, although binary and categorical variables can be translated into dummies and used as explanatory variables such as regression (see box 8.5). For the reasons stated in the previous section, and because non-metric dependent variables violate both the normality and the homoskedasticity assumptions, an alternative approach is used to estimate discrete choice models.

To show the basic idea behind discrete choice models, let us consider the binary case, where y can assume the discrete values zero or one. In order to model y as a function of x one can exploit the concept of latent variables, which has already been particularly useful for factor analysis (see chapter 10) and structural equation models (see chapter 15). More specifically, y may assume either value zero or one depending on the threshold value δ of a metric and continuous latent variable z.

So, one can rewrite the above regression model as:

$$\begin{cases} y_i = 0 \text{ if } z_i \leq \delta \\ y_i = 1 \text{ if } z_i > \delta \end{cases}$$

which states that the dependent variable y is one when a latent continuous variable z is above the threshold δ, and zero otherwise. The model is completed by the following regression equation

$$z_i = \alpha + \beta x_i + \varepsilon_i$$

which is the usual regression model since the dependent variable z is metric and continuous and after some assumptions the distribution of ε is known. There are two problems here, (a) we do not observe z (it is latent) and (b) we do not know δ. The latter issue is easily resolved, since as long as the intercept α is included in the regression equation one may arbitrarily choose δ (the easiest way is to fix it at zero) and the only result which changes is the estimate of the intercept α. The other problem is finding a way to create z for each observation as a function of y, taking into account the available information, that is the proportions of zero and 1 for the y variable. For this purpose, one needs to make an assumption on the probability distribution for this latent variable and how it is linked to y. To put it in technical words, it is necessary to specify the *link function* between y and z. The link function defines the relationship between z and y through the expected value of the appropriate distribution function for the generic observation y_i.

For example, with binary data, one can assume that the probabilities of each observation y_i follow a *binomial distribution* (see appendix). Without getting too much into the math (see box 16.1), there are a number of transformations of y which create a z variable compatible with the binomial distribution. One of them is the *logistic function*, which can be understood from figure 16.1.

The usefulness of the logistic function is clear; probabilities that $y = 1$ (on the vertical axis) tend to concentrate around zero for values of x below a certain threshold, then they go quickly toward 1 when x is above the threshold. The function fits well with the need for approximating the probabilities of a binary outcome as a function of the explanatory variable. The logistic transformation of y into z is obtained by applying the logit link function to the expected value of y. See box 16.1 for the math.

Now there is everything. In short, the logit transformation is the link function for logistic regression and allows one to transform the binary variable y into a continuous variable z. The final equation is a regression with the continuous variable (z) on the left-hand side. The empirical observations for z can be computed by applying the logit transformations of π, which is the expected value for the probability that $y = 1$. The only difference with the standard regression model is that the distribution of the error is not normal, but logistic. This is not a problem, since estimation of α and β can be obtained by *maximum likelihood*, which works with any known probability distribution for errors and returns the maximum likelihood estimates (to put it roughly, the most probable values for the parameters).

What we have described so far is *logistic regression* (since the starting point was the logistic transformation of y, hence the assumption that the error ε follows a logistic distribution). Note that for simplicity we have used a single explanatory variable, but the above result is also valid for multiple explanatory variables. Conventionally, the term logistic regression refers to the case where at least one of the explanatory variables is metric and continuous. When all of the variables on the right-hand side are non-metric

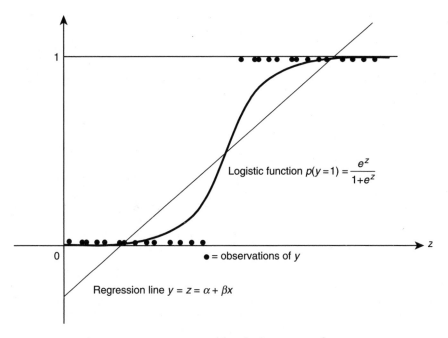

Figure 16.1 *From binary to continuous variables: the logistic transformation*

(binary or categorical), the model is named *logit model* (see 'logit models' in Upton and Cook, 2006), which are closely related to log-linear models discussed in chapter 9. With the logit model, having a categorical or binary x variable, the coefficient β is mathematically related to the *odds ratio* (with respect to the baseline category of x) of having a positive outcome.

For example, if the dependent variable is 1 when the consumer buys a specific brand and x measures whether the consumer has children or not, one can compute with e^{β} the odds ratio of buying the brand for consumers with children as compared to consumers without.

Having explained logistic regression and the logit model, generalization to other discrete choice models is straightforward.

The *probit model* is also applied to binary dependent variables, but with different assumptions on the link function and the error distribution. The link function (called *probit*) is the inverse of the standard normal cumulative distribution function and guarantees that the distribution of the model which is finally estimated is still normal. The choice between the probit and the logit distribution depends on the type of dependent variable one is dealing with. If the dependent variable can be reasonably assumed to be a proxy for a true underlying variable which is normally distributed, then the probit model should be chosen. Instead, if the dependent variable is considered to be a truly qualitative and binomial character, then logit modeling should be preferred. Generally, the two models lead to very similar results, unless cases are concentrated to the tails of the distributions, in which case the logit link function should be chosen.

Logit and probit models can be generalized to *ordered logit* and *ordered probit* models when the dependent variable is not binary but categorical and the categories are ordered. If the categories cannot be ordered, the adopted specifications are the *multinomial logit* or *multinomial probit*. The multinomial forms refer to the case where there is a single dependent variable with multiple categories and should not be

---

BOX 16.1 *Binary logistic regression and the logit link*

The logistic function in figure 16.1 represents the relationship between the probabilities of each observation of y assuming value 1 and the values of the explanatory variable x and can be written as:

$$p_i = P(y_i = 1 \,|\, \alpha + \beta x_i) = \frac{e^{\alpha + \beta x_i}}{1 + e^{\alpha + \beta x_i}}$$

Using the above expression, it is easy to make z_i enter the picture, since we know that $y_i = 1$ when $z_i > 0$ and $z_i = \alpha + \beta x_i + \varepsilon_i$:

$$p_i = P(y_i = 1 \,|\, \alpha + \beta x_i) = P(z_i > 0) = P(\alpha + \beta x_i + \varepsilon_i > 0) = \frac{e^{\alpha + \beta x_i}}{1 + e^{\alpha + \beta x_i}} + \varepsilon_i$$

which means that the probability that the i-th observation is equal to one (given the values of x_i, α and β), is a logistic function plus the random error.

Note that the above discussion is still valid for any binary coding of y, because the left-hand side of the equation is simply the probability of a given outcome for y. Thus, the coding is purely conventional, the y outcomes could be one and two or zero and ten and the results would be the same. Now, what is needed to exploit the logistic function as a link function is simply to link the expected (mean) value of the binomial distribution of y_i to the values of x. However, in binomial distributions, $E(y_i) = P(y_i = 1)$, that is the expected value of the observation y_i is equal to the probability that the observation is 1, while the expected value of the error ε_i is 0. Thus, by using expected values, the error term on the right-hand side disappears and again:

$$E(y_i) = P(y_i = 1) = p_i = \pi = \frac{e^{\alpha + \beta x_i}}{1 + e^{\alpha + \beta x_i}} = \frac{e^{z_i}}{1 + e^{z_i}}$$

which basically means that the expected (average) value of y_i corresponds to the (unconditional) probability that $y_i = 1$ (which is equal to π for all observations) and is a logistic function of x. This is a big step, because now there are all elements to link y to a continuous (logistic) latent variable z. This can be obtained by referring to the *odds ratio*, the ratio between the probability of success ($y_i = 1$), equal to p_i, and the probability of failure ($y_i = 0$), equal to $1 - p_i$:

$$\frac{p_i}{1 - p_i} = \frac{e^{z_i}}{1 + e^{z_i}} \bigg/ \frac{1 + e^{z_i} - e^{z_i}}{1 + e^{z_i}} = e^{z_i}$$

We are almost there. The logarithm of the left-hand side of the above equation is the so-called *logit* and allows to write that:

$$\log \left(\frac{p_i}{1 - p_i} \right) = z_i = \alpha + \beta x_i + \varepsilon_i$$

Since p_i is the binary outcome for observation i and the distribution of the observations is binomial, we can estimate the expected value π as $\sum_{i=1}^{n} y_i / n$ (that is the proportion of cases where $y = 1$).

---

confused with the *multivariate* form, where several discrete choice models are estimated simultaneously (there are multiple dependent variables). For an exhaustive review of discrete choice specifications, the reader is referred to Greene (2003).

16.3 Discrete choice models in SPSS

The terminology for discrete choice models and the approaches used in SPSS can be slightly confusing in some circumstances. Here we provide a brief introduction to the

Table 16.1 *SPSS logistic regression, summary output*

Model summary

Step	−2 Log-likelihood	Cox & Snell R-Square	Nagelkerke R-Square
1	467.079[a]	0.157	0.217

[a]Estimation terminated at iteration number 5 because parameter estimates changed by less than 0.001.

Hosmer and Lemeshow Test

Step	Chi-square	df	Sig.
1	3.030	8	0.932

estimation in SPSS of some of the discrete choice models explored in this chapter, using the Trust data-set for examples.

For comparison purposes, it is useful to exploit the same examples as in chapter 11 for discriminant analysis. First, consider the choice of buying chicken at the butcher's shop as explained by weekly expenditure on chicken, age, safety of butcher's chicken, trust in supermarkets.

For the case of binary logistic regression, the easier way is to start from the menu ANALYZE/REGRESSION/BINARY LOGISTIC, select q8d as the dependent variable and q5 (weekly expenditure on chicken), q21b (perceived safety of butcher's chicken), q43d (trust in supermarkets) and q51 (age) as the covariates and simply choose ENTER as the method. This dialog box also allows us to run *stepwise logistic regression* which exploits procedures for the selection of meaningful covariates (see stepwise regression in chapter 8). Through the button STATISTICS we can ask for some additional output, like the Hosmer and Lemeshow test and the confidence intervals for the odds ratios.

The model summary output in table 16.1 reports the value of the log-likelihood function, together with two goodness-of-fit statistics that can be interpreted in a similar way to R-Square in regression.

The *Hosmer and Lemeshow* test compares the expected frequencies with those actually observed, after dividing the subject in ten equal groups according to their predicted probabilities. The above result (with a *p*-value largely above the 0.05 threshold) is quite satisfactory and tells us that the model fits the data well.

The classification table 16.2 compares model predictions with the observed values. The values on the diagonal are the correctly predicted ones, while those outside are the incorrect classifications. Overall, the model is able to predict correctly 70.7% of the cases. This percentage should be compared with the proportion of cases in the largest category, as a rough model which allocates all cases to that category would still correctly predict 66% of cases (the percentage of those who do not buy chicken at the butcher's shop). On balance the gain of using logistic regression is not very large, yet we are able to predict correctly 38% of the individuals shopping for chicken at the butcher's shop. Finally, table 16.3 shows the estimated coefficients, the odds ratios (that is e^b, where b is the estimated coefficient) and the relative confidence interval.

All covariates are highly significant (*Sig.* below 0.01) and the odds ratios tell us the change in the odds of purchasing chicken at the butcher's shop associated with a unit increase in the explanatory variable. Thus, if age increases by one year (q51),

Table 16.2 *Classification table*

Classification Table[a]

Observed			Predicted		
			Butcher		Percentage correct
			no	yes	
Step 1	Butcher	No	243	34	87.7
		Yes	89	54	37.8
	Overall percentage				70.7

[a]The cut value is 0.500.

Table 16.3 *Coefficient estimates and diagnostics*

Variables in the Equation

		B	S.E.	Wald	df	Sig.	Exp(B)	95.0% C.I. for EXP(B)	
								Lower	Upper
Step 1[a]	q51	0.022	0.007	8.988	1	0.003	1.022	1.008	1.037
	q43b	−0.269	0.074	13.327	1	0.000	0.764	0.661	0.883
	q21d	0.441	0.077	32.888	1	0.000	1.554	1.337	1.807
	q5	0.085	0.028	8.975	1	0.003	1.088	1.030	1.150
	Constant	−3.169	0.615	26.539	1	0.000	0.042		

[a]Variable(s) entered on step 1: q51, q43b, q21d, q5.

the odds of purchasing chicken at the butcher's shop (which is the ratio between the probability of doing it and the probability of not doing it) increase by 2.2%. Instead, a unit increase in trust in supermarket (q43b) decreases the same odds ratio by 23.6%. The safety perception is the most relevant covariate, as a unitary increase improves the odds ratio by 55.4%.

Note that logit and probit models could also be estimated by using the menu ANALYZE/REGRESSION/PROBIT. In this case, however, SPSS requires data to be structured as counts of 'success' cases (*response frequency*), with an additional column for the total number of cases. This is barely a complication, since we can create the *total observed* variable by creating a new variable of ones (for example *tot*). If we do that, we can repeat the above analysis by selecting q8d as the response frequency and *tot* as the *total observed* variable. The *covariates* are then chosen as before and the analysis yields very similar results as the above, provided that *logit* is chosen. Finally, logistic regression, logit and probit models can be estimated using the *generalized linear model* approach. To start the analysis, click on ANALYZE/GENERALIZED LINEAR MODELS/GENERALIZED LINEAR MODEL. The window shown refers to the RESPONSE variable. Select q8d as the dependent variable and specify that the variable represents a binary response by clicking on the appropriate box and defining the *reference category* (which is useful to interpret the output) as the lowest one, which means those who do not purchase chicken at the butcher's. The distribution for the dependent variable is automatically chosen as the binomial distribution, but we can complete the specification of the link function by stating that the *logit link function* should be used (one might later try with the *probit link function* to explore differences). Once the response variable has been defined, one can move to the PREDICTORS window. In SPSS *factors* correspond to non-metric (categorical) predictor, while *covariates* refer to metric (continuous) variables. According to the convention mentioned in section 16.2, we have a logit model if we only include factors, while with covariates we get a logistic regression. In our example we go for a logistic regression and we can consider all explanatory variables to be covariates. Once the covariates have been selected, they need to be included in the model by opening the MODEL menu. For our introductory view of discrete choice models, we can select all the variables as *main effects*, but those interested in the full complexity of discrete choice models are advised to read Louviere et al. (2000) to appreciate the possibilities of interaction terms and random effect specifications. Here we ignore all other opportunities for adjusting the model and we simply run the analysis by clicking on OK. The results are identical to those obtained with binary logistic regression.

Finally, let us see how SPSS deals with the ordered (ordinal) and multinomial dependent variables. Consider trust in food safety information provided by government (q43j), measured on a scale from one to seven. Do income and age status influence that ordinal variable? The model we employ is the *ordered logit* which can be estimated by clicking on ANALYZE/REGRESSION/ORDINAL, selecting the appropriate dependent variable (q43j), the factor (income range q60) and the covariate (age q51).

The output is shown in table 16.4.

The initial warning is fundamental to evaluating problems due to the presence of empty cells. A large proportion of empty cells (zero frequencies in the dependent variable levels conditional to the predictors' values) invalidate some of the goodness-of-fit statistics.

The model-fitting information compares the ordered model with an intercept only model. Since the final (ordered) model shows a significant chi-square statistic (at least at the 5% significant level), there is an improvement in using the ordered model.

The goodness-of-fit statistic based on Pearson's chi-square indicate a good fit, intended as the similarity between the predicted and observed data. However, these statistics are sensitive to the number of empty cells. Furthermore, the Pseudo R-Square statistics are quite low, suggesting that the model could be improved by the inclusion of other covariates and factors.

Finally, there are two sets of parameter estimates. The first (*threshold*) determines the cut-off points for allocating an observation to a given value of the dependent variable, according to the value of the latent variable. The second set contains the *location* parameters, which translate the predictors into a value for the latent variable. The Wald test (corresponding to the *t*-test in regression) shows that the predictors do not actually contribute significantly. This is consistent with the poor Pseudo R-Square statistics. What could be interesting (at least for a model with a better fit) is the computation of the *marginal effects*, which tell us the change in the probability of an observation being classified in each specific category of the dependent variable according to the values of the predictors. Unfortunately, SPSS does not provide marginal effects. Finally, multinomial logit models can be estimated by clicking on ANALYZE/REGRESSION/MULTINOMIAL LOGISTIC. The process is similar to the one leading to the estimation of ordered logistic regression and the output should also be interpreted accordingly.

Table 16.4 *Ordered logit output*

Warnings

There are 975 (77.0%) cells (i.e. dependent variable levels by combinations of predictor variable values) with zero frequencies.

Model fitting information

Model	−2 Log-Likelihood	Chi-square	df	Sig.
Intercept Only	1004.627			
Final	987.352	17.276	7	0.016

Link function: Logit.

Goodness-of-fit

	Chi-square	df	Sig.
Pearson	1088.968	1073	0.360
Deviance	816.931	1073	1.000

Link function: Logit.

Pseudo R-Square

Cox and Snell	0.051
Nagelkerke	0.052
McFadden	0.014

Link function: Logit.

Parameter estimates

		Estimate	Std. Error	Wald	df	Sig.	95% Confidence interval Lower bound	95% Confidence interval Upper bound
Threshold	[q43j = 1]	−2.861	0.941	9.250	1	0.002	−4.705	−1.017
	[q43j = 2]	−2.389	0.935	6.528	1	0.011	−4.222	−0.556
	[q43j = 3]	−1.738	0.930	3.492	1	0.062	−3.560	0.085
	[q43j = 4]	−0.672	0.925	0.527	1	0.468	−2.486	1.142
	[q43j = 5]	0.051	0.925	0.003	1	0.956	−1.761	1.864
	[q43j = 6]	1.302	0.930	1.960	1	0.162	−0.521	3.126
Location	q51	−0.009	0.006	1.961	1	0.161	−0.021	0.004
	[q60 = 0]	−0.845	0.905	0.873	1	0.350	−2.618	0.928
	[q60 = 1]	−0.158	0.898	0.031	1	0.861	−1.917	1.602
	[q60 = 2]	−0.069	0.912	0.006	1	0.940	−1.857	1.719
	[q60 = 3]	0.121	0.918	0.017	1	0.895	−1.678	1.920
	[q60 = 4]	−0.081	0.952	0.007	1	0.933	−1.947	1.786
	[q60 = 5]	0.774	1.111	0.486	1	0.486	−1.403	2.952
	[q60 = 6]	0[a]			0			

Link function: Logit.
[a]This parameter is set to zero because it is redundant.

BOX 16.2 *Statistical packages and discrete choice models*

Although SPSS allows estimation of the most common discrete choice models, it lacks some useful features like the estimation of *marginal effects*, which are the changes in the probability of falling in a specific category of the dependent variable induced by a unity change in each of the predictors.

In SAS discrete choice models can be estimated with several procedures of SAS/STAT. The procedure *CATMOD* is employed for estimating logistic regression when the data are structured as a frequency table. Otherwise, binary and ordered logistic regression can be obtained through the procedure *LOGISTIC*. The same models can be estimated with the *PROBIT* procedure, which also enables estimation of probit models. Finally, the procedure *GENMOD* allows to specify a variety of link functions for generalized linear models, thus enabling estimation of all models described in this chapter.

The software *LIMDEP* (see Greene, 2002) was specifically created by William Greene for the estimation of *limited dependent variable* models, which include discrete choice models. Although its interface is less immediate than SPSS, it is extremely flexible and contains all the required features and the most up-to-date diagnostics. Those familiar with *STATA* can easily estimate discrete choice models using that software, which has also commands (STATA, 2005) for generating marginal effects. Finally, discrete choice models can be estimated using another very popular econometric software, *Econometric Views*, but the availability of diagnostics is rather limited as compared to LimDep and no marginal effects are displayed.

16.4 Choice modeling and conjoint analysis

Conjoint analysis is a very popular research technique in marketing, closely associated with stated preference analysis and mainly exploited for the development of new products and the modification of product characteristics (see Green et al., 2001). It should be made clear that conjoint analysis is not a model or an estimation technique, but rather a methodology for constructing the data collection instrument when the objective is choice modeling.

To appreciate the potential uses of conjoint analysis, we can consider the most common application in consumer research, that is the analysis of consumer evaluations of different combinations of product attributes. Suppose that a car manufacturer needs to take some decision about some options to be provided for car configuration, like the range of colors, the model of car stereo, presence of air conditioning, etc. Rather than asking consumers about their evaluation of these attributes on a one-by-one basis, conjoint analysis starts by creating all potential combinations of the product attributes. For example, one potential combination is a red car, with an *mp3* stereo player and no air conditioning, another could still relate to a red car, but with a standard CD player and air conditioning and so on. Respondents are asked to choose among these alternative potential products defined by the combination of attributes. Obviously, when several attributes are considered simultaneously, the number of potential combinations is quite high. Thus, conjoint analysis creates many different choice sets, each one containing a limited number of options. The way choices are allocated in the sample and the distribution of attributes are controlled statistically, so that the collected data enable inference on preferences and evaluations for the individual attributes. The underlying theory for conjoint analysis is based on the economic concept of *utility*, which assumes that each individual has a specific set of preferences for bundles of products (and attributes) and takes decisions in a way to maximize the level of satisfaction from consumption (the utility level). Through this theory, by observing many individuals, it is possible to go back from stated choices to preferences.

Conjoint analysis is inspired to scientific *experimental designs* and the terminology reflects this association. Attributes are called *factors* (like car color) and the different values they can assume are the *levels* (red, blue, yellow, etc.). Using the same example, note that an additional factor could be the price of the car. By including price levels in the choice set, it becomes possible to evaluate how much consumers would be willing to pay for the car they prefer. Furthermore, it is quite likely that different levels for other factors influence the price level, thus nonsense choices (like including all car options but setting a very low price) can be excluded by the researcher, who has control on the overall *choice set*. In the actual questionnaire, respondents must choose the preferred combination of attributes or possibly rank all possible choices according to their preferences. To make a parallel with the discussion in chapter 13 on multidimensional scaling, conjoint analysis is a *decompositional method*, since it starts from an overall evaluation to infer preferences for the product attributes.

The experimental design and the modeling of preferences also depend on some theory which links the evaluation of single attributes to the final choice. Consumer behavior offers some alternatives (Green and Srinivasan, 1990), like:

- the *part-worth* model, which assumes that total utility of a choice is equal to the sum of utilities of the attributes of that specific choice
- the *vector linear* model, only applicable when all attributes are measured on a metric (continuous) scale, which assumes that the utility of individual attributes is related to total utility through a linear relationship
- the *ideal point* model, which assumes that the consumer has an ideal level for all factors and the total utility depends from the distance between the actual levels and the ideal levels

A key problem in running conjoint analysis is the large number of alternative combinations of attributes which arises when there are many factors and levels to be taken into account. Consider that a product with six attributes, each with three levels, potentially allows for 729 different combinations and it would be quite unrealistic to assume that respondents can choose among so many alternatives. This problem is faced through an appropriate experimental design, aimed at knowing the relationship between the factors and the potential choice with a number of observations as small as possible. Since consumer researchers are generally not interested at eliciting the preferences of a single specific individual, but they rather target a group of homogeneous consumers, the experimental design sets the criteria to obtain the preference information from more than one respondent, using an aggregate analysis. Thus, one can distinguish between *full factorial designs*, where all potential products (729 in the example above) are compared and *fractional factorial designs* which exploit the experimental design to reduce the number of choices, still guaranteeing that the sample will produce meaningful aggregate results. For a detailed discussion on the optimal designs, see for example Green (1974) or Louviere et al. (2000).

Traditional conjoint analysis is based on the constraint that each respondent is faced with the whole set of attributes. Thus, it requires either a full factorial design or a fractional factorial design, where all attributes appear in the choice set of each respondent (although not necessarily for all levels). Traditional conjoint analysis becomes quickly inapplicable as the number of factors or levels increases.

An alternative approach is *adaptive conjoint analysis* where these design issues are dealt with and each respondent only faces a sub-set of potential choices which can be defined in different ways. For example, respondents could be asked to rank the factors first, and then their ranking is exploited to 'adapt' data collection.

In this direction, the evolution of computer programs has been particularly helpful, since the software 'learns' from the earlier responses and builds the data-sets accordingly.

The decomposition of the observed choices into weights and preferences for single attributes is generally obtained for an aggregate of consumers or for homogeneous groups of consumers. Several techniques can be employed to this purpose (see Hair et al., 1998), but the evolution of the discrete choice models discussed in this chapter has given relevance to a specific type of adaptive conjoint analysis, called *choice-based conjoint*. The idea of choice-based conjoint is to give the respondent the possibility of evaluating all attributes, not in a single (often too complex) choice, but rather within a sequence of smaller choice sets, where the possibility of choosing none of the alternatives is also given.

To make a very trivial example, suppose that a car can be red or blue, with and without air conditioning. This implies four combinations. Instead of using a single choice set (red with air conditioning, red without, blue with and blue without), the respondent might be asked to choose first among red with air conditioning, blue without or none of them, then between blue with and blue without or none of them, etc. While advantages might not be apparent with a small number of factors and levels as in the above example, this method is especially desirable for more complex cases. Respondents are not asked to compare too many stimuli at once, but rather they face a more realistic choice. Furthermore, with many factors and levels, each respondent can be asked to deal with limited number of choice sets. As a matter of fact, it is sufficient that an homogeneous group of respondents (that is respondents that are similar in terms of characteristics that can influence the choice) is confronted with the whole range of alternatives, then the estimation technique will do the rest. The relevance of the experimental design in running a successful choice-based conjoint is clear and there is an evolving research effort to guarantee the quality of the analysis. Once the data has been collected, the natural estimation technique is the *multinomial logit*, where choices are related to the attributed levels (see for example Louviere and Woodworth, 1983).

There are computer packages specifically developed for conjoint analysis. The SPSS Conjoint module deals with the experimental design and provides estimates based on an orthogonal decomposition of the design matrix (see Golub, 1969). In SAS/STAT, the *TRANSREG* procedure is a useful support to define the experimental design.

Summing up

The extension of regression model to the cases where the dependent variable is non-metric leads to discrete choice models. This class of models is especially useful for analyzing stated preference and stated choice data, where respondents choose among a set of alternatives differing for a set of explanatory factors. The basis for estimating discrete choice models is the definition of a link function between the observed discrete outcome and a latent (non-observed) outcome which is continuous. The most common discrete choice model is the logistic regression for binary outcomes, which is based on the assumption that the dependent variable, conditional on the values of the predictors, follows the logistic distribution. Alternatively one could assume the normal distribution of the latent variable (as in the probit model). When the dependent variable is ordinal, the choice is between the ordered logit and ordered probit model, while categorical dependent variables are modeled with the multinomial logit and probit models. Discrete choice models are also applied in choice-based conjoint analysis. Conjoint analysis is a decompositional technique aimed at explaining consumer choices on the basis of the preferences for alternative combinations of attributes. It is particularly useful for new product development. The key step in conjoint analysis is the definition of an appropriate experimental design, which determines the choice set for each respondent. The purpose of conjoint analysis is to allocate the choice sets in a way that allows the elicitation of preferences on the aggregate sample, still considering all potential combinations of attribute levels but without requiring that the individual respondent is faced with extremely large and unrealistic choice sets.

EXERCISES

1. Open the EFS data-set
   a. Estimate a binary logistic regression where the possession of a computer (a1661) is related to income (incanon), age of the household reference person (p396) and household size (a049)
   b. Which of the variables is the most important in predicting computer possession?
   c. How many cases are correctly predicted by the model?
   d. Estimate the model with ordinary linear regression and compare the results
   e. Estimate the model with a probit regression and compare the results
   f. Replicate the results using the generalized linear model
   g. Include NS-SEC 8 Class of Household Reference Person (a094) as a categorical predictor and the interaction between age and income. Do results improve?
2. Open the EFS data-set
   a. Recode the tenure-type variable (a121) into three categories, owned (7), owned with mortgage (5), rented (1 to 4) and treat other values as missing
   b. Explain the variable with a multinomial logistic regression using the following predictors (covariates):
      i. Income (incanon)
      ii. Age of the household reference person (p396)
      iii. Household size (a049)
      iv. Recreation expenditure (p609)
   c. Use the stepwise method to maintain the relevant variables
   d. Is the model satisfactory in terms of correct prediction rate?
3. Open the Trust data-set
   a. Estimate an ordered probit model to explain how the likelihood to purchase chicken (q7) is explained by the determinants of the Theory of Planned Behavior (see box 15.3) – HINT: include them in the model as covariates:
      i. Attitudes toward chicken (q9)
      ii. Subjective norm (SN)
      iii. Perceived behavioral control (PBC)
   b. Which variable is the most relevant in explaining the likelihood to purchase?
   c. What is the rate of correct prediction?
   d. Add average price (PRICE) and income (INC) and check the model performance.

Further readings and web-links

❖ **Revealed versus stated preference theories** – For revealed preference theory see Samuelson (1948) or Varian (2006), for stated preference theory see Louviere et al. (2000), Louviere and Hensher (1983) or Hensher et al. (1999). Train (2003, chapter 7) deals with both methods and a combination between the two data sources is discussed together with further issues in Ben Akiva et al. (2002).

❖ **Generalized Linear Model, all the link functions** – For further details on the various models available within the generalized framework, see Liao (1994), the classic textbook by McCullagh and Nelder (1989) or the more recent book by Dunteman and Ho (2006).

❖ **Designing the experimental plan for conjoint analysis** – To learn more on how choice sets can be designed for multinomial logit analysis and conjoint analysis see Kessels et al. (2006), Green and Srinivasan (1990), Louviere and Woodworth (1983), Sandor and Wedel (2001 and 2005) and references therein.

Hints for more advanced studies

☞ How are marginal effects computed?

☞ A further extension of the regression model which exploits the link between an observed variable and a latent variable is the threshold model, where only values above a certain threshold are observed. Find out how the model can be specified and estimated.

☞ Can data from conjoint analysis be analyzed by multidimensional scaling? Find the relationship between the two approaches.

CHAPTER 17

The End (and Beyond)

T HIS CHAPTER draws some (personal) conclusions from all the efforts of previous chapters, but more importantly points out some directions in consumer and marketing research which definitely cannot be defined 'new,' but are increasingly employed. This chapter recalls useful quotes from Socrates, provides some initial references to learn about data meaning and tells the story of an encounter with Sir Thomas Bayes.

Section 17.1 adds some concluding lines to the textbook
Section 17.2 points the readers to the world of data mining
Section 17.3 shortly contributes to the cause of proselytism for Bayesian statistics

THREE LEARNING OUTCOMES

This chapter enables students to attain the three most important achievements:

➡ Learn that there are further and evolving topics in the field of statistics for consumer and marketing research
➡ Finish this textbook (at last)
➡ Hopefully passing the exam, do something else in case satisfaction was low, move to application and further study if satisfaction was high.

PRELIMINARY KNOWLEDGE: The whole book, the exercises, the appendix, the glossary, possibly all suggested further readings as well. And, probably, it is not enough.

All movement is accomplished in six stages
And the seventh brings return.
The seven is the number of the young light
It forms when darkness is increased by one.
Change returns success
Going and coming without error.
Action brings good fortune.
Sunset, sunrise.

Syd Barrett and Pink Floyd, chapter 24, The Piper at the Gates of Dawn

17.1   Conclusions

Among the various sayings that made Socrates[1] a great thinker, there is the well-known sentence which sounds like 'I know nothing except the fact of my ignorance.' Apparently, he left nothing written and other people took care of handing down his philosophy to posterity. I should have thought of the above before writing this book. First, after 16 chapters, one has the uneasy feeling that they are not so successful to convey the knowledge in an effective and helpful way. I can still hope I am wrong, since I know nothing… The second lesson is not to leave anything written down, but it is too late for me now.

Socrates also said that 'Education is the kindling of a flame, not the filling of a vessel.'[2] It is not false modesty, but I doubt this book kindled any flame (unless you tried to burn it). But I fully agree with the sentence. This book cites something like 300 references and in putting things together, many times I was astonished to see the genius of those who created some of the powerful statistical techniques described throughout the chapters. Most of them came out in their current shape in less than a century, but they all rely on millenary knowledge. Still, after collecting 300 references and looking back at the achievements of mathematical and statistical thinking over the last few thousands of years, one must admit that it is quite difficult (if not impossible) to appreciate the uses of a statistical method and use it properly without applying it several times and learning from errors. This is also to say, if you are not able to do statistics for consumer and marketing research after reading every page of this book, it is not necessarily my fault. Provided that some flame has been kindled, it is your turn now to get crazy on SPSS or any other software and get really familiar with the methods.

Like it or not, we are now in the computer age and this apparently makes things quicker (sorry, I have no time now to check whether Socrates said anything about computers and modern lifestyles). This means that complex methods, once discarded because they were time consuming, can now be recovered and implemented by anyone who can find the right button on a keyboard. On the one hand, this inflates the risk of careless use of statistics. On the other hand, there are 'new fashions' in consumer research that deserve at least an introduction. This is the scope for this chapter, besides allocating some room to my mental wanderings after completing the manuscript in the middle of an August night when everybody else was on holiday.

17.2   Data mining

Long ago, the term 'data miner' had a very offensive meaning. It was directed to those that, rather than learning proper statistical theories, preferred to crunch numbers without thinking too much about how they came out that way from the PC screen. Now *data mining* has gained popularity and even respect.

Data mining, according to the Oxford Dictionary of Statistics, is 'the exploration of a large set of data with the aim of uncovering relationships between variables' (Upton and Cook, 2006). Data mining is also known with the less direct title of *knowledge discovery in databases* (KDD) and makes no apology of making full use of information technology, through the automation of data analysis procedures. Press (2004) discusses the historical roots of data mining and cites Aristotle and Francis Bacon as precursors of modern data mining.

Note that the statistical knowledge is all but ignored by data mining (see Hand, 1998). Simply, it is adapted to deal with (very) large data-sets, it valorizes computer intensive methods and it is merged into a discipline which also learns from Computer Science, Machine Learning, Artificial Intelligence, Database Technology, and Pattern Recognition (Press, 2004). Luckily, practically all of the multivariate techniques described in this book are relevant to data mining. Unfortunately, they only represent a sample of the population of techniques in data mining, which also includes other techniques like genetic algorithms and neural networks. The common denominator among the techniques is always the use of very large databases. These databases are the outcome of *data warehousing*, which consists in organizing all of the data available to a company into a common format, which allows integration of different data types and their analysis through data mining. The organization of company information in data warehouses requires one to recognize linkages of data which relate to the same objects and the explicit recognition of the time dimension to monitor changes.

Data mining is widely applied in marketing research (see Drozdenko and Drake, 2002; Shaw et al., 2001). A typical application is *market basket analysis* (see Chen et al., 2004), where customer purchasing patterns are discovered by looking at the databases of transactions in one or more stores of the same chain (think how loyalty cards can help...) and the contents of the trolley are analyzed to detect repeated purchases and brand switching behaviors.

Data mining is a complex and automated process, which faces many risks (see Hand, 1998), like the fact that data-sets may be contaminated (to use the terminology of this book, they are affected by error), the possibility that they are affected by selection biases and non-independent observations, the possibility that automated data analysis could find spurious relationships (see spurious regression in box 8.2). Thus, successful data mining (or rather knowledge discovery in databases which includes data mining) needs to follow all the necessary (iterative and interactive) steps to provide meaningful solutions, namely (Fayyad, 1997):

1. data warehousing
2. target data selection
3. data cleaning
4. preprocessing
5. transformation and reduction
6. data mining
7. model selection (or combination)
8. evaluation and interpretation
9. consolidation and use of the extracted 'knowledge.'

For a discussion of these steps, see Fayyad et al. (1996). A comprehensive and recent treatment for data mining is provided in the book by Han and Kamber (2006). Many data mining techniques are based on the Bayesian approach, briefly introduced in the next paragraph.

17.3 The Bayesian comeback

Not long ago, at the University of Reading, I happened to have a friend and office neighbor (fully acknowledged in the preface) who made a daily effort to convert me (and most of the people around the Department) into Bayesian thinking. I have postponed my conversion on pure laziness grounds, but since then I have seen several 'frequentists' converting to the Bayes religion. As time passes, it is probably becoming less of a field choice and many former frequentist statisticians have started using Bayesian techniques which are undoubtedly more effective in some situations. Furthermore, when I went to SAGE Publications to deliver my first draft manuscript by hand, I was early. I wandered in the neighborhood of the SAGE building and I had a walk in Bunhill Fields, just on the other side of the street. There you can see the graves of William Blake the poet, and Daniel Defoe, the author of Robinson Crusoe.

Without any prior information (…) I ended up just in front of the grave of Thomas Bayes.

I felt guilty and – while my laziness to switch the paradigm has not succumbed yet – I decided to add at least a few lines to introduce readers to the realm of Bayesian thinking. Destiny insisted and an anonymous reviewer who read the first draft also suggested I should include some paragraphs on the Bayesian way.

Not an easy task and I hope my Bayesian friends (and Bayes) will forgive me for this rough attempt, which is inevitably very rough and simplistic, but has the only objective of showing this different perspective on estimation. Those who are ready for a more rigorous treatment of Bayesian statistics can refer to Gelman et al. (2004).

To put things in context, a good start is a definition of the two competing parties. The Frequentist school (also called the Classical school) owes its popularity in the twentieth century to the work of Ronald Aylmer Fisher (yes, the guy of the F-test in chapters 6 and 7), while there are claims that in the previous century every statistician, including Gauss, was a Bayesian (see Efron, 1986).[3] The methods described in this book are essentially based on the Classical or Frequentist view, which assumes that 'true' and fixed population parameters exist, albeit unknown, and statistics can exploit sampling to estimate them. Furthermore, Frequentists associate probabilities with observations, so that the probability of a given outcome for a random event can be proxied by the frequency of that outcome and – as usual – the larger is the sample, the closer is the estimated probability to the true probability. Thus, a linear regression model tries to estimate the 'true' coefficients which link the explanatory variables and the dependent variable using a sample of observations. A key concept of the Frequentist approach is the confidence interval of chapter 8, where a range of values contains the true and fixed value with a confidence level and the confidence level is nothing else than the frequency with which an interval contains the true and fixed value considering different random samples.

Bayesians look at the problem of estimation from a different perspective.[4] The unknown parameters in the population are not fixed, but treated as random variables with their own probability distribution (see the appendix for a definition of random variable). Besides this, one is allowed to exploit knowledge or beliefs about the shape of the probability distribution which exist *prior* to estimation. Once data are collected, Bayesian methods exploit this information to update this *prior probability distribution* and the final outcome is a *posterior distribution*, which depends on the data and the prior knowledge. The estimation of the posterior distribution opens the way to Bayesian statistical operations and is based on the Bayes rule, which relates the probability of the outcomes of two random events in the following way:

$$P(A|B) = \frac{P(A, B)}{P(B)} = \frac{P(B|A)P(A)}{P(B)}$$

The above equation is easier than it might seem at a first glance, but first one should become familiar with the concept of *conditional probability* as expressed by the vertical bars between brackets. $P(A|B)$ is the probability of the first random event to generate the outcome A when the second random event has generated the outcome B, thus it is the probability of A conditional on B, which means that a different outcome for the second event leads to different probabilities for a given outcome of the first event. This probability can be computed as the ratio between $P(A, B)$ – that is the joint probability that both events A and B happen – and $P(B)$, the *unconditional probability* of the event B. The Bayes theorem shows that the joint probability that the events A and B happen can be also expressed as the product $P(B|A)P(A)$, that is the product between the probability that the event B happens conditional on the outcome A and the probability of the event A.

To understand the use of the Bayes rule, the two random events could be respectively the value of unknown parameter (A), which in Bayesian statistics is determined by a random variable, and the available data (B), which is also the outcome of a random variable since it was obtained through sampling. Thus, if A is a particular value of the parameters and B is a particular sample, the Bayes theorem says that the probability to obtain the parameter estimate A given the observed sample B (the *posterior probability*) can be computed through the above equation as a function of the probability of observing sample B when the parameter estimate is A and the unconditional probabilities of the parameter estimate A. The unconditional probability of the parameter estimate A is the *prior probability*.

The Bayes rule is very helpful when it is easier to estimate $P(B|A)$ than $P(A|B)$. In this case, if one can compute the probability of having the sample B conditional on the unknown parameter A, using some prior information on the probability of the parameter A and the unconditional probability of the sample B it becomes possible to find the probability distribution of the parameter A conditional on the data, which is the final objective of estimation. Now, considering q possible outcomes (estimates) for the parameter A, note that the denominator of the Bayes rule can be rewritten as:

$$P(B) = \sum_{i=1}^{q} P(B|A_i)P(A_i)$$

which means that the unconditional probability of the sample B can be seen as the sum of probabilities of the sample B conditional on all of the possible estimates A_i, weighted by the probability of each estimate A_j.

To take the final step into Bayesian estimation, two elements have to be considered – (1) $P(B|A)$ is the *likelihood function* of A, that is the probability of a given set of observations depending on a set of parameters and its generally known (Frequentists use it in maximum likelihood methods as well); and (2) the denominator of the Bayes rule is a constant and it is generally not necessary to estimate it, so that estimation can be based on the following result:

$$P(A|B) \propto P(B|A)P(A)$$

where the sign which substitutes the equal sign means that the left-hand side is *proportional* to the right-hand side. Now there is everything and we can make an example of Bayesian estimation.

Consider the estimation of a single regression coefficient in a bivariate regression which relates caviar expenditure (c) to income (i), using data from a random sample which generates a set of observations contained in the vectors **c** (for simplicity consider **i** as the observations of a fixed exogenous variable). The equation is, $c = \beta i$. Frequentists start from some assumptions on the probability distribution of the data and the error term, then get point estimates that are the most likely given the observed sample. Building on the randomness of sampling, confidence intervals can be associated to the coefficient estimate.

Bayesians face the same problem in the following way. They start with the assumption that caviar expenditure follows a (say) normal probability distribution (the *prior distribution*) around its mean, which is equal to $\beta \bar{i}$. Second, they assume a given standard deviation for this normal distribution. For example, one may assume that the standard deviation of caviar expenditure is 0.02.

Consider the value $\beta = 0.05$. If the prior distribution holds, we should have that c is normally distributed around 0.05i. Now it becomes necessary to evaluate the probability to get the observed sample c given that $\beta = 0.05$.

This is not too difficult, one may generate c^* by multiplying i by 0.05. Then, considering that c is a random sample from a normal distribution, one can get the likelihood of c^* conditional on $\beta = 0.05$ using the known *likelihood function*. The unconditional (prior) probability that $\beta = 0.05$ is also known, given that we have assumed that the distribution is normal, with a 0.05 mean and a standard deviation of 0.02 (which means that the probability of $\beta = 0.05$ is about 20%). Note that with a computer and given the prior distribution of β, we can easily compute the unconditional probabilities for all possible values of β and the probabilities of all possible values of c^*. Using a slightly different notation of the Bayes rule which defines $L(\beta|c)$ as the likelihood function of the sample c, one may write:

$$P(\beta|c) \propto L(\beta|c)P(\beta)$$

where the left-hand side is the (unknown) posterior probability of β. As mentioned, for any fixed value of β it is possible to compute both the likelihood function and the unconditional probability using the prior. Suppose that for $\beta = 0.05$ the likelihood of observing the collected data-set is 10%. Then, one may compute:

$$P(0.05|c) \propto 0.10 \cdot 0.20 = 0.02$$

The above result does not mean that the probability is 2%, since there is a proportionality relationship. However, repeating the experiment for the whole range of values that β may assume, allows to compute the probability distribution for β conditional on the observed sample (the posterior distribution), which allows to determine the most likely estimate for β. This estimate will be different from 0.05 unless we had an excellent prior. The posterior distribution might also differ from the normal distribution (although not in this case). From the posterior distribution it is possible to compute the *percentiles* (see appendix). Thus a 95% Bayesian confidence interval can be obtained by considering the values of β corresponding to the 2.5th percentile and the 97.5th one from the posterior distribution.

Obviously, the final result depends on the quality of the prior. However, Bayesian statistics have extended the above founding concepts very much and there are many ways to relax the relevance of the prior assumptions and check for their robustness. For example, there are *non-informative priors*, which do not assume particular knowledge on the parameters, basically only that they are uniformly distributed around the maximum range of possible values. One of the reasons for the Bayesian statistics comeback in the twenty-first century is the fact that the Bayes rule can be applied iteratively (that is the prior distribution can be updated) and the progress in automated computing power has led to brilliant results in estimating complex models. For example, modern Bayesian methods exploit the posterior distribution to generate a larger number of draws from which estimates are actually computed.

Rossi and Allenby (2003) have explored the major role that Bayesian methods can play in marketing. They include a long and annotated list of Bayesian studies in the marketing area, including hypothesis testing with scanner data (Allenby, 1990), extensions of conjoint analysis (Andrews et al., 2002), Bayesian multidimensional scaling (DeSarbo et al., 1999), the multinomial probit (McCulloch and Rossi, 1994) and many other Bayesian alternatives to Frequentist multivariate statistics.

Finally, Socrates also said 'The hour of departure has arrived, and we go our ways' (mind you, I am not going to drink any hemlock after I spent so much time writing this textbook).

Summing up

In extreme synthesis, this was a book on statistics for and consumer marketing research. It has presented a multitude of multivariate techniques and in this final chapter it has been acknowledged that more techniques exist and even different philosophies of data processing.

Among the modern evolutions, a special mention is given to data mining techniques, which extend and adapt statistical data analysis to work with large company databases in an automated and integrated way and to the comeback of Bayesian statistics, an approach to statistical analysis which is proving very effective in some marketing research areas.

Notes

1. Just to avoid the risk of confusion in the Wikipedia age, I am referring to the Greek philosopher, not the famous Brazilian football player of the Eighties, although there are rumors that the latter also holds a doctorate in philosophy. By the way, I must confess here that Wikipedia (wikipedia.org) has been a useful support in several instances, although the long list of references which follows this chapter is still the most incredible source of knowledge on the topics covered in this book.
2. Yes, again, this was found on Wikiquotes...
3. The comments and reply on Efron's article in the same issue of *The American Statistician* are amusing and it is interesting to see the 1986 predictions about the proportion of Bayesian statisticians in the twenty-first century.
4. See for example the discussion in Rupp, Dey and Zumbo, 2004.

Fundamentals of Matrix Algebra and Statistics

A.1 Getting to know x and y

One of the challenges of this book is to make complex techniques accessible without the need for advanced knowledge in mathematics and statistics. Still, the use of some elementary definitions may be a valuable help to understanding some of the concepts of statistics. This appendix helps by recalling the basic notions of algebra and statistics and serves as an immediate reference for interpreting the meaning of equations.

A.1.1 Variables and observations

First of all it is important to be familiar with the concept of *variable*, which is a symbol that represents a measurable characteristic or attribute of an object or person. A variable represents a quantity which is not specified explicitly. It is called variable, because one may reasonably expect this measurable characteristic to change across individuals (or over time). Variables are generally indicated by Latin letters, with a preference for those toward the end of the alphabet (x, y, z). Once characteristics are observed, variables assume numeric values and can be referred to as *observations*.

In statistics, *random variables* play a key role. As explained in the next sections, the main strength of statistics is to deal with 'randomness,' as probability laws allow one to make reasonable guesses on the outcomes of random situations. As for other variables, random variables can also assume different values, but each of the values that a random variable can assume depends on a precise probability law.

For example, a variable x could be the price of books in marketing research. There are many books and they are likely to have different prices. If prices are recorded for 100 different books, there are 100 observations. Formally, observations for the x variable can be denoted with x_i where the suffix i refers to a specific observation and ranges from 1 to 100 in this example. For example, $x_{50} = 20$ would mean that the price of the 50th book in our set of observation is £20.

There are no reasons to think that the price of a book is a random variable, but consider a different example. Let y be the daily temperature of central England on 20 May of a given year. One can reasonably think that the temperature on that day in different years oscillates around a mean value, with oscillations driven by random factors (unless disturbing events like global warming are generating a systematic upwards trend). As further discussed in chapter 1 in relation to random errors, factors acting on temperature may be regarded as random if they lead to fluctuations that do not follow a systematic direction. This includes the fact that on 20 May there might be sunshine or rain. Thus, y_j might denote the temperature observed on 20 May of a given year j, with j ranging from 1772 to 2006. Figure A.1 shows the observed temperatures over the 234 years of the sample, with an horizontal line indicating the average value.[1]

Observed temperatures are below or above the average value, with no apparent systematic pattern. The temperature on 20 May in central England can be regarded as the observed outcome of a random effect which causes deviation from the average (expected) value of 11.6°C. These deviations can be very large (for example $y_{1916} = 17.8$ and $y_{1906} = 5.5$), but large deviations are more unlikely than small deviations. More details on random distributions are provided in the section A.2.

A.1.2 Functions

Variables become more interesting when we explore linkages among them through a *function*, that is a math relationship which links some *independent variables* to *dependent variables*. For example, it is reasonable to assume that beer consumption of an individual (y) can be expressed as a function of price (p), available budget (b) and temperature (t):

$$y = f(p, b, t)$$

If the relationship is strictly mathematical, the function is said to be *deterministic*. For example, one could express total expenditure on food (z) of a given household with the following equation:

$$z = \sum_{i=1}^{n} x_i p_i$$

where the household purchases n food items, x_i represents the quantity consumed of the i-th food item and p_i is the price of the same food item. The symbol $\sum_{i=1}^{n}$ is the *summation operator* and corresponds to writing:

$$z = x_1 p_1 + x_2 p_2 + x_3 p_3 + \ldots + x_n p_n$$

However, most of the time relationships are not deterministic. Take the beer consumption example again; while we can be comfortable with the idea that a relationship exists between the amount of beer, its price, the available budget and temperature, it is quite likely that more factors exist to explain consumption choices, for example the price of other drinks or discriminating whether the individual is driving or not. Since it would be impossible to account for all potential factors, so it can be expected to measure this relationship with some degree of error. If one indicates the error with ε, the function becomes:

$$y = f(p, b, t) + \varepsilon$$

As shown for regression analysis (see section A.2 and chapter 8), even without knowing all of the factors determining beer consumption, it would be possible to learn something about the relationship with the help of statistics, especially when the characteristics of the error term comply with a few statistical requirements. More precisely, if the error term is random, we can define the relationship as a *stochastic function*.

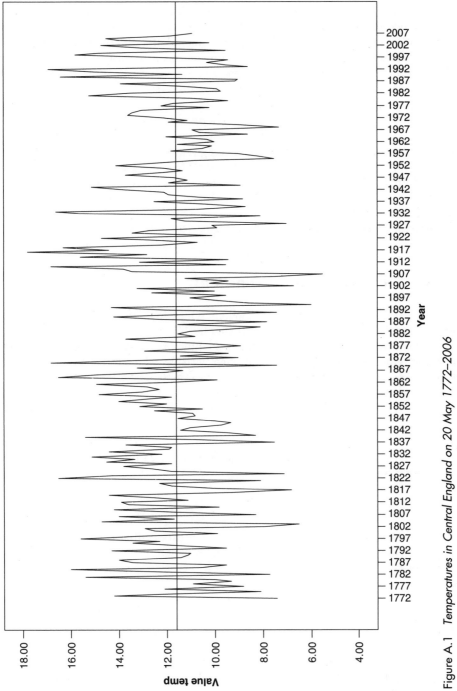

Figure A.1 Temperatures in Central England on 20 May 1772–2006

A.1.3 Parameters and regression

Obviously what matters is the quantification of the relationship in a function. It could be useful to know how much beer consumption would decrease with a 10% price increase or how much consumption would react to a temperature increase of 1°C. In other words, it becomes relevant to know the *parameters* which define the function and quantify constant characteristics. Parameters are conventionally indicated by Greek letters. Thus, assuming that beer consumption has a linear relationship with the explanatory variables, the consumption function could be written as:

$$y = f(p, b, t) + \varepsilon = \alpha + \beta p + \gamma b + \delta t + \varepsilon$$

The objective of statistical estimation is to assign numbers to these unknown parameters, at least with some degree of approximation. More specifically, the above equation can be configured as a *linear regression model*, discussed in chapter 8. Note that the term *parameter* does not only refer to the coefficients of a regression model, but it is also employed to refer to the statistical properties of a random variable or distribution, for example the mean or the variance. This is made clearer in section A.2.

A.1.4 Some special math functions

There are a few widely used math functions which help sorting out problems with non-linear relationships (see for example chapter 16).

The exponential function of a number is the function which allows one to express a variable as the non-linear function of another variable, which is the exponent (power) of a number, where the number is the *base*. For example, one might write:

$$y = 10^x$$

so that $y = 10$ for $x = 1$, $y = 100$ for $x = 2$, and so on.[2] The equation above is an exponential function with base 10 and the same function can be written as $y = \exp_{10}(x)$, or more generally:

$$y = \exp_b(x)$$

where b is the base. The inverse function of the exponential is the *logarithm*. Thus, if $y = 10^x$, one can write x as the logarithmic function of y as $x = \log_{10}(y)$, or more generally as:

$$x = \log_b(y)$$

As before, logarithms can assume various bases (a different b) and the above example refers to a logarithm with base 10. There is one base which is particularly used in mathematics, to an extent that it is called the *natural base*. Its value is a quite peculiar one: 2.718 (actually with many more decimals). Exponential functions with the natural base are written as:

$$y = e^x$$

while the inverse logarithmic function, the *natural logarithm*, is written as

$$x = \ln y$$

Among the many mathematical properties of logarithms, there is one which is especially helpful in statistics because it allows one to transform multiplicative relationships into linear (additive) ones:

$$\log_b xy = \log_b x + \log_b y$$

A.1.5 Vectors

So far variables have been treated in their individuality. However, in many circumstances – and especially when complex and multi-faceted phenomena like consumer behavior are investigated – it is quite convenient to consider several elements simultaneously. In particular, it might be convenient to refer to a set of variables. We might combine the price, budget and temperature variable in a vector $\mathbf{x}$ (usually vectors of variables are denoted with small bold Latin letters) as follows:

$$\mathbf{x} = \begin{bmatrix} p \\ b \\ t \end{bmatrix}$$

Once the variables are observed, the object becomes a vector of *scalars* (numbers). For example:

$$\mathbf{x} = \begin{bmatrix} 1.5 \\ 15 \\ 22 \end{bmatrix}$$

refers to a consumer facing a beer price of £1.5, a budget constraint of £15 and a temperature of 22°C. Similarly to vectors of variables, it is possible to define vectors of parameters (in which case a Greek bold letter should be used):

$$\boldsymbol{\theta} = \begin{bmatrix} \alpha \\ \beta \\ \gamma \\ \delta \end{bmatrix}$$

With a slight modification of the vector of explanatory variables $\mathbf{x}$ (an additional unity element to account for the intercept term), it becomes possible to rewrite the beer consumption equation using the vector notation:

$$\mathbf{x} = \begin{bmatrix} 1 \\ p \\ b \\ t \end{bmatrix}$$

This makes it possible to write the beer equation as:

$$y = f(p, b, t) + \varepsilon = \alpha + \beta p + \gamma b + \delta t + \varepsilon = \boldsymbol{\theta}'\mathbf{x}$$

where the final term $\boldsymbol{\theta}'\mathbf{x}$ on the right-hand side of the equation is the product of two vectors (discussed later in this appendix).

Conventionally, vectors are structured as a single column. However, it is sometimes useful to have their elements arraned in a row. The operation which transforms a *column vector* into a *row vector* is called *transposition* and is commonly denoted by an elongated apostrophe (or a T superscript) after the vector's symbol. For example

$$x' = \begin{bmatrix} 1 & p & b & t \end{bmatrix}$$

is the transpose of **x**.

A.1.6 Matrices

With multivariate problems it is quite useful to have objects with two dimensions. Consider the key starting point of any statistical analysis, the data-set. For example, a consumer data-set could be organized with a bi-dimensional structure, where each row represents a consumer and each column a variable. As a generalization, imagine a data-set with 100 individuals, which records the price faced by each of these individuals, their individual budget and the temperature they are subject to. Thus, there is a vector of variables for each of the observations. This data-set can be conveniently arranged into a *matrix* of variables, conventionally denoted by Latin capital letters:

$$\mathbf{W} = \begin{bmatrix} 1 & p_1 & b_1 & t_1 \\ 1 & p_2 & b_2 & t_2 \\ 1 & p_3 & b_3 & t_3 \\ \dots & \dots & \dots & \dots \\ 1 & p_i & b_i & t_i \\ \dots & \dots & \dots & \dots \\ 1 & p_{100} & b_{100} & t_{100} \end{bmatrix}$$

The matrix **W** has 100 rows (one for each observation) and four columns (one for the intercept, plus one for each variable). When a matrix is transposed, its rows become its columns (thus **W'** has 4 rows and 100 columns). Matrices are said to be *square matrices* when the number of rows equals the number of columns, while square matrices are *symmetric* when the transpose operation does not change the matrix. If 100 observations on beer consumption (one per individual) are available, these can be arranged into a column vector with 100 elements:

$$\mathbf{y} = \begin{bmatrix} y_1 \\ y_2 \\ y_3 \\ \dots \\ y_i \\ \dots \\ y_{100} \end{bmatrix}$$

One can generalize the beer consumption relationship for all of the observations using a matrix notation:

$$\mathbf{y} = \mathbf{W}\boldsymbol{\theta} + \boldsymbol{\varepsilon}$$

where $\boldsymbol{\varepsilon}$ is a vector of 100 error terms and $\boldsymbol{\theta}$ is the previous parameter vector.

A.1.7 Basic operations on vectors and matrices

Operations on vectors and matrices can be executed, provided that their dimensions conform to some rule. The *dimensions of vectors and matrices* are indicated by the notation $r \times c$ where r is the number of row elements and c is the number of column elements. Thus, **y** and **ε** are a 100×1 vector, **W** is a 100×4 matrix and **θ** is a 4×1 vector.

Below some basic operations on vectors and matrices are illustrated. These should suffice to understand the equations in this book, but for more rigorous notions of algebra one is strongly recommended to refer to a good math textbook.

A.1.8 Summation

When matrices or vectors have the same dimensions, their elements can be summated one by one (*elementwise summation*). Consider two 4×1 vectors, **a** and **b**:

$$
\mathbf{a} = \begin{bmatrix} 5 \\ 2 \\ 4 \\ 3 \end{bmatrix}, \ \mathbf{b} = \begin{bmatrix} 3 \\ 3 \\ 6 \\ 2 \end{bmatrix}, \ \mathbf{a} + \mathbf{b} = \begin{bmatrix} 5+3 \\ 2+3 \\ 4+6 \\ 3+2 \end{bmatrix} = \begin{bmatrix} 8 \\ 5 \\ 10 \\ 5 \end{bmatrix}
$$

or two 2×2 matrices, **A** and **B**:

$$
\mathbf{A} = \begin{bmatrix} 2 & 4 \\ 5 & 7 \end{bmatrix}, \ \mathbf{B} = \begin{bmatrix} 5 & 2 \\ 1 & 3 \end{bmatrix}, \ \mathbf{A} + \mathbf{B} = \begin{bmatrix} 2+5 & 4+2 \\ 5+1 & 7+3 \end{bmatrix} = \begin{bmatrix} 7 & 6 \\ 6 & 10 \end{bmatrix}
$$

A.1.9 Multiplication

Multiplication of two vectors, a vector and matrix or two matrices is less straightforward. First, contrary to scalar multiplications, the order of object matters. The key condition is that the number of columns of the first object needs to be equal to the number of rows of the second object. The resulting object has the number of rows of the first object and the number of columns of the second object.

Consider vector multiplication first. If a row vector is multiplied by a column vector (provided they have the same number of elements), then the result is necessarily scalar. If a column vector is multiplied by a row vector, then the result is necessarily a matrix. This follows from the definition of vector multiplication. Consider the first case, where a 1×3 row vector $\mathbf{a}'$ is multiplied by a 3×1 vector **b** (remember the apostrophe/transposition sign which transforms standard column vectors into row vectors):

$$
\mathbf{a}' = \begin{bmatrix} a_1 & a_2 & a_3 \end{bmatrix}, \ \mathbf{b} = \begin{bmatrix} b_1 \\ b_2 \\ b_3 \end{bmatrix}, \ \mathbf{a}'\mathbf{b} = [a_1 b_1 + a_2 b_2 + a_3 b_3]
$$

Each element of the row vector is multiplied by the corresponding element in the column vector, and the terms obtained through these multiplications are summated. Using the summation operator, vector multiplication can be written as $\mathbf{a}'\mathbf{b} = \sum_{i=1}^{n} a_i b_i$

where n is the size of the vectors and a_i and b_i denote the i-th element of the vectors **a** and **b**, respectively.

Instead, consider the product **ba'**. The product of a column vector with r elements with a column vector with c columns results in a matrix having r rows and c columns. The elements of the resulting matrix are computed as follows: the value in the i-th row and j-th column of the resulting matrix is obtained by multiplying the i-th element of **b** and the j-th element of **a'**, for example:

$$\mathbf{b} = \begin{bmatrix} b_1 \\ b_2 \\ b_3 \end{bmatrix}, \ \mathbf{a'} = \begin{bmatrix} a_1 & a_2 & a_3 \end{bmatrix}, \ \mathbf{ba'} = \begin{bmatrix} b_1a_1 & b_1a_2 & b_1a_3 \\ b_2a_1 & b_2a_2 & b_2a_3 \\ b_3a_1 & b_3a_2 & b_3a_3 \end{bmatrix}$$

The above procedures for multiplication are easily generalized to the case of a matrix and a vector or two matrices. Examples are

$$\mathbf{A} = \begin{bmatrix} a_{11} & a_{12} & a_{13} \\ a_{21} & a_{22} & a_{23} \end{bmatrix}, \ \mathbf{B} = \begin{bmatrix} b_{11} & b_{12} \\ b_{21} & b_{22} \end{bmatrix}, \ \mathbf{c} = \begin{bmatrix} c_1 \\ c_2 \\ c_3 \end{bmatrix}$$

$$\mathbf{Ac} = \begin{bmatrix} a_{11}c_1 + a_{12}c_2 + a_{13}c_3 \\ a_{21}c_1 + a_{22}c_2 + a_{23}c_3 \end{bmatrix}$$

where the matrix resulting from multiplication of the 2×3 matrix **A** with the 3×1 vector **c** is a 2×1 vector where the first element is the product of the first row of **A** with the vector **c** and the second element is the product of the second row of A with the vector **c**.

Instead, the product **BA** results from the multiplication of a 2×2 matrix (**B**) with a 2×3 matrix (**A**). Again, the resulting matrix is a 2×3 matrix with each (i, j) element obtained by multiplying the i-th row of **B** and the j-th column of **A**:

$$\mathbf{BA} = \begin{bmatrix} b_{11}a_{11} + b_{12}a_{12} & b_{11}a_{12} + b_{12}a_{22} & b_{11}a_{13} + b_{12}a_{23} \\ b_{21}a_{11} + b_{22}a_{21} & b_{21}a_{12} + b_{22}a_{22} & b_{21}a_{13} + b_{22}a_{23} \end{bmatrix}$$

A.1.10 Other algebraic notions

Throughout this book there are equations with other slightly more complicated matrix operators. For those not familiar with the joys of matrix algebra, it may suffice to give a brief (and possibly rough) interpretation of their meaning.

An *identity matrix* **I** is a square matrix where all elements on the main diagonal are 1 and any other elements is zero. Multiplying a matrix by **I** returns the same matrix (similarly to multiplication of a number by 1).

The *inverse* of a matrix **A** is denoted with $\mathbf{A}^{-1}$ and the multiplication of a matrix by its inverse generates the identity matrix **I**:

$$\mathbf{AA}^{-1} = \mathbf{A}^{-1}\mathbf{A} = \mathbf{I}$$

Note that not all matrices are invertible, only those that are termed as *non-singular*. A matrix is non-singular when its *determinant* is different from zero. The *determinant* of a square matrix is a unique scalar value associated with that matrix and can be computed from the matrix elements. It has a series of desirable properties that are quite useful in matrix algebra. For example, the determinant of a matrix obtained as the product of two matrices is the product of the respective determinants.

The *eigenvalues* of a square matrix are particular relevant to statistics and are often cited in this book. To understand their meaning, consider a given $n \times n$ square matrix **A** and an $n \times 1$ vector **x** of unknown terms. If it is possible to find a scalar λ so that the following relationship holds:

$$\mathbf{Ax} = \lambda \mathbf{x}$$

Then λ is an eigenvalue of **A**. Every $n \times n$ square matrix has at most n eigenvalues.

The product of all eigenvalues is the determinant of the matrix. It follows that eigenvalues of non-singular matrices cannot be zero.

Some methods discussed in this book (especially those described in chapter 10) refer to orthogonality. Two vectors are orthogonal when their product returns a vector of 0.

A matrix is orthogonal when $\mathbf{A}'\mathbf{A} = \mathbf{I}$ (which means that $\mathbf{A}' = \mathbf{A}^{-1}$) and the determinant of an orthogonal matrix is either -1 or 1). An *orthogonal transformation* of a vector **b** is obtained by pre-multiplying the vector by an orthogonal matrix.

A.2 First steps into statistical grounds

After familiarizing with the key notation and operations in matrix algebra, it is helpful to briefly review the basic elements of statistics, with concepts that are exploited (and explained) throughout the book.

A.2.1 Frequencies and distributions

The absolute frequency of a data value is the number of times it appears within the given sample. Consider a data-set containing 500 responses to the question: 'How many days of vacation have you had last year?'

If 157 respondents have answered '28 days,' then the *absolute frequency* for that category is 157. It is usually helpful to think in terms of *relative frequency*, where each category is represented as a proportion of the sampled units, so that '28 days of vacation' represent 31.4% of the sample (157 divided by the sample size of 500). For quantitative variables, it is possible to compute frequencies for sets of values (classes). For example, if days of holidays are measured, a table of frequencies can be built on categories obtained after grouping individual days into classes, as in table A.1.

A *frequency distribution* is the complete set of values of the data-set, possibly grouped into classes, together with their absolute or relative frequencies. The *cumulative frequency* for a given value v of the variable being analyzed is the total number of observations that are less or equal to v.

If there are the conditions to generalize the observed frequencies into a distribution of probabilities (see chapters 5 and 6), then it becomes possible to refer to a *probability distribution*. Considering the above example, if our 500 respondents are representative of all workers in Europe, it becomes possible to associate each class of holidays to a probability level, so that 'a worker who gets a job in Europe has a 39.4% probability of having a number of days of vacations between 22 and 28.'

Take the example of the previous section, where temperatures on 20 May have been measured every year from 1772 to 2006. If temperatures oscillate randomly around

Table A.1 *Holidays, absolute and relative frequencies*

Days of holidays	Absolute frequency	Relative frequency	Cumulative frequency
Less than 7	18	0.036	0.036
7–14	49	0.098	0.134
15–21	109	0.218	0.352
22–28	197	0.394	0.746
29–35	92	0.184	0.930
More than 35	35	0.070	1.000
Total	500	1.000	

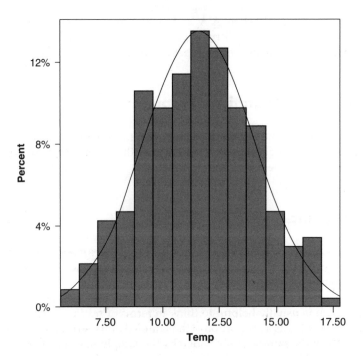

Figure A.2 *Frequency distribution of temperatures between 1772 and 2006*

their mean, then one should obtain something similar to a normal distribution. The histogram below (where observations are grouped into equal-width classes) shows that this is indeed the case in figure A.2.

A.2.2 *Statistical distributions and inference*

The aspects of measurement errors (see chapter 1) and probability distributions open the way to the concept of *statistical* or *probability distributions*. The Gaussian or normal distribution of chapter 1 is the key distribution for randomness of a metric variable, but it is not the only one. There are many other distributions to deal with other forms of randomness. For example, when an experiment with two outcomes is repeated several times (flipping a coin 30 times and counting heads and tails), the *binomial distribution* is the preferred distribution. Instead, the *uniform distribution* describes a set of events which have exactly the same probability of occurring. For example, when throwing a dice, one would expect that each outcome has the same (1/6) probability. Probability distributions cover a wide range of potential outcomes and one should refer to a statistics textbook to learn more about the variety of probabilistic situations, which include curious ones. For example, the *Poisson distribution* for very rare events in large populations (sometimes called 'the law of small numbers'), named after the mathematician Siméon-Denis Poisson, gained recognition with the application to the yearly number of deaths by horse-kicks in the Prussian cavalry.[3]

The advantages of associating a random event of a particular type to a known theoretical probability distribution are multi-fold. First of all, this allows us to get an easy and reliable estimate of the *parameters* of a distribution. Each theoretical distribution is characterized by a set of parameters which determine its shape, like the mean value, the median value or the variance, statistical concepts briefly discussed over the next pages. Another crucial application of statistical distributions is central to *sampling theory* (see chapter 5) or *hypothesis testing* (see chapter 6). In rough terms, when we have the guarantee that the observed (measured) variables are the outcome of some random data generating process (as the one determined by probabilistic sampling) it becomes possible to draw statistical conclusions (that is with some margin of uncertainty) on the 'true' variables.

This allows us, for example, to make an estimate of the mean expenditure for chips of the whole US population, despite measuring chips expenditure on 5,000 US consumers, provided the selection of the consumers is based on randomness. Hypothesis testing exploits probability distributions to test the likelihood of a given event. Consider the data on temperature in central England again. If one can safely assume that temperature on 20 May follows a normal distribution with a given mean (11.6°C in this example) and variability, since the characteristics of the normal distribution are known, it becomes possible to associate each possible temperature with a probability level. The most likely temperature is the average one (11.6°C), while probability levels decrease symmetrically as temperature departs from the mean (less or more). Thus, a temperature of 18°C is very unlikely, but not impossible.

The 'statistical way of thinking' is as follows. If some events occur which are very unlikely since they are well below a *significance level* (for example an event that has less than 1% of probability of occurring), while one cannot fully exclude randomness it is advisable to check for non-random factors that may have provoked that event. For example, between 1997 and 2007, temperature on 20 May has been above the average in 8 years out of 11. Does this mean that there is some factor (global warming) that raises temperatures? Without going into technical details, since the initial assumption is that the temperature is randomly distributed around the mean value, the probability of having 8 values above the average in 11 years is about 8%. Thus, observing 8 temperatures above the average is still quite possible under the randomness condition, albeit less likely than having 7 years above the average (about 16%). The statistician proceeds by setting an arbitrary but fixed probability threshold (the *significance level*), which is used to discriminate between random events and events that are unlikely to be random. In social science and marketing a conservative significance level is 1%, while a less conservative one is 5%. In this example, even using the less conservative threshold, the hypothesis that the eight observations above the average are due to randomness cannot be rejected.

A.2.3 Averaging and central tendency measures

Averaging is probably the most known statistical activity. We can now exploit the mathematical notation developed in this appendix to define how a mean can be computed:

$$\overline{x} = \frac{\sum\limits_{i=1}^{n} x_i}{n}$$

The above equation considers a situation where n observations of the variable x, that is $x_1, x_2, ..., x_n$ are taken. To compute the mean, one must sum all of the observations and divide by n. Conventionally, when the mean is denoted by a barred Latin letter, it refers to a mean computed (estimated) on a sample of observations. However, what really matters is the 'true' mean of the variable (which does not necessarily correspond to the estimated mean, as it is explained in chapters 5 and 6). Suppose we are in a situation where we can measure all of the outcomes of the variable x. For example, we have the age of all individuals living in New York from the population registry, which makes it possible to compute the true mean of the population. In that case, the notational convention requires the use of a Greek letter (generally μ), but the equation does not change.

As mentioned, statistical distributions can be characterized by a small number of parameters. One of them is certainly the mean. With a discrete frequency distribution, the mean can be computed as follows. If $f_1, f_2, ..., f_k$ indicate the (relative or absolute) frequencies for the values $z_1, z_2, ..., z_k$, the mean value is given by:

$$\overline{z} = \frac{\sum\limits_{i=1}^{k} z_i f_i}{\sum\limits_{i=1}^{k} f_i}$$

where the variable z can assume k different values and each of them is associated with an absolute or relative frequency, or, for random variables, to a probability level. When relative frequencies or probabilities are used, the denominator of the above equation disappears, since $\sum\limits_{i=1}^{k} f_i = 1$.

Take the example of a dice; the possible outcomes are 1, 2, 3, 4, 5 and 6, each of them with a probability of 1/6. The average value is 3.5. Sometimes, when working with probability distributions and means, averaging produces the so-called *expected value*. It is 'expected' in the sense that it represents the mean value which one expects to compute as the number of observations increases. Thus, if one throws the dice only a few times, it is possible that the computed mean differs from 3.5, but as one increases the number of independent observations (or realizations of the random process) the computed mean should get closer to the expected value.

There are other statistics that are quite useful to learn about the characteristics of a variable. Together with the mean, they are sometime called *central tendency measures*, because they refer to the tendency of quantitative data to concentrate around some central value within the range of possible values.

The *median* is the value which splits the sample of observations in two equally-sized halves when the data are sorted in ascending or descending order. Considering the relative cumulative frequency described earlier in this chapter, the median is the value of the variable corresponding to a cumulative relative frequency of 50%. If the sample size is an odd number, then the median value will correspond to one of the data values in the sample. If the sample size is an even number, the median is taken as the midpoint between the two values splitting the sample in two halves. The median value can be also computed for qualitative variables, if they are measured on an ordinal scale.

For example, the median value of the temperatures observed in central England on 20 May over the years between 1772 and 2006 is 11.7°C, which means that in 50% of the years in our data-set the temperature was less or equal to 11.7 degrees. When data are grouped into classes as in table A.1, the median can be computed by identifying the median class first, then choose the appropriate value within the range of the median class. The cumulative frequency of the distribution reaches 50% for some value within the class 22–28. The class starts from a cumulative frequency of 0.352 and ends with a cumulative frequency of 0.746. Assuming that values are uniformly distributed within a class, one could estimate the relative position of the value 0.5 within the class as $(0.5 - 0.352)/(0.746 - 0.352) = 0.376$.

Since the class goes from 22 to 28, still assuming uniform distribution within the class the median value is computed as $22 + [0.376 \cdot (28 - 22)] = 24.25$, that is the left end of the class plus the proportion of values within the class that are estimated to be below the 0.5 threshold.

A third measure of central tendency is the *mode,* also known as the 'most frequent value,' which is the value with the highest relative frequency. Note that the mode refers to a data value and not to its frequency, which is called the 'modal frequency.' The modal value can be extracted from both qualitative and quantitative data, although it is generally difficult to identify a meaningful mode for metric variables without classifying them into classes.

For example, the data on temperatures in central England are grouped into 15 equally-sized classes. The modal class is the one with the highest peak in the histogram, which corresponds to the class ranging between 11.24°C and 12.06°C, with a modal frequency of 13.6%. Without grouping the temperature values into classes, the modal value is 11.8, with a very low modal frequency, since only 3.6% of cases (years) register exactly that temperature. Note that observed frequency distributions can have more than one modal class or value.

The median and the mode can also be regarded as parameters of probability distributions. For example, the normal distribution of a continuous variable has the characteristic of showing the same value for the mean, the median and the mode. Instead, the uniform distribution, where all of the values have the same probability, has no modal value (or all values are modal, as one prefers to look at it).

Another useful statistic which provides information on the distribution of values is based on *quartiles* (and *percentiles*). Quartiles are the three values splitting the (ordered) data-set into four parts, each with 25% of the total number of observations. Considering the cumulative frequency, the lower (first) and higher (third) quartiles are the values corresponding to a cumulative frequency of 25% and 75%, respectively. The lower quartile can be viewed (and computed) as the median of the lower half of the data and the higher quartile as median of the higher half. The second is the median itself. The difference between the third and first quartile represents the *interquartile range* and is a measure of dispersion. A generalization of the quartiles is provided by the concept of *percentiles*, where the r-th percentile is the value of the variable corresponding to a cumulative relative frequency of $r\%$.

For example, the lower quartile of the temperature distribution is 9.9°C, the higher quartile is 13.35°C and the 95th percentile is 15.6°C.

A.2.4 Dispersion and variability

Knowing the central tendency of a variable is generally not enough to draw meaningful conclusions. For example, suppose you need to decide what clothes to take with you in a trip which will keep you abroad for three months. The information that the mean temperature of that place in the target period is 20°C is not very helpful. Whether the temperature oscillates between 19 and 21°C or between 0 and 40°C makes a big difference on clothing. That is why central tendency measures are generally accompanied by measures of variability (or dispersion) which tell how the data spread around a central tendency value (usually the mean). As before, variability measures can be applied to both empirically observed data or to theoretical probability distributions. The variability of a normal distribution indicates the shape of the curve. The larger the variability, the flatter is the normal curve. The study of variability is fundamental in statistics, because the main purpose of statistical analysis is to explain the reasons why individual observations are different, that is the factors leading to variability.

The most known measure of variability is the *variance*, which is a (quadratic) measure of the variability of the observed data around their mean. The variance of a variable over a complete data-set is computed as:

$$\sigma^2 = \frac{\sum_{i=1}^{n}(x_i - \bar{x})^2}{n}$$

As shown in chapter 5, when the mean computed on a sample is used instead of the population mean to compute the sample variance, the denominator should be replaced by $n-1$.

The *standard deviation* is the square root of the variance and provides a more intuitive interpretation since the measurement scale is the same used for the observed values and the mean. For example, the average temperature of central England is 11.61°C, with a standard deviation of 2.42, which can be interpreted as an average measure of dispersion in °C.

Sometimes it could be useful to eliminate the influence of measurement units and/or variability to compare very different variables. Suppose a data-set contains information on income (measured in thousands of dollars) and on age (measured in years). Obviously the former variable will have a much larger mean and variance than the latter, just because of the measurement unit.

If one simply creates two additional variables by dividing all of the original values by the variable mean, then the influence of the measurement unit is eliminated, since both the resulting variables will have a mean equal to one. This opens the way to compare the variances of two variables with different measurement units, since the two rescaled varaibales maintain their variability. In other cases it could be useful to eliminate the differences in variability as well and create variables with the same mean and variance. This might be useful to compare the relative position of an individual observation with reference to the mean properties of the data-set. The process which creates variables with zero mean and variance equal to 1 is called *standardization* and a *standardized variable* can be obtained from a given variable by subtracting its mean from each observation and dividing for its standard deviation as follows:

$$\hat{x}_i = \frac{x_i - \mu}{\sigma}$$

Finally, the *coefficient of variation* is a synthetic measure of the degree of variability in a given variable. It is computed as the ratio between the standard deviation and the mean, it is independent from the measurement unit and can be expressed as a percentage. In our case, the coefficient of variation of temperature on 20 May in central England is 0.208 or 20.8% (that is 2.42/11.61).

A.2.5 Some generalizations to the multivariate case: covariance and correlation

In every chapter of this book dealing with multivariate statistical methods (that is dealing with several variables) there is some reference to *covariance* and *correlation* matrices.

While a more exhaustive discussion is left for chapter 8, here it may be useful to recall some of the matrix algebra and statistical concepts to 'test' their understanding and define covariance and its matrix.

Covariance is a measure looking at the joint variability of two variables. In rough terms it provides an indication on how two variables move together, on average. For example, one may reasonably think that height and weight of individuals tend to move in the same direction, as in general taller (shorter) individuals weigh more (less). If we indicate with x the height variable and with y the weight variable, one could compute, for each observation, the following quantity:

$$(x_i - \bar{x})(y_i - \bar{y})$$

which is nothing else than the product of distances from the mean. This product is negative for any observation which is above the average for one variable and below for the other, positive when both variables are above or below the average. The larger the distances from the means, the larger the product. If there is no link between the two variables, considering all of the observations, one would expect negative values to compensate positive values. Instead, if the observations 'move together,' that is observations with values of x above the mean tend also to have values of y above the mean and the other way round, the output of the above multiplication across the observations will be a large number of positive values. Similarly, if in most cases where x is above the average y is below the average, then most of the products will be negative. Thus, by averaging the value of the above product across all of the observations, one gets a good indication of the average co-movement; if it's close to zero, then the variables are weakly related, if it is positive and large, then the variables move together, if it is negative and large, then the variables move in opposite directions. Thus, one can compute the covariance as follows:

$$Cov(x, y) = \frac{\sum\limits_{i=1}^{n} (x_i - \bar{x})(y_i - \bar{y})}{n}$$

Obviously, the value of covariance depends on the measurement units and the degree of variability within each variable. Thus, one could think of standardizing the variables as described earlier in this appendix. Covariance between standardized variables is called correlation; it is independent from the measurement unit and ranges between −1 (perfect negative correlation) and +1 (perfect positive correlation). There are other ways to look at correlations. These are discussed in chapter 8.

Now consider the case of more than two variables. This could be a good excuse to make some use of the matrix algebra concepts introduced earlier in this appendix. Let **x** be a vector of variables and $\bar{\mathbf{x}}$ the vector of scalars containing the means of the variables in **x**. In a three variable case (y, z and w) one can write:

$$\mathbf{x} = \begin{bmatrix} y \\ z \\ w \end{bmatrix}, \bar{\mathbf{x}} = \begin{bmatrix} \bar{y} \\ \bar{z} \\ \bar{w} \end{bmatrix}, (\mathbf{x} - \bar{\mathbf{x}}) = \begin{bmatrix} y - \bar{y} \\ z - \bar{z} \\ w - \bar{w} \end{bmatrix}, (\mathbf{x} - \bar{\mathbf{x}})' = \begin{bmatrix} y - \bar{y} & z - \bar{z} & w - \bar{w} \end{bmatrix}$$

Now, with some basic vector multiplication, one can obtain the *covariance matrix* as:

$$\Sigma = \frac{1}{n}(\mathbf{x} - \bar{\mathbf{x}})(\mathbf{x} - \bar{\mathbf{x}})' = \frac{1}{n}\begin{bmatrix} y - \bar{y} \\ z - \bar{z} \\ w - \bar{w} \end{bmatrix} \begin{bmatrix} y - \bar{y} & z - \bar{z} & w - \bar{w} \end{bmatrix}$$

We leave to the reader as an exercise to show that:

(1) The matrix Σ is a 3×3 symmetric matrix.
(2) The elements on the diagonal of Σ are the variances of y, z and w.
(3) The elements off the diagonal are the covariances between the original variable.

Notes

1. Data available from the Met Office Hadley Centre, hadobs.metoffice.com/hadcet/data/download.html
2. Remember that any number with power 0 returns one, so $10^0 = 1$, $5^0 = 1$, etc.
3. The example was used by Bortkiewicz (1898) in his treaty 'Das Gesetz der kleinen Zahlen' (which means 'The Law of Small Numbers'), Leipzig: Teubner.

Glossary

Accuracy: the degree to which the empirical measurement reflects the true value of the object being measured, not to be confused with *precision* or *validity*.

Adjusted *R*-Square: see *R*-Square.

Administration method: the method for interviewing people in a survey (see chapter 3).

Alternative hypothesis: the hypothesis which works as an alternative when the null hypothesis is rejected. See *hypothesis testing*.

Analysis of covariance (ANCOVA): *analysis of variance* where some of the factors are metric variables and others are categorical (see chapter 7). See also *general linear model*.

Analysis of variance (ANOVA): a multivariate statistical method (see chapter 7) which allows one to test whether one or more explanatory *factors* which appear in different levels (*treatments*) induce significant variations in the target variable.

Association: in general, it is the strength of the relationship between two variables (without assuming any causation). More specifically, the association between *metric* variables is measured through *correlation*, while *associations measures* commonly refer to the relationship between categorical variables (see chapter 9).

Binary variable (dummy): see *dummy variable*.

Burt table: a tabulation approach which allows one to show the frequencies of more than two categorical variables in the same contingency table. Especially useful for *correspondence analysis* and *log-linear analysis*.

Canonical correlation: the bivariate correlation between two sets of variables. See *canonical correlation analysis*.

Canonical correlation analysis: a statistical model which explores the correlation between two sets of metrics and non-metric variables (see chapter 9).

CAPI: Computer-assisted personal interview, a survey administration method which associates personal interviews with the use of computer software (see chapter 3).

Categorical variable: a.k.a. qualitative or non-metric variable. See variable.

CATI: Computer-assisted telephone interview, a survey administration method which associates telephone interviews with the use of computer software (see chapter 3).

CAWI: Computer-assisted web interview, a survey administration method which associates electronic interviews with the use of computer software (see chapter 3).

Census: a survey to collect data on all units in a population.

Chi-square test: a statistical test based on the chi-square distribution. It is commonly used to compare an empirical (observed) frequency distribution with a theoretical (probability) distribution (see chapter 6) or to test the association between two categorical variables (see chapter 9). See *hypothesis testing*. In a *structural equation model* it is used to test the difference between the estimated covariance matrix and the observed one.

Cluster analysis: a classification technique which allows one to group observations into clusters, where observations within the same cluster are as similar as possible according to a given set of variable, while observations in different cluster are as different as possible (see chapter 12).

Coefficient of determination: see *R*-Square.

Collinearity: the correlation between the explanatory variables in a statistical model, which can induce problems in estimation (see chapter 8).

Communality: in *factor analysis* and *principal component analysis*, the proportion of the original variability of a variable which is maintained by the extracted factor or principal components.

Compositional methods: methods for analyzing consumer evaluations of products or brands starting from a set of attributes. In *multidimensional scaling*, compositional MDS methods generate *perceptual maps* which allow a better interpretation of the dimensions and the relative positioning of objects and subjects in the map.

Confidence interval: a range of values estimated on a sample that is expected to contain the true (population) parameter with a probability equal to the *confidence level* (see chapter 6).

Confidence level: in *hypothesis testing* (see chapter 6), the degree of confidence associated with non-rejection of a null hypothesis. It is arbitrarily set by the researcher (usually at 95% or 99%) and represents the probability that the non-rejected hypothesis is actually true. It is complementary to the *significance level*. See *type I error*.

Confirmatory factor analysis: see *factor analysis*. See also *structural equation model*.

Conjoint analysis: a *decompositional method* to explain consumer choices on the basis of the preferences for alternative combinations of attributes (see chapter 16). Through a *factorial design*, conjoint analysis drives the data collection process and the collected responses are generally analyzed through a *multinomial logit model* or other *discrete choice models*.

Contingency table: a table summarizing the joint frequencies for two or more variables (see chapter 9).

Contrasts: within an *analysis of variance* design, they allow one to test hypotheses on specific sub-sets of the treatments (see chapter 7). They open the way to explore further the sources of variability when the null hypothesis of mean equality is rejected. The comparisons are usually based on a theory and planned before the analysis, thus they are also called *planned comparisons*. See also *post-hoc tests*.

Correlation: the strength of the relationship between two *metric* variables (see chapter 8). See also *association*.

Correspondence analysis: multivariate statistical technique which allows one to look into the *association* of two or more categorical variables and display them jointly on a bivariate graph (see chapter 14). It can be used to apply *multidimensional scaling* to categorical variable.

Covariates: although this is not a rigorous statistical definition, SPSS terms as covariates the metric explanatory variables in a *general linear model*. See also *analysis of variance*.

Critical value: in *hypothesis testing*, the values of a theoretical probability distribution which set the borders between rejection and non-rejection of the null hypothesis (see chapter 6).

Cronbach's Alpha: a measure of *reliability* (see box 1.2).

Decompositional methods: methods for analyzing consumer evaluations of products or brands starting from a ranking of the objects in their integrity. In *multidimensional scaling*, decompositional MDS methods generate *preference maps*.

Degrees of freedom: they count the number of 'independent' observations available for estimation. Basically, it is a measure of the discrepancy between the available number of observations and the constraints associated with the estimation of unknown parameters. Every estimator is associated with a given number of degrees of freedom. For example, a sample variance estimator uses all of the observations, but one parameter needs to be estimated (the mean), so there are $n - 1$ degrees of freedom, that is $n - 1$ pieces of information available for estimation. In *structural equation models* the degrees of freedom indicate if a model is under-identified (negative degrees of freedom), just-identified (degrees of freedom equal to zero) or over-identified (positive degrees of freedom). See also *identification*.

Dependent variable: in a statistical model, it represents the target variable which is expected to be influenced by the *explanatory variables*.

Discrete choice model: regression model with categorical dependent variables (see chapter 16). See also *general linear model*.

Discriminant analysis: a classification technique which explains the allocation of observations into two or more groups on the basis of a set of predictors. When more than two groups are considered, it is termed as *multiple discriminant analysis* (see chapter 11). Note that the technique is used to explain a classification rather than producing one, contrarily to *cluster analysis*.

Discriminant function: in *discriminant analysis*, the function of the predictors which allows to allocate an observation into one of the groups.

Dummy variable (binary): a categorical variable which can only assume two values and is especially useful to model the presence or absence of an attribute.

Effect size: in *hypothesis testing*, when the null hypothesis is rejected, the effect size measures the intensity of the outcome. This is especially useful when rejection of the null hypothesis implies that a *factor* is influential, so that individual effect sizes allow comparisons among different influential factors.

Eigenvalues: a.k.a. characteristic values (see the math explanation in the appendix), they are values which can be computed out of a square matrix and have quite a few desirable uses in multivariate statistics. For example, in *principal component analysis*, the eigenvalues of the covariance (or correlation) matrix measure the amount of variability explained by each component compared to the average component (see chapter 10).

Endogenous variables: in a statistical model, they are the variables that are determined internally to the model as they depend on the *exogenous variables* of the model and possibly on other endogenous variable. *Dependent variables* in regression models are always endogenous (as they depend on the explanatory variables), while sometime explanatory variables might be endogenous if they are influenced by the dependent variable, which pose statistical problems. *Structural equation models* can accommodate multiple endogenous variables, provided the model allows for *identification*.

Exogenous variables: in a statistical model, they are variables that are determined externally to the model and cannot be influenced by the endogenous variables or by other exogenous variables. In the *regression model*, the explanatory variables are assumed to be exogenous. See also *endogenous variables*.

Explanatory variable: in a statistical model, it is one of the variables which is expected to explain the *dependent variable*.

Exploratory factor analysis: see *factor analysis*.

External preference mapping: *preference mapping* where the *proximity matrix* contains objective (analytic) measures of product characteristics or evaluations from expert panels, so that the map contains information external to the set of consumers which provide their evaluation of the products. The final map shows products as they are evaluated by the external source and consumers according to their preferences. See also *internal preference mapping*.

Factor analysis: a technique which can be used for the identification of latent variables and their relationship with a set of manifest indicators and is also exploited for data-reduction purposes (see chapter 10). See also *principal component analysis*. Factor analysis can be *exploratory* when no prior assumption on the *factor loadings* is made (i.e. all *manifest variables* can potentially load on all factors), while *confirmatory factor analysis* starts from an hypothesized relationship between the *manifest variables* and the factors and tests it statistically.

Factorial design: is an approach to produce an *experimental design* for data collection with two or more *factors*. A *full factorial design* produces all of the possible combinations of all of factors and their levels (*treatments*). It is usually helpful, for example in *conjoint analysis*, to reduce the number of combinations, as it happens with a *fractional factorial* design (see chapter 16). See also *analysis of variance*.

Factors: in analysis of variance, the factors are the categorical variables whose variation is expected to induce a variation in the dependent variable. See *analysis of variance*.

Frequency distribution: the complete set of values of the data-set, possibly grouped into classes, together with their absolute or relative frequencies (see appendix).

F-test: a statistical test based on the F probability distribution, commonly used in ANOVA and in multiple regression (see chapters 7 and 8). See *hypothesis testing*.

General linear model (GLM): a comprehensive modeling approach based on multivariate multiple linear regression which incorporates as special cases a set of multivariate methods (especially those of analysis of variance) like *one-way ANOVA, factorial ANOVA, MANOVA, ANCOVA, MANCOVA, t-tests, multiple regression* and simultaneous regression equations (see chapter 7). It is especially helpful for more complex ANOVA designs. Not to be confused with the *generalized linear model*.

General log-linear model: a model where the dependent variable is the logarithm of the frequencies in a *contingency table* and the explanatory variables are categorical variables and their interactions. It allows one to evaluate the associations between two or more categorical variables (see chapter 9). See also *log-linear analysis*.

Generalized linear model: a comprehensive modeling approach for discrete choice modeling, where one or more *dependent* categorical variables are modelled as the outcome of one or more *explanatory variables*, which can be metric or non-metric. Depending on the type of *link function*, the generalized linear model collapses into *logistic regression, logit or probit models*, and *multinomial* or *multivariate logistic regression, logit or probit models* (see chapter 16).

Goodness-of-fit: a measure of the ability of a statistical model to replicate the original data-set. In *regression analysis*, a goodness-of-fit measure is the *R-Square*. In a *structural equation model*, there is a range of goodness-of-fit tests which compare the original covariance matrix with the one estimated through the model. In *multidimensional scaling* a measure of goodness-of-fit is the *STRESS function*.

Heteroskedasticity: two or more (random) variables are said to be heteroskedastic when they have different variances. See also *homoskedasticity*.

Hierarchical clustering methods: *cluster analysis* methods which proceed hierarchically, either through *divisive algorithms* which start from a single cluster and progressively divide it into as many clusters as the number of observations or through *agglomerative algorithms*, where each observation initially represents a cluster and the algorithm progressively groups them into a unique cluster. All potential numbers of clusters are considered, a rule is needed to decide the optimal number of clusters (see chapter 12).

Hierarchical log-linear analysis: a procedure for *log-linear analysis* which proceeds hierarchically to exclude the non-relevant interaction terms (see chapter 9). See also *general log-linear model*.

Homoskedasticity: two or more (random) variables are said to be homoskedastic when they have the same variance. Since elsewhere in this book this term has been set as the distinction mark between those who can claim to know statistics and those who cannot, it may be useful to know that sometimes it is written with a *c* instead of the *k*. See also *heteroskedasticity*.

Hypothesis testing: a statistical procedure which allows one to decide whether the *null hypothesis* should be rejected or not, based on probability grounds (see chapter 6).

Ideal point model: especially useful in *multidimensional scaling*, it is the statistical model used to estimate the point corresponding to the ideal product or brand for consumers, according to their stated preferences. See *multidimensional unfolding*.

Identification (problem): a model is *under-identified* when the unknown parameters are too many compared to the available observations, *just-identified* when the model allows a single solution since there is a single set of estimates compatible with the available observations, or *over-identified* when a single *structural model* is compatible with various sets of estimates. In the latter case, estimates might be not unique nor optimal (see box 15.2). *Structural equation models* allow one to test over-identified models and find a solution which is statistically acceptable. See also *degrees of freedom*.

Imputation: the process of assigning values to missing data in a data-set, generally using statistical techniques (statistical imputation).

Inertia: a measure of association between two categorical variables, based on the chi-square statistic. In *correspondence analysis*, the proportion of inertia explained by each of the dimension can be regarded as a measure of *goodness-of-fit*.

Inference: the generalization process which allows one to project characteristics observed in a sample to the whole population of interest.

Internal preference mapping: *preference mapping* where the *proximity matrix* for the objects (products or brands) is obtained exclusively on the basis of evaluations from consumers. The final map shows products as they are perceived by the consumers according to their preferences. In some circumstances external data (i.e. objective measures or expert evaluations) are used to interpret the dimensions, but not to draw the map. See also *external preference mapping*. It is complementary to the *confidence level*.

Latent variable: a variable which cannot be observed or directly measured, as it quantifies an object or construct which cannot be defined and measured in an unequivocal and verifiable way.

Likelihood function: is a statistical function which allows one to measure the probability of a given set of observations depending on a set of parameters. If parameters are known, the likelihood function defines the probability of the set of observations. If the parameter are not known, by maximizing the likelihood function (the *maximum likelihood* method) it is possible to estimate the most likely values for the parameters, conditional on the observations.

Likelihood ratio test: is a test which compares two *likelihood functions* and generally follows the *Chi-square distribution*. Usually, it compares the observed likelihood with the likelihood under the null hypothesis.

Linear regression: a statistical model which assumes a linear relationship between the *explanatory variables* and the *dependent variable*.

Link function: in *discrete choice models*, the function which allows one to transform the discrete dependent variable into a latent continuous variable and opens the way to estimation (see chapter 16). See *generalized linear model*.

List-wise (case-wise) deletion: when the statistical processing of two or more variables encounters one or more missing data for a given observation, the whole observation (all variables) is omitted from the analysis.

Loadings: in *factor analysis* and *principal component analysis*, the loadings are the correlations between the original variables and the factors or components (see chapter 10). They can be exploited to assign a meaning to the latent factors or components. See also *orthogonal rotation*.

Logistic regression: a *discrete choice model* with a categorical dependent variable and one or more metric or non-metric explanatory variables (see chapter 16). See also *general linear model*.

Logit model: a *discrete choice model* with a binary dependent variable, which exploits the *logit link function* (see chapter 16). Conventionally, one refers to logistic regression when there are metric explanatory variables and to the logit model when all of the explanatory variables are categorical, but these terms are commonly used interchangeably. See also *generalized linear model*.

Log-linear analysis: a statistical procedure which allows one to explore the association among two or more categorical variables (see chapter 9). See *general log-linear model*.

Manifest variables: variables that can be observed and measured, contrarily to *latent variables*. A.K.a indicators.

Maximum likelihood estimation: see *likelihood function*.

Measurement model: see *structural equation model*.

Measurement scale: the rule for assigning numerical values to measure attributes or objects. The various types of measurement scales are discussed in chapter 1. See *measurement*.

Measurement: the action of assigning numerical values to the dimension of an object, an event, or their attributes according to rules. See *measurement scale*.

Metric variable: a.k.a. quantitative or scale variable. See variable.

Missing data: observations with no recorded measurement for a given variable, as a result of non-responses or other omissions (see chapter 4).

Multidimensional scaling: a set of statistical techniques which allow one to translate consumer *preferences* or *perceptions* toward products or brands into a reduced number of dimensions (usually two or three), so that they can be graphically represented into a *preference map* or *perceptual map* (see chapter 13). It is also possible to show both objects and subjects (the consumers) in the same graph through *multidimensional unfolding*, a technique which unfolds the co-ordinates for consumers (or groups of consumers) on the basis of their preferences or perceptions through an *ideal point model*.

Multidimensional unfolding: a method which allows one to estimate the co-ordinates of the *ideal point* or *ideal vector* for each consumer on the basis of their rankings. *Ideal points* can be used to represent a consumer's ideal product (or brand) in bi-plots, together with the actual positioning of brands (see chapter 13).

Multinomial logit (probit) model: a *discrete choice model* which extend the logit (probit model) to account for a dependent variable with more than two categories (see chapter 16). See also *generalized linear model*.

Multinomial models: discrete choice models are said to be multinomial when the dependent variable has more than two categories (see chapter 16).

Multiple discriminant analysis: see *discriminant analysis*.

Multiple regression model: a regression model with two or more explanatory variables (see chapter 8).

Multivariate analysis of variance (MANOVA): *analysis of variance* with two or more target variables (see chapter 7). See also *general linear model*.

Multivariate methods: in general, methods or models are said to be *multivariate* when they have more than one target (*dependent*) *variable*. For example, multivariate regression refers to models with several dependent variables. Note the difference with *multiple regression*, which still refer to a single dependent variable, but with two or more explanatory variables.

Multi-way (factorial) analysis of variance: *analysis of variance* with more than one factor (see chapter 7). See also *general linear model*.

Nominal variable: a type of categorical variable with non-ordered values. See *variable*.

Non-hierarchical clustering methods: *cluster analysis* methods which allocate the observations into a pre-determined number of clusters and contrarily to *hierarchical algorithms* allow reallocation of units into different cluster throughout the work of the algorithm (see chapter 12).

Non-metric variable: a.k.a. categorical or non-metric variable. See variable.

Non-parametric test: a typology of statistical tests for *hypothesis testing* which is free from the distributional assumptions for the population of interest, contrarily to *parametric tests* (see chapter 6).

Non-probability sampling: see sampling.

Non-response error: the portion of error in a survey due to the fact that some of the surveyed units do not respond (due to refusal, inability or missed contact). It generates missing data.

Non-sampling error: the portion of survey error which does not depend on sampling and includes non-response and response errors, together with errors by interviewers and researchers.

Null hypothesis: in *hypothesis testing*, the null hypothesis is the hypothesis to be tested, which defines the probability distribution of the test *statistic*. The null hypothesis is rejected in favor of the *alternative hypothesis* when the probability of the test *statistic* is below the *significance level*. See *hypothesis testing*.

Observation: each of the measurements of a variable.

Odds ratio: a measure (expressed as a ratio) of the probability of an outcome compared to an alternative outcome. For example, an odds ratio of 1 means that two outcomes are equally probable, while 2 indicates that the odds of one event double those of the alternative one. In *discrete choice models* odds ratio can be computed from the model coefficients (see chapter 16).

Ordered logit (probit) model: a *discrete choice model* which extends the logit (probit model) to account for an *ordinal* dependent variable (see chapter 16). See also *generalized linear model*.

Ordinal variable: a type of categorical variable with ordered values. See variable.

Orthogonal rotation: in *factor analysis* and *principal component analysis*, it is a matrix transformation of the factor or component *loadings* which facilitates the interpretation of the factors or components (see chapter 10 and the appendix).

Outlier: an observation for a variable which strongly differs from other observations, thus might be the outcome of a response error or other non-sampling errors (see chapter 4).

Pair-wise deletion: when the statistical processing of more than two variables encounters missing data for a given observation, only the variables with missing data for that observation are omitted from the analysis, while the same observation enters the analysis for those variables without missing data.

Parametric test: a typology of statistical tests for *hypothesis testing* which stems from the assumption of a *probability distribution* for the population being investigated.

Partial correlation: the *correlation* between two variables after accounting for correlations with other influential variables (see chapter 8).

Path analysis: is an extension of multiple regression which allows for correlation between explanatory variables and assumes a causal relationship, although there is quite a debate on the difference between causality in statistical models and cause-effect relationships in reality (see chapter 15). Path analysis is associated with a diagrammatic representation of the relationship through boxes for variables and arrows indicating causality (one-sided arrow) or correlations (two-sided arrows), called *path diagram*. See also *structural equation models*.

Path diagram: see *path analysis*.

Perception: see *perceptual map*.

Perceptual map: a plot of products (brands) into a graph of two or more dimensions according to *consumer perceptions*, defined as the subjective evaluations of the attributes of the objects as stated by the respondents (usually measured through measurement scales). See also *preference mapping* and *multidimensional scaling*.

Planned comparisons: see *contrasts*.

Population: the set of units which are the object of the research.

Post-hoc tests: within an *analysis of variance* design, they allow running multiple comparisons among sub-sets of the treatments (see chapter 7). They open the way to further explore the sources of variability when the null hypothesis of mean equality is rejected. Contrarily to planned *contrasts* these comparisons are not planned before data collection and adjust the error to account for the fact that they are not carried out independently from each other. See also *contrasts*.

Power (of a test): the probability for a test of not committing a *type II error*.

Precision: the variability in repeated measurements, not to be confused with *accuracy*.

Preference map: a plot of products (brands) into a graph of two or more dimensions according to *consumer preferences*, defined as the ranking of the objects as stated by the consumers through a survey. See also *perceptual mapping* and *multidimensional scaling*.

Preference: see *preference map*.

Primary data: data explicitly collected for the purposes of the research (see chapter 3).

Principal component analysis: a data-reduction technique which exploits correlation between multiple variables to summarize them into a reduced number of components (see chapter 10). These principal components are linear combinations of the original variables and are not correlated between each other. See also *factor analysis*.

Probability distribution: the complete set of values that a random variable can assume, possibly grouped into classes, together with the probability of their occurrence (see appendix).

Probability sampling: see *sampling*.

Probit model: a discrete choice model with a binary dependent variable, which exploits the probit link function (see chapter 16). See also *generalized linear model*.

Proximity matrix: especially useful as an input matrix for *multidimensional scaling*, it is a symmetric matrix containing some measure of similarity or dissimilarity between objects (brands, products or their attributes) or subjects (consumers).

*p***-value:** in *hypothesis testing*, the probability associated with a given value of the test statistic (see chapter 6).

Random error: fluctuations in measurement which do not follow any systematic direction, but are due to factors that act in a random fashion.

Random variable: a *variable* whose values depend on a random variation or are the outcomes of random extractions, thus can be associated to a probability law (see appendix).

Reliability: consistency of the measurement across several questionnaire items measuring the same latent construct or over time.

Representativeness: the extent to which observations in a sample reflect the key (targeted) characteristics of the *population*.

Response error: the portion of error in a survey which is related to inaccurate or false responses, collection or recording. It generates incorrect data.

R-Square: a.k.a. *coefficient of determination*, a measure of goodness-of-fit in *linear regression* (see chapter 8), measured by the ratio between the variability explained by the regression model and the total variability in the dependent variable. This measure is bounded between zero and 1, but it increases as the number of explanatory variables increases. Thus, to value the *parsimony* of a model, it can be adjusted to account for the number of explanatory variables and takes the name of *adjusted R-Square*.

Sample: a sub-set of the units in a population, usually chosen to represent the whole population.

Sampling error: the portion of survey error which is due to the fact that data are collected from samples rather than the whole population.

Sampling frame: the complete list of all elements in a population, which can be used to extract a sample.

Sampling: the process which allows one to extract a sample from a population. In *probability sampling*, extraction is random and based on probability distributions. In *non-probability sampling*, extraction of sample units is not based on probability rules. See *sample*.

Scale variable: a.k.a. quantitative or metric variable. See *variable*.

Scores: in *factor analysis*, factor scores are the estimates of each of the extracted latent factor for each of the observations in the data-set. In *principal component analysis*, component scores are not estimates, but are computed exactly (see chapter 10).

Scree diagram: a bi-dimensional graph used in *factor, principal component* and *cluster analysis* to plot the explained variance (or other similar statistics) against the number of factors, components or clusters. When an elbow is encountered, this may help in identifying the optimal number of factors, components or clusters (see chapters 10 and 12).

Secondary data: data originally collected for purposes other than those of the research, usually available through external data sources (see chapter 2).

Significance level: in *hypothesis testing* (see chapter 6), a threshold probability level, below which the null hypothesis is rejected. It represents the probability of rejecting the null hypothesis when it is actually true and it is arbitrarily set by the researcher (usually at 5% or 1% level). It is denoted by α and its complementary $1 - \alpha$ is the *confidence level* test.

Standard deviation: a measure of variability (see appendix).

Standard error: a measure of precision for an estimator (see appendix).

Standardization: a variable with zero mean and variance equal to 1. Any metric variable can be standardized by subtracting the mean and dividing by the *standard deviation*.

Statistic: generally, a *statistic* is the function of one or more random variables. A *test statistic* is a statistic which allows one to test an hypothesis (see *hypothesis testing*), since its probability distribution is known when the *null hypothesis* is true.

Stepwise methods: an estimation strategy for statistical models, which allow one to select among a set of explanatory variables those who are relevant to explain the dependent variable (see chapter 8).

STRESS function: a measure of goodness-of-fit for *multidimensional scaling* methods which compares the dissimilarities obtained after scaling with the observed ones.

Structural equation model (SEM): a *multivariate* statistical model which allows one to estimate and test multiple and simultaneous relationships involving several *dependent* and *explanatory* variables and allows the inclusion of *latent variables* (see chapter 15). SEMs are made by a *measurement model*, which relate *manifest variables* to *latent variables* through *confirmatory factor analysis* and a *structural model*, which represents all of the relationships among the *endogenous* and *exogenous* variables. The structural equation is an extension of *path analysis* to allow for latent variables. SEMs start with *path diagrams* which incorporate circles for latent variables beside squares for manifest variables and causality/correlation arrows (see *path analysis*). SEMs are particularly useful to test alternative theories.

Structural model: see *Structural equation model*.

Survey: a systematic collection of data on a population or sample. Commonly the term refers to data collection on people through questionnaires.

Systematic error: a bias in measurement which makes each of the measures systematically too high or too low.

Treatments: the various levels (categories) of a *factor*. See *analysis of variance*.

t-test: a statistical test based on the Student-t probability distribution, commonly used to test equality between the means of two independent sample or whether a regression coefficient is different from zero (see chapters 6 and 8). See *hypothesis testing*.

Type I error: in *hypothesis testing* (see chapter 6), the probability of rejecting the null hypothesis when it is actually true. The probability of a Type I error is equal to the significance level of a test.

Type II error: in *hypothesis testing* (see chapter 6), the probability of non-rejecting the null hypothesis when it is actually false (the alternative hypothesis is true). The probability of a type II error is denoted by β and its complementary $1 - \beta$ is the power of a test.

Validity: the extent to which measurement reflects the 'true' phenomenon under study.

Variable: a symbol that represents a measurable characteristic or attribute of an object or person. It represents a quantity which is not specified explicitly and can be of different types. *Non-metric* (also *qualitative* or *categorical*) *variables* assume values that are not strictly numeric, while *metric* (also *quantitative* or *scale*) variables are associated with numerical values. Non-metric variables can be *ordinal*, when the categories follow an objective order, or *nominal* when there is no order. Metric variables can be distinguished in *interval scales* (when the zero point is arbitrary and depends on the unit of measurement) or *ratio scales* (when there is an objective zero reference point). Furthermore, metric variables can be *continuous* when they can assume all values in the numeric continuum or *discrete*, when they can only assume some of the values in the numeric continuum, for example integers.

Z-test: the statistical test based on the normal distribution (see chapter 8). See *hypothesis testing*.

References

Aaker, J. L. (1997) 'Dimensions of brand personality,' *Journal of Marketing Research,* 34(3): 347–356.

Abdi, H., Valentin, D. and Edelman, B. (1998) *Neural networks,* Thousand Oaks, CA: Sage Publications.

Acito, F. and Anderson, R. D. (1986) 'A Simulation Study of Factor Score Indeterminacy,' *Journal of Marketing Research,* 23(2): 111–118.

Agresti, A. (Ed., 2002) *Categorical Data Analysis.* Hoboken, NJ: John Wiley & Sons.

Ajzen, I. (1991) 'The Theory of Planned Behavior,' *Organizational Behavior and Human Decision Processes,* 50(2): 179–211.

Ajzen, I. (2002) 'Perceived behavioral control, self-efficacy, locus of control, and the theory of planned behavior,' *Journal of Applied Social Psychology,* 32(4): 665–683.

Ajzen, I. and Fishbein, M. (1980) *Understanding attitudes and predicting social behaviour,* Englewood Cliffs, NJ: Prentice-Hall.

Allenby, G. M. (1990) 'Hypothesis testing with scanner data – the advantage of Bayesian methods,' *Journal of Marketing Research,* 27(4): 379–389.

Alpert, M. I. and Peterson, R. A. (1972) 'Interpretation of canonical analysis,' *Journal of Marketing Research,* 9(2): 187–192.

Alves, S. G., Neto, N. M. O., and Martins, M. L. (2002) 'Electoral surveys influence on the voting processes: A cellular automata model,' *Physica A-Statistical Mechanics and Its Applications,* 316(1-4): 601–614.

Anderson, R. D. and Rubin, H. (Ed., 1956) *Statistical inference in factor analysis* (5th ed.), Berkeley, CA: University of California Press.

Andrews, R. L., Ansari, A., and Currim, I. S. (2002) 'Hierarchical Bayes versus finite mixture conjoint analysis models: A comparison of fit, prediction, and partworth recovery,' *Journal of Marketing Research,* 39(1): 87–98.

Angelis, V. A., Lymperopoulos, C., and Dimaki, K. (2005) 'Customers' perceived value for private and state-controlled Hellenic banks,' *Journal of Financial Services Marketing,* 9(4): 360–374.

Arnold, S. J. (1979) 'Test for Clusters,' *Journal of Marketing Research,* 16(4): 545–551.

Assael, H. and Keon, J. (1982) 'Non-sampling vs sampling errors in survey research,' *Journal of Marketing,* 46(2): 114–123.

Baker, K. (1991) 'Using Geodemographics in Market Research Surveys,' *The Statistician,* 40(2): 203–207.

Bartlett, M. S. (1937) 'The statistical conception of mental factors,' *British Journal of Psychology,* 28: 97–104.

Baumgartner, H. and Steenkamp, J. B. E. M. (2006) 'An extended paradigm for measurement analysis of marketing constructs applicable to panel data,' *Journal of Marketing Research,* 43(3): 431–442.

Beh, E. J. (2004) 'Simple correspondence analysis: A bibliographic review,' *International Statistical Review,* 72(2): 257–284.

Ben Akiva, M., McFadden, D., Train, K., Walker, J., Bhat, C., Bierlaire, M., Bolduc, D., Boersch-Supan, A., Brownstone, D., Bunch, D. S., Daly, A., De Palma, A., Gopinath, D., Karlstrom, A., and Munizaga, M. A. (2002) 'Hybrid choice models: Progress and challenges,' *Marketing Letters*, 13(3): 163–175.

Bentler, P. M. (1980) 'Multivariate analysis with latent variables – Causal modeling,' *Annual Review of Psychology*, 31: 419–456.

Berk, D. R. A. and Berk, R. (2004) *Regression Analysis: A Constructive Critique*, Thousand Oaks, CA: Sage Publications.

Bernard, H. R. (2000) *Social Research Methods: Qualitative and Quantitative Approache*, Thousand Oaks, CA: Sage Publications.

Bethel, J. W. (1989) 'Sample allocation in multivariate surveys,' *Survey Methodology*, 15(1): 47–57.

Biemer, P. B., Groves, R. M., Lyberg, L. E., Mathiowetz, N. A., and Sudman, S. (Eds.) (2004) *Measurement Errors in Surveys*, Hoboken, NJ: John Wiley & Sons.

Blattberg, R. C. and Dolan, R. J. (1981) 'An assessment of the contribution of log linear models to marketing research,' *Journal of Marketing*, 45(2): 89–97.

Bogue, J. and Ritson, C. (2006) 'Integrating consumer information with the new product development process: The development of lighter dairy products,' *International Journal of Consumer Studies*, 30(1): 44–54.

Bollen, K. A. (1989) *Structural Equations with Latent Variables*, New York, NY: Wiley.

Bollen, K. A. (2002) 'Latent variables in psychology and the social sciences,' *Annual Review of Psychology*, 53(1): 605.

Bollen, K. A. and Long, J. S. (1993) *Testing Structural Equation Models*, Newbury Park, CA: Sage Publications.

Borg, I. and Groenen, P. J. F. (2005) *Modern Multidimensional Scaling: Theory and Application*, New York, NY: Springer.

Braunsberger, K., Wybenga, H., and Gates, R. (2007) 'A comparison of reliability between telephone and web-based surveys,' *Journal of Business Research*, 60(7): 758–764.

Bray, J. E. and Maxwell, S. E. (1985) *Multivariate Analysis of Variance*, Newbury Park, CA: Sage Publications.

Breiman, L., Friedman, J. H., Olshen, R. A., and Stone, C. J. (1984) *Classification and Regression Trees*, Boca Raton, FL: Chapman & Hall/CRC.

Brown, M. B. and Forsythe, A. B. (1974) 'Robust tests for equality of variances,' *Journal of the American Statistical Association*, 69(346): 364–367.

Brown, R. A. (1974) 'Robustness of studentized range statistic' *Biometrika*, 61(1): 171–175.

Brunso, K. and Grunert, K. G. (1998) 'Cross-cultural similarities and differences in shopping for food,' *Journal of Business Research*, 42(2): 145–150.

Bryman, A. (Ed., 2007) *Qualitative Research I*, Thousand Oaks, CA: Sage Publications.

Burnham, K. P. and Anderson, D. R. (2004) 'Multimodel inference – understanding AIC and BIC in model selection,' *Sociological Methods & Research*, 33(2): 261–304.

Busing, F. M. T. A., Groenen, P. J. K., and Heiser, W. J. (2005) 'Avoiding degeneracy in multidimensional unfolding by penalizing on the coefficient of variation,' *Psychometrika*, 70(1): 71–98.

Byrne, B. M. (2001a) 'Structural equation modeling with AMOS, EQS, and LISREL: Comparative approaches to testing for the factorial validity of a measuring nstrument,' *International Journal of Testing*, 1(1): 55–86.

Byrne, B. M. (2001b) *Structural Equation Modeling With Amos: Basic Concepts, Applications, and Programming*, Mahwah, NJ: Lawrence Erlbaum Associates.

Carroll, J. D. (1972) 'Individual differences and multidimensional scaling,' *Multidimensional Scaling: Theory and Applications in the Behavioral Sciences*. Shepard, R.N., Romney, A.H., and Nerlove S.B.(Eds.), New York, NY: Seminar Press.

Carroll, J. D. and Chang, J. J. (1970) 'Analysis of individual differences in multi-dimensional caling ia An N-way generalization of Eckart-Young decomposition,' *Psychometrika*, 35(3): 283–319.

Carroll, J. D. and Green, P. E. (1997) 'Psychometric methods in marketing research 2. Multidimensional scaling,' *Journal of Marketing Research*, 34(2): 193–204.

Chen, Y. L., Tang, K., Shen, R. J., and Hu, Y. H. (2004) 'Market basket analysis in a multiple store environment,' *Decision Support Systems*, 40(2): 339–354.

Cheung, G. W. and Rensvold, R. B. (2002) 'Evaluating goodness-of-fit indexes for testing measurement invariance,' *Structural Equation Modeling*, 9(2): 233–255.

Cochran, W. G. (1977) *Sampling Techniques*, New York, NY: John Wiley & Sons.

Churchill, G. A. and Peter, J. P. (1984) 'Research design effects on the reliability of rating-scales – A meta-analysis,' *Journal of Marketing Research*, 21(4): 360–375.

Clark, T. (1989) 'Managing outliers: Qualitative issues in the handling of extreme observations in marketing research,' *Marketing Research*, 1(2): 31–48.

Cohen, J. (1988) *Statistical Power Analysis for the Behavioral Sciences* (2nd ed.), Hillsdale, NJ: Lawrence Erlbaum.

Commandeur, J. J. F. and Heiser, W. J. (1993) *Mathematical derivations in the proximity scaling (PROXSCAL) of symmetric data matrices*, Tech. Rep. No. RR- 93–03 Leiden, NL: University of Leiden.

Cooper, L. G. (1983) 'A Review of multi-dimensional scaling in marketing research,' *Applied Psychological Measurement*, 7(4): 427–450.

Cortina, D. J. M. and Nouri, D. H. (2000) *Effect Size for Anova Designs*, Thousand Oaks, CA: Sage Publications.

Couper, M. P. (2000) 'Web surveys – A review of issues and approaches,' *Public Opinion Quarterly*, 64(4): 464–494.

Cox, T. F. and Cox, M. A. A. (2001) *Multidimensional Scaling* (2nd ed.), New York, NY: Chapman and Hall.

Crask, M. R. and Perreault, W. D. (1977) 'Validation of discriminant-analysis in marketing research,' *Journal of Marketing Research*, 14(1): 60–68.

Creswell, J. W. (2003) *Research Design: Qualitative, Quantitative, and Mixed Method Approaches*, Thousand Oaks, CA: Sage Publications.

Cronbach, L. J. (1951) 'Coefficient alpha and the internal structure of tests,' *Psychometrika*, 16(3): 297–334.

Danaher, P. J. (1988) 'A log-linear model for predicting magazine audiences,' *Journal of Marketing Research*, 25(4): 356–362.

Davis, W. R. (1993) 'The Fc1 rule of identification for confirmatory factor-analysis – A general sufficient condition,' *Sociological Methods & Research*, 21(4): 403–437.

Deming, W. E. (1990) *Sample design in business research*, New York, NY: John Wiley & Sons.

Denzin, N. K. and Lincoln, Y. S. (Ed., 2005). *The Sage Handbook of Qualitative Research* (3rd ed.) Thousand Oaks, CA: Sage Publications.

DeSarbo, W. S. and Hildebrand, D. K. (1980) A marketers guide to log-linear models for qualitative data-analysis,' *Journal of Marketing*, 44(3): 40–51.

DeSarbo, W. S., Fong, D. K. H., Liechty, J., and Saxton, M. K. (2004) 'A hierarchical Bayesian procedure for two-mode cluster analysis,' *Psychometrika*, 69(4): 547–572.

DeSarbo, W. S., Kim, Y., and Fong, D. (1999) 'A Bayesian multi-dimensional scaling procedure for the spatial analysis of revealed choice data,' *Journal of Econometrics*, 89(1–2): 79–108.

Devlin, S. J., Dong, H. K., and Brown, M. (1993) 'Selecting a scale for measuring quality,' *Marketing Research*, 5(3): 12–17.

Dillon, W. R. (1979) 'Performance of the linear discriminant function in non-optimal situations and the estimation of classification error rates – review of recent findings,' *Journal of Marketing Research*, 16(3): 370–381.

Dillon, W. R., White, J. B., Rao, V. R., and Filak, D. (1997) ' "Good Science" – Use structural equation models to decipher complex customer relationships,' *Marketing Research*, 9(Winter): 22–31.

Drozdenko, R. G. and Drake, P. D. (2002) *Optimal Database Marketing: Strategy, Development, and Data Mining*, Thousand Oaks, CA: Sage Publications.

Dunteman, G. H. (1989) *Principal Component Analysis*, Newbury Park, CA: Sage Publications.

Dunteman, G. H. and Ho, M. H. R. (2006) *An Introduction to Generalized Linear Models*, Thousand Oaks, CA: Sage Publications.

East, R. (1997) *Consumer Behaviour: Advances and Applications in Marketing*, London, UK: Prentice-Hall.

Efron, B. (1986) 'Why Isn't Everyone A Bayesian?,' *American Statistician*, 40(1): 1–5.

Eisend, M. (2005) 'The role of meta-analysis in marketing and consumer behavior research: Stimulator or inhibitor?,' *Advances in Consumer Research*, 32: 620–622.

Everitt, B. S. (1979) 'Unresolved problems in cluster-analysis,' *Biometrics*, 35(1): 169–181.

Everitt, B. S. (1995) 'The analysis of repeated-measures - A practical review with examples,' *Statistician*, 44(1): 113–135.

Everitt, B. S., Landau, S., and Leese, M. (2001) *Cluster Analysis*, London, UK: Arnold.

Farley, J. U. and Lehmann, D. R. (1986) *Meta-Analysis in Marketing: Generalization of Response Models*, Lexington, MA: Lexington Books.

Fava, J. L. and Velicer, W. F. (1992) 'The Effects of over-extraction on factor and component analyses,' *Multivariate Behavioral Research*, 27(3): 387–415.

Fava, J. L. and Velicer, W. F. (1996) 'The effects of underextraction in factor and component analyses,' *Educational and Psychological Measurement*, 56(6): 907–929.

Fayyad, U. M. (1997) 'Editorial,' *Data Mining and Knowledge Discovery*, 1(1): 5–10.

Fayyad, U. M., Piatetsky-Shapiro, G., and Smyth, P. (1996) 'From data mining to knowledge discovery: an overview,' *Advances in knowledge discovery and data mining*. U. M. Fayyad, G. Piatetsky-Shapiro, P. Smyth and R. Uthurusamy. (Eds), pp. 1–34.

Field, A. (2005) *Discovering Statistics Using SPSS* (2nd ed.), London, UK: Sage Publications.

Flick, U. (2006) *An Introduction to Qualitative Research* (3rd ed.), London, UK: Sage Publications.

Fox, J. (1997) *Applied Regression Analysis, Linear Models, and Related Method*, Thousand Oaks, CA: Sage Publications.

Frank, R. E., Massy, W. F., and Morrison, D. G. (1965) 'Bias in multiple discriminant analysis,' *Journal of Marketing Research*, 2(3): 250–258.

Fuller, W. A. (1987) *Measurement Error Models*, New York, NY: John Wiley & Sons.

Galton, F. (1869) *Hereditary Genius: An Inquiry into Its Laws and Consequence*, London, UK: MacMillan.

Garcia-Escudero, L. A. and Gordaliza, A. (1999) 'Robustness properties of k means and trimmed k means,' *Journal of the American Statistical Association*, 94(447): 956–969.

Garson, G. D. (1998) *Neural Networks: An Introductory Guide for Social Scientist*, London, UK: Sage Publications.

Gelb, B. D. and Gelb, G. M. (1986) 'New coke's fizzle – lessons for the rest of us,' *Sloan Management Review*, 28(1): 71–76.

Gelman, A., Carlin, J. B., Stern, H. S., and Rubin, D. B. (2004) *Bayesian data analysis*, Boca Raton, FL: Chapman & Hall/CRC Press.

Gentle, J. E. (2006) *Optimization Methods for Applications in Statistic*, New York, NY: Springer Verlag.

Gibbons, J. D. (1993) *Nonparametric Statistics: An Introduction*, Newbury Park, CA: Sage Publications.

Golub, G. H. (1969) 'Matrix decompositions and statistical calculations,' *Statistical Computation*. R. C. Milton and J. A. Nelder (Eds.), New York, NY: Academic Press, pp. 1–34.

Gorsuch, R. L. (1983) *Factor Analysis* (2nd ed.), Hillsdale, NJ: Lawrence Erlbaum.

Gorsuch, R. L. (1990) 'Common factor-analysis versus component analysis – some well and little known facts. *Multivariate Behavioral Research*, 25(1): 33–39.

Green, P. E. (1974) 'Design of choice experiments involving multifactor alternatives,' *Journal of Consumer Research*, 1(2): 61–68.

Green, P. E. and Carroll, J. D. (1978) *Mathematical Tools for Applied Multivariate Analysi*, New York, NY: Academic Press.

Green, P. E. and Rao, V. R. (1969) 'A note on proximity measures and cluster analysis,' *Journal of Marketing Research*, 6(3): 359–364.

Green, P. E. and Srinivasan, V. (1990) 'Conjoint-analysis in marketing – new developments with implications for research and practise,' *Journal of Marketing*, 54(4): 3–19.

Green, P. E., Krieger, A. M., and Wind, Y. (2001) 'Thirty years of conjoint analysis: Reflections and prospects,' *Interfaces*, 31(3): S56–S73.

Greenacre, M. J. and Blasius, J. (2006) *Multiple Correspondence Analysis and Related Methods*, Boca Raton, FL: Chapman & Hall/CRC Press.

Greenberg, B. G., Kuebler R. R. Jr, Abernathy, J. R., and Orvitz, D. G. (1971) 'Application of the randomized response technique in obtaining quantitative data,' *Journal of the American Statistical Association*, 66(334): 243–250.

Greene, W. H. (2002) *Econometric Modelling Guide (Volumes 1 and 2): Version 8.0.*, Plainview, NY: Econometric Software.

Greene, W. H. (2003) *Econometric Analysis* (5th ed.), Upper Saddle River, NJ: Prentice-Hall.

Grosh, M. E. and Glewwe, P. (2000) *Designing household survey questionnaires for developing countries: lessons from 15 years of the living standards measurement study*, Washington, DC: The World Bank.

Groves, R. M. (2006) 'Non-response rates and non-response bias in household surveys,' *Public Opinion Quarterly*, 70(5): 646–675.

Grunert, K. G. and Bech-Larsen, T. (2005) 'Explaining choice option attractiveness by beliefs elicited by the laddering method,' *Journal of Economic Psychology*, 26(2): 223–241.

Gutman, J. (1982) 'A means-end chain model based on consumer categorization processes,' *Journal of Marketing*, 46(2): 60–72.

Guttman, L. (1953) 'Image theory for the structure of quantitative variate,' *Psychometrika*, 18(4): 277–296.

Guttman, L. (1977) 'What is not what in statistics,' *Statistician*, 26(2): 81–107.

Hair, J. F. Jr., Anderson, R. E., Tatham, R. L., and Black, W. C. (1998) *Multivariate Data Analysis* (5th ed.), Upper Saddle River, NJ: Prentice-Hall.

Hamilton, J. D. (1994) *Time Series Analysis*, Princeton, NJ: Princeton University Press.

Han, J. and Kambers, M. (2006) *Data Mining: Concepts and Techniques*, San Francisco, CA: Morgan Kaufmann.

Hand, D. J. (1998) 'Data mining: Statistics and more,' *American Statistician*, 52(2): 112–118.

Hansen, M. H. and Hauser, P. M. (1945) 'Area sampling – some principles of sample design,' *The Public Opinion Quarterly*, 9(2): 183–193.

Harris, C. W. (1962) 'Some Rao-Guttman relationships,' *Psychometrika*, 27(3): 247–263.

Hauser, J. R. and Koppelman, F. S. (1979) 'Alternative perceptual mapping techniques – relative accuracy and usefulness,' *Journal of Marketing Research*, 16(4): 495–506.

Heckman, J. J. (1979) 'Sample selection bias as a specification error,' *Econometrica*, 47(1): 153–161.

Hennig, C. (1998) 'Clustering and outlier identification: Fixed point cluster analysis,' *Advances in Data Science and Classification*. Rizzi, A., Vichi, M., and H. H. Bock (Eds.), Berlin, GE: Springer, pp. 37–42.

Hennig, C. (2003) 'Clusters, outliers, and regression: fixed point clusters,' *Journal of Multivariate Analysis*, 86(1): 183–212.

Hensher, D., Louviere, J., and Swait, J. (1999) 'Combining sources of preference data,' *Journal of Econometrics*, 89(1–2): 197–221.

Hodge, V. J. and Austin, J. (2004) 'A survey of outlier detection methodologies,' *Artificial Intelligence Review*, 22(2): 85–126.

Hoeppner, F., Klawonn, F., Kruse, R., and Runkler, T. (1999) *Fuzzy Cluster Analysis: Methods for Classification, Data Analysis and Image Recognition*, Chichester, UK: John Wiley & Sons.

Hoffman, D. L. and Franke, G. R. (1986) 'Correspondence analysis – graphical representation of categorical-data in Marketing research,' *Journal of Marketing Research*, 23(3): 213–227.

Horton, N. J. and Lipsitz, S. R. (2001) 'Multiple imputation in practice: Comparison of software packages for regression models with missing variables,' *American Statistician*, 55(3): 244–254.

Horton, R. L. (1986) *The general linear model: data analysis in the social and behavioral sciences*, Malabar, FL: Krieger Publishing Company.

Houston, M. B. (2004) 'Assessing the validity of secondary data proxies for marketing constructs,' *Journal of Business Research*, 57(2): 154–161.

Hoyle, R. H. (Ed., 1995) *Structural Equation Modeling: Concepts, Issues and Applications*, Thousand Oaks, CA: Sage Publications.

Iacobucci, D. and Henderson, G. (1997) 'Log linear models for consumer brand switching behavior: What a manager can learn from studying standardized residuals,' *Advances in Consumer Research*, 24: 375–380.

Ip, B. and Jacobs, G. (2005) 'Segmentation of the games market using multivariate analysis,' *Journal of Targeting, Measurement and Analysis for Marketing*, 13(3): 275–287.

Jackson, S. A. and Brashers, D. E. (1994) *Random factors in ANOVA*, Thousand Oaks, CA: Sage Publications.

Johnstone, D. J. (1989) 'On the necessity for random sampling,' *British Journal for the Philosophy of Science*, 40(4): 443–457.

Jolliffe, I. T. (1972) 'Discarding variables in a principal component analysis .1. Artificial Data,' *Journal of the Royal Statistical Society Series C-Applied Statistics*, 21(2): 160–173.

Jolliffe, I. T. (1989) 'Rotation of ill-defined principal components,' *Journal of the Royal Statistical Society, Series C-Applied Statistics*, 38(1): 139–147.

Jolliffe, I. T. (1995) 'Rotation of principal components - choice of normalization constraints,' *Journal of Applied Statistics*, 22(1): 29–35.

Jolliffe, I. T. (2002) *Principal Component Analysis*, New York, NY: Springer.

Jones, J. P. (Ed., 1998) *How Advertising Works: The Role of Research*, Thousand Oaks, CA: Sage Publications.

Joreskog, K. G. (1962) 'On the statistical treatment of residuals in factor analysis,' *Psychometrika*, 27(4): 335–354.

Joreskog, K. G. (1967) 'Some contributions to maximum likelihood factor analysis,' *Psychometrika*, 32(4): 443–482.

Joreskog, K. G. (1969) 'A general approach to confirmatory maximum likelihood factor analysis,' *Psychometrika*, 34(2): 183–202.

Kagel, J. H. and Roth, A. E. (Ed., 1995) *The Handbook of Experimental Economics*, Princeton, NJ: Princeton University Press.

Kaiser, H. F. (1960) 'The application of electronic computers to factor analysis,' *Educational and Psychological Measurement*, 20: 141–151.

Kaiser, H. F. and Caffrey, J. (1965) 'Alpha factor analysis,' *Psychometrika*, 30(1): 1–14.

Kang, S. K. and Hsu, C. H. C. (2004) 'Spousal conflict level and resolution in family vacation destination selection,' *Journal of Hospitality & Tourism Research*, 28(4): 408–424.

Kaplan, D. W. (2000) *Structural Equation Modeling: Foundations and Extension*, Thousand Oaks, CA: Sage Publications.

Kessels, R., Goos, P., and Vandebroek, M. (2006) 'A comparison of criteria to design efficient choice experiments,' *Journal of Marketing Research*, 43(3): 409–419.

Khattree, R. and Naik, D. N. (1999) *Applied Multivariate Statistics with SAS Software* (2nd ed.). Cary, NC: SAS Institute Inc.

Kiang, M. Y. (2003) 'A comparative assessment of classification methods,' *Decision Support Systems*, 35(4): 441–454.

King, B. F. (1985) 'Surveys combining probability and quota methods of sampling,' *Journal of the American Statistical Association*, 80(392): 890–896.

Klecka, W. R. (1980) *Discriminant Analysis*, Newbury Park, CA: Sage Publications.

Kline, R. B. (2005) *Principles and practice of structural equation modeling* (2nd ed.), New York, NY: Guildford Press.

Knoke, D. and Burke, P. J. (1980) *Log-Linear Models*, Newbury Park, CA: Sage Publications.

Knoke, J. D. (1982) 'Discriminant analysis with discrete and continuous variables,' *Biometrics*, 38(1): 191–200.

Kruskal, J. B. (1964) 'Multidimensional scaling by optimizing goodness-of-fit to a non-metric hypothesis,' *Psychometrika*, 29(1): 1–28.

Kruskal, J. B. (1964) 'Nonmetric multi-dimensional scaling: A numerical method,' *Psychometrika*, 29(2): 115–129.

Kruskal, J. B. and Wish, M. (1978) *Multidimensional Scaling*, Newbury Park, CA: Sage Publications.

Krzanowski, W. J. (2000) *Principles of Multivariate Analysis* (2nd ed.), New York, NY: Oxford University Press.

Kuk, A. Y. C. (1990) 'Asking sensitive questions indirectly,' *Biometrika*, 77(2): 436–438.

Lastovicka, J. L. and Thamodaran, K. (1991) 'Common factor score estimates in multiple-regression problems,' *Journal of Marketing Research*, 28(1): 105–112.

Lavrakas, P. J. (1996) 'To err is human,' *Marketing Research*, 8(1): 30–36.

Lawley, D. N. and Maxwell, A. E. (1962) 'Factor analysis as a statistical method,' *The Statistician*, 12(3): 209–229.

Lawrence, F. R. and Hancock, G. R. (1999) 'Conditions affecting integrity of a factor solution under varying degrees of over-extraction,' *Educational and Psychological Measurement*, 59(4): 549–579.

Le Cam, L. (1986) 'The central limit theorem around 1935,' *Statistical Science*, 1(1): 78–91.

Lee, C. E. (1965) 'Measurement and the development of science and marketing,' *Journal of Marketing Research*, 2(1): 20–25.

Lehmann, E. L. and Romano, J. P. (2005) 'Generalizations of the familywise error rate,' *Annals of Statistics*, 33(3): 1138–1154.

Levine, M. S. (1977) *Canonical Analysis and Factor Comparison*, Newbury Park, CA: Sage Publications.

Levy, P. S. and Lemeshow, S. (1999) *Sampling of Populations : Methods and Applications*, New York, NY: John Wiley & Sons.

Lewis-Beck, D. M. S. (1980) *Applied Regression: An Introduction*, Newbury Park, CA: Sage Publications.

Liao, F. T. (1994) *Interpreting Probability Models: Logit, Probit, and Other Generalized Linear Model*, Thousand Oaks, CA: Sage Publications.

Liao, Z. Q. and Cheung, M. T. (2001) Internet-based e-shopping and consumer attitudes: an empirical study,' *Information & Management*, 38(5): 299–306.

Linder, R., Geier, J., and Koelliker, M. (2004) 'Artificial neural networks, classification trees and regression: Which method for which customer base?,' *Journal of Database Marketing and Customer Strategy Management*, 11(4): 344–356.

Little, R. J. A. and Rubin, D. B. (1989) 'The analysis of social-science data with missing values,' *Sociological Methods & Research*, 18(2-3): 292–326.

Little, R. J. A. and Rubin, D. B. (2002) *Statistical Analysis with Missing Data* (2nd ed.), Hoboken, NJ: John Wiley & Sons.

Louviere, J. J. and Hensher, D. A. (1983) 'Using discrete choice models with experimental design data to forecast consumer demand for a unique cultural event,' *Journal of Consumer Research*, 10(3): 348–361.

Louviere, J. J. and Woodworth, G. (1983) 'Design and analysis of simulated consumer choice or allocation experiments – an approach based on aggregate data,' *Journal of Marketing Research*, 20(4): 350–367.

Louviere, J. J., Hensher, D. A., and Swait, J. D. (2000) *Stated Choice Methods: Analysis and Application*, Cambridge, UK: Cambridge University Press.

Lundahl, D. S. and McDaniel, M. R. (1988) 'The panelist effect – fixed or random,' *Journal of Sensory Studies*, 3(2): 113–121.

Lusk, J. L., Feldkamp, T., and Schroeder, T. C. (2004) 'Experimental auction procedure: Impact on valuation of quality differentiated goods,' *American Journal of Agricultural Economics*, 86(2): 389–405.

Mahmood, M. A., Bagchi, K., and Ford, T. C. (2004) 'On-line shopping behavior: Cross-country empirical research,' *International Journal of Electronic Commerce*, 9(1): 9–30.

Marks, S. and Dunn, O. J. (1974) 'Discriminant functions when covariance matrices are unequal,' *Journal of the American Statistical Association*, 69(346): 555–559.

Marsh, C. and Scarbrough, E. (1990) 'Testing 9 hypotheses about quota sampling,' *Journal of the Market Research Society*, 32(4): 485–506.

Maruyama, G. M. (1998) *Basics of Structural Equation Modeling*, Thousand Oaks, CA: Sage Publications.

Maxwell, J. A. (2005) *Qualitative Research Design: An Interactive Approach* (2nd ed.), Thousand Oaks, CA: Sage Publications.

McCullagh, P. P. and Nelder, J. A. (1989) *Generalized Linear Models*, Boca Raton, FL: CRC Press.

McCulloch, R. and Rossi, P. E. (1994) 'An exact likelihood analysis of the multinomial probit model,' *Journal of Econometrics*, 64(1-2): 207–240.

McCullough, B. D. (1998) 'Assessing the reliability of statistical software: Part I,' *American Statistician*, 52(4): 358–366.

McCullough, B. D. (1999) 'Assessing the reliability of statistical software: Part II,' *American Statistician*, 53(2): 149–159.

McQuarrie, A. D. R. and Tsai, C. L. (1998) *Regression and Time Series Model Selection*, Singapore: World Scientific Publishing.

Mead, A. (1992) 'Review of the development of multi-dimensional-scaling methods,' *Statistician*, 41(1): 27–39.

Mehdizadeh, M. (1990) 'Loglinear models and student course evaluations,' *Journal of Economic Education*, 21(1): 7–21.

Milligan, G. W. and Cooper, M. C. (1985) 'An examination of procedures for determining the number of clusters in a data-set,' *Psychometrika*, 50(2): 159–179.

Morrison, D. G. (1969) 'On the interpretation of discriminant analysis,' *Journal of Marketing Research*, 6(2): 156–163.

Moschis, G. P. and Churchill, G. A. (1978) 'Consumer socialization – theoretical and empirical analysis,' *Journal of Marketing Research*, 15(4): 599–609.

Moser, C. A. and Stuart, A. (1953) 'An experimental study of quota sampling,' *Journal of the Royal Statistical Society Series A-General*, 116(4): 349–405.

Moses, L. W. (1952) 'Non-parametric statistics for psychological research,' *Psychology Bulletin*, 49: 122–143.

Mulaik, S. A. and McDonald, R. P. (1978) 'Effect of additional variables on factor indeterminancy in models with a single common factor,' *Psychometrika*, 43(2): 177–192.

Naes, T., Kubberod, E., and Sivertsen, H. (2001) 'Identifying and interpreting market segments using conjoint analysis,' *Food Quality and Preference*, 12(2): 133–143.

Narula, S. C. and Wellington, J. F. (2002) 'Multiple objective optimization problems in statistics,' *International Transactions in Operational Research*, 9(4): 415–425.

Neal, W. D. and Wurst, J. (2001) 'Advances in market segmentation,' *Marketing Research*, 13(1): 14–18.

Neslin, S. A. (1981) 'Linking product features to perceptions – self-stated versus statistically revealed importance weights,' *Journal of Marketing Research*, 18(1): 80–86.

Netemeyer, R., Bentler, P. M., Bagozzi, R., Cudeck, R., Cote, J., Lehmann, D. R., McDonald, R. P., Heath, T., Irwin, J., and Ambler, T. (2001) 'Structural equations modeling and statements regarding causality,' *Journal of Consumer Psychology*, 10(1–2): 83–100.

Niedrich, R. W., Sharma, S., and Wedell, D. H. (2001) 'Reference price and price perceptions: a comparison of alternative models,' *Journal of Consumer Research*, 28(3): 339–354.

Ofir, C., Reddy, S. K., and Bechtel, G. G. (1987) 'Are semantic response scales equivalent,' *Multivariate Behavioral Research*, 22(1): 21–38.

Olejnik, S. and Algina, J. (2003) 'Generalized eta and omega squared statistics: measures of effect size for some common research designs,' *Psychological Methods*, 8(4): 434–447.

Ostrom, C. W. (1990) *Time Series Analysis: Regression Techniques*, Thousand Oaks, CA: Sage Publications.

Parsons, L. J. (1976) 'Rachet model of advertising carryover effects,' *Journal of Marketing Research*, 13(1): 76–79.

Pearl, J. (2000) *Causality: Models, Reasoning, and Inference*, Cambridge, UK: Cambridge University Press.

Peter, J. P. (1981) 'Construct-Validity – A review of basic issues and marketing practices,' *Journal of Marketing Research*, 18(2): 133–145.

Peterson, R. A. (1994) 'A meta-analysis of Cronbach coefficient-alpha,' *Journal of Consumer Research*, 21(2): 381–391.

Plassmann, H., Ambler, T., Braeutigam, S., and Kenning, P. (2007) 'What can advertisers learn from neuroscience?,' *International Journal of Advertising*, 26(2): 151–175.

Press, S. J. (2004) 'The Role of Bayesian and Frequentist multivariate modeling in statistical data mining,' *Statistical Data Mining and Knowledge Discovery*, H. Bozdogan (Ed.) Boca Raton, FL: CRC Press, pp. 1–14.

Punj, G. and Stewart, D. W. (1983) 'Cluster-analysis in marketing research – review and suggestions for application,' *Journal of Marketing Research*, 20(2): 134–148.

Rabianski, J. S. (2003) 'Primary and secondary data: concepts, concerns, errors and issues,' *Appraisal Journal*, 71(1): 43–55.

Reed, J. F. and Stark, D. B. (2004) 'Robust two-sample statistics for testing equality of means: a simulation study,' *Journal of Applied Statistics*, 31(7): 831–854.

Rencher, A. C. (1992) 'Interpretation of canonical discriminant functions, canonical variates and principal components,' *American Statistician*, 46(3): 217–225.

Rentz, J. O. (1987) 'Generalizability theory – a comprehensive method for assessing and improving the dependability of marketing measures,' *Journal of Marketing Research*, 24(1): 19–28.

Reynolds, T. J. and Gutman, J. (1988) 'Laddering theory, method, analysis, and interpretation,' *Journal of Advertising Research*, 28(1): 11–31.

Reynolds, T. J. and Olson, J. C. (2001) *Understanding Consumer Decision Making: The Means-end Approach to Marketing and Advertising Strategy*, Mahwah, NJ: Lawrence Erlbaum.

Richman, M. B. (1986) 'Rotation of principal components,' *Journal of Climatology*, 6(3): 293–335.

Rigdon, E. E. (1995) 'A necessary and sufficient identification rule for structural models estimated in practise,' *Multivariate Behavioral Research*, 30(3): 359–383.

Ripley, B. D. (1994) 'Neural networks and related methods for classification,' *Journal of the Royal Statistical Society Series B-Methodological*, 56(3): 409–437.

Rossi, P. E. and Allenby, G. M. (2003) 'Bayesian statistics and marketing,' *Marketing Science*, 22(3): 304–328.

Rubin, D. B. (2003) 'Discussion on multiple imputation,' *International Statistical Review*, 71(3): 619–625.

Rudas, D. T. (1998) *Odds Ratios in the Analysis of Contingency Tables*, Thousand Oaks, CA: Sage Publications.

Rupp, A. A., Dey, D. K., and Zumbo, B. D. (2004) 'To Bayes or not to Bayes, from whether to when: Applications of Bayesian methodology to modeling,' *Structural Equation Modeling – A Multidisciplinary Journal*, 11(3): 424–451.

Rutherford, A. (2001) *Introducing ANOVA and ANCOVA – A GLM Approach,* London, UK: Sage Publications.

Samuelson, P. A. (1948) 'Consumption theory in terms of revealed preference,' *Economica,* 15(60): 243–253.

Sanchez, P. M. (1974) 'The unequal group size problem in discriminant analysis,' *Journal of the Academy of Marketing Science,* 2(4): 629–633.

Sandor, Z. and Wedel, M. (2001) 'Designing conjoint choice experiments using managers' prior beliefs,' *Journal of Marketing Research,* 38(4): 430–444.

Sandor, Z. and Wedel, M. (2005) 'Heterogeneous conjoint choice designs,' *Journal of Marketing Research,* 42(2): 210–218.

SAS Institute Inc. (2004) *BASE SAS 9.1 Procedures Guide,* Chicago, IL: SAS Publishing.

SAS Institute Inc. (2004) *SAS/STAT 9.1 User's Guide,* Cary, NC: SAS Institute.

Sawyer, A. G. and Ball, A. D. (1981) 'Statistical power and effect size in marketing research,' *Journal of Marketing Research,* 18(3): 275–290.

Sawyer, A. G. and Peter, P. J. (1983) 'The significance of statistical significance tests in marketing research,' *Journal of Marketing Research,* 20(2): 122–133.

Schafer, J. L. and Graham, J. W. (2002) 'Missing data: Our view of the state of the art,' *Psychological Methods,* 7(2): 147–177.

Scheffe, H. (1959) *The Analysis of Variance,* New York, NY: John Wiley & Sons.

Schervish, M. J. (1987) 'A review of multivariate analysis,' *Statistical Science,* 2(4): 396–413.

Schleifer A. Jr., and Bell, D. E. (Eds., 1995) *Data Analysis, Regression and Forecasting,* Cambridge, MA: Course Technology Inc.

Schlesinger, B. (1962) A survey of methods used to study decision making in the family,' *The Family Life Coordinator,* 11(1): 8–14.

Schmitt, N. and Stults, D. M. (1985) 'Factors defined by negatively keyed items – the result of careless respondents,' *Applied Psychological Measurements,* 9(4): 367–373.

Schriesheim, C. A. and Eisenbach, R. J. (1995) 'An exploratory and confirmatory factor-analytic investigation of item wording effects on the obtained factor structures of survey questionnaire measures,' *Journal of Management,* 21(6): 1177–1193.

Shaffer, J. P. (1995) 'Multiple hypothesis testing,' *Annual Review of Psychology,* 46: 561–584.

Shaw, M. J., Subramaniam, C., Tan, G. W., and Welge, M. E. (2001) 'Knowledge management and data mining for marketing,' *Decision Support Systems,* 31(1): 127–137.

Shepard, R. N. (1962a) 'The analysis of proximities: Multidimensional scaling with an unknown distance function I,' *Psychometrika,* 27(2): 125–140.

Shepard, R. N. (1962b) 'The analysis of proximities: Multidimensional scaling with an unknown distance function II,' *Psychometrika,* 27(3): 219–246.

Smith, T. M. F. (1983) 'On the validity of inferences from non-random samples,' *Journal of the Royal Statistical Society Series A-Statistics in Society,* 146: 394–403.

Solomon, H. and Zacks, S. (1970) 'Optimal design of sampling from finite populations: a critical review and indication of new research areas,' *Journal of the American Statistical Association,* 65(330): 653–677.

Solomon, M. R. (2007) *Consumer Behavior* (7th ed.), Upper Saddle, NJ: Prentice-Hall.

Spearman, C. E. (1904) '"General intelligence" objectively determined and measured,' *American Journal of Psychology,* 5: 201–293.

Spector, P. E., VanKatwyk, P. T., Brannick, M. T., and Chen, P. Y. (1997) 'When two factors don't reflect two constructs: How item characteristics can produce artifactual factors,' *Journal of Management,* 23(5): 659–677.

SPSS Inc. (2006) *SPSS Base 15.0 User's Guide,* Chicago, IL: SPSS Inc.

Squire, P. (1988) 'Why the 1936 literary digest poll failed,' *The Public Opinion Quarterly,* 52(1): 125–133.

STATA (2005) *STATA User's Guide: Release 9,* College Station, TX: STATA Press.

Stevens, J. P. (2002) *Applied Multivariate Statistics for the Social Sciences* (4th ed.), Mahwah, NJ: Lawrence Erlbaum.

Stevens, S. S. (1946) 'On the theory of scales of measurement,' *Science*, 103(2684): 677–680.

Takane, Y., Young, F. W., and Deleeuw, J. (1977) 'Non-metric individual differences multi-dimensional scaling – aternating least-squares method with optimal scaling features,' *Psychometrika*, 42(1): 7–67.

Tellis, G. J. (2004) *Effective Advertising: Understanding When, How, and Why Advertising Works*, Thousand Oaks, CA: Sage Publications.

Tenenhaus, M. and Young, F. W. (1985) 'An analysis and synthesis of multiple correspondence-analysis, optimal-scaling, dual scaling, homogeneity analysis and other methods for quantifying categorical multivariate data,' *Psychometrika*, 50(1): 91–119.

Thompson, B. (1984) *Canonical Correlation Analysis: Uses and Interpretation*, Thousand Oaks, CA: Sage Publications.

Thompson, M. E. (1997) *Theory of Sample Surveys*, London, UK: Chapman & Hall.

Thomson, M. (2006) 'Human brands: investigating antecedents to consumers' strong attachments to celebrities,' *Journal of Marketing*, 70(3): 104–119.

Thurstone, L. L. (1927) 'A law of comparative judgment,' *Psychological Review*, 34: 273–286.

Thurstone, L. L. (1935) *The Vectors of Mind*, Chicago, IL: University of Chicago Press.

Thurstone, L. L. (1946) 'Theories of Intelligence,' *The Scientific Monthly*, 62(2): 101–112.

Tibshirani, R., Walther, G., and Hastie, T. (2001) 'Estimating the number of clusters in a data set via the gap statistic,' *Journal of the Royal Statistical Society Series B-Statistical Methodology*, 63(2): 411–423.

Timm, N. H. and Mieczkowski, T. A. (1997) *Univariate & Multivariate General Linear Models: Theory and Applications Using SAS Software*, Cary, NC: SAS Publishing.

Torgerson, W. S. (1958) *Theory and Methods of Scaling*, New York, NY: John Wiley & Sons.

Train, K. E. (2003) *Discrete Choice Methods with Simulation*, Cambridge, UK: Cambridge University Press.

Upton, G. and Cook, G. (2006) *A Dictionary of Statistics*, Oxford, UK: Oxford University Press.

van Kleef, E. (2006) *Consumer Research in the Early Stages of New Product Development: Issues and Applications in the Food Domain*, PhD thesis, Wageningen University.

van Kleef, E., van Trijp, H. C. M., and Luning, P. (2006) 'Internal versus external preference analysis: An exploratory study on end-user evaluation,' *Food Quality and Preference*, 17(5): 387–399.

Van der heijden, P. G. M., Defalguerolles, A., and Deleeuw, J. (1989) 'A combined approach to contingency table analysis using correspondence analysis and log-linear analysis,' *Journal of the Royal Statistical Society Series C-Applied Statistics*, 38(2): 249–292.

Varian, H. R. (2006). Revealed Preference. *Samuelsonian Economics and the Twenty-First Century*. M. Szenberg, L. Ramrattan, and A. A. Gottesman (Eds.), Oxford, UK: Oxford University Press, pp. 99–115.

Velicer, W. F. and Jackson, D. N. (1990a) 'Component analysis versus common factor-analysis – someissues in selecting an appropriate procedure,' *Multivariate Behavioral Research*, 25(1): 1–28.

Velicer, W. F. and Jackson, D. N. (1990b) 'Component analysis versus common factor-analysis – some further observations,' *Multivariate Behavioral Research*, 25(1): 97–114.

Velicer, W. F., Peacock, A. C., and Jackson, D. N. (1982) 'A comparison of component and factor patterns – a Monte-Carlo approach,' *Multivariate Behavioral Research*, 17(3): 371–388.

Viswanathan, M. (2005) *Measurement Error and Research Desig*, Thousand Oaks, CA: Sage Publications.

Welsch, R. E. (1977) 'Stepwise multiple comparison procedures,' *Journal of the American Statistical Association*, 72(359): 566–575.

Wilcoxon, F. (1945) 'Individual comparisons by ranking methods,' *Biometrics Bulletin*, 1(6): 80–83.

Wilkerson, M. and Olson, M. R. (1997) 'Misconceptions about sample size, statistical significance, and treatment effect,' *Journal of Psychology*, 131(6): 627–631.

Wind, Y. and Denny, J. (1974) 'Multivariate analysis of variance in research on effectiveness of Tv commercials,' *Journal of Marketing Research*, 11(2): 136–142.

Winer, R. S. (1999) 'Experimentation in the 21st century: the importance of external validity,' *Journal of the Academy of Marketing Science*, 27(3): 349–358.

Wiseman, F. and Rabino, S. (2002) 'The effects of sample size and the specification of hypotheses in the substantiation of superiority and parity Claims,' *The Marketing Review*, 2(4): 427–439.

Wong, M. A. and Lane, T. (1983) 'A kth nearest neighbor clustering procedure,' *Journal of the Royal Statistical Society Series B-Methodological*, 45(3): 362–368.

Wright, B. D. (1977) 'Solving measurement problems with Rasch model,' *Journal of Educational Measurement*, 14(2): 97–116.

Wu, C. and Hsing, S. (2006) 'Less is more: How scarcity influences consumers' value perceptions and purchase intents through mediating variables,' *Journal of the American Academy of Business*, 9(2): 125–132.

Yan, M. and Ye, K. (2007) 'Determining the number of clusters using the weighted gap statistic,' *Biometrics*.

Young, F. W. (1981) 'Quantitative analysis of qualitative data,' *Psychometrika*, 46(4): 357–388.

Zeithaml, V. A. (1988) 'Consumer perceptions of price, quality, and value – a means-end model and synthesis of evidence,' *Journal of Marketing*, 52(3): 2–22.

Zhang, T., Ramakrishnon, R., and Livny, M. (Ed., 1996) *BIRCH: An Efficient Data Clustering Method for Very Large Databases*.

Index

Page numbers in **bold** indicate entries with a concise definition of the subject, while those in *italics* refer to the main discussion about the topic.

Accuracy 9–10, 103, 106, 106–9, *114–117*, 125, 182, **375**
ACNielsen 35, 41, 43, 45
Adjusted R^2 *see* R^2
Administration Method 11, 29, 32, 34, 40, 46, 48, 51–53, *55–58*, 60, 62–64, 67, 74–80, 122, 124, **375**
  See also Computer-assisted Interviews, Electronic Interviews, Mail Interviews, Mall-intercepts, Personal Interviews, Telephone Interviews
Agglomeration Schedule **272**, 278
Akaike Information Criterion *see Information Criteria*
Allocation of Sample Size (in stratified sampling) 110, *112–113*, 118
  Neyman (Optimal) Allocation 110, *113*
  Proportional Allocation *113*
  See also Stratified Sampling
Alpha Factoring **225**
ALSCAL 290, 295, **299**
AMOS 316, 322, *325–333*, **335**
Analysis of Covariance (ANCOVA) *164–165*, 168, 191, 199, 249, 259, **375**, 378
  Multivariate ANCOVA (MANCOVA) 165, 199, 378
  See also General Linear Model
Analysis of Variance (ANOVA) 93, 103, 106, 145, *150–168*, 173–174, 179, 181, 183–184, 187–189, 191, 199, 201, 249, 253–254, 259–260, 270–271, 274, **375**, 376–378, 381–383, 385
  Effects *see Effects (in ANOVA)*
  Multi-way ANOVA *160–162*, 164–165, 167, 181, 191, 199, 378–379, **382**
  Multivariate ANOVA (MANOVA) 161, *163*, 165, 167–168, 191, 199, 378–379, **381**

Non-parametric ANOVA *159–160*, 165–167
One-way ANOVA *151–160*, 161–162, 165, 167, 253–254, 259
  See also General Linear Model
Anderson-Rubin Method **226**
Annual Population Survey *64–68*
Area Sampling 11, 56, 66, **111**, 114, *119*, 128
  See also Cluster Sampling
Aristotle 353
Arnold's Criterion **270**
Arrow, Kenneth Joseph 76
Association 92, 99, 167, 171, 175, 178, 193, *196–207*, 214, 216, 247, 300–301, 303–309, **375**, 376–377, 379–381
  Association Measures *196–207*, 214, 216, 301, 303–309, 312, **375**
  Chi-square Association Test *197–199*, 303–304, **376**
  Partial Association Table 203–204
  See also Contingency Coefficient, Cramer's V, Goodman and Kruskal's Lambda, Uncertainty Coefficient, Gamma Statistic, Somer's d Statistic, Kendall's Tau b and Tau c Statistics
Attitudes 3–4, 8, 32–34, 35, 44, 50, 63–66, 69–70, 76, 174, 208, 219, 318, 324, 333
Attitudinal Survey 33, **35**, 66, 286
Attribute-based Methods *see Compositional Methods*
Attribute-free Methods *see Decompositional Methods*
Average *see Mean*
Average Linkage Method **267**

Bacon, Francis 353
Banding *see Frequency Classes*
Bar Chart **86**, *86–91*, 99

Bartlett's
  Likelihood Ratio Test 227
  Scores 226
  Test of Sphericity 227, 236–237
Bayes Information Criterion (BIC) *see*
  Information Criteria
Bayes Rule **355**, *355–357*
  See also Bayesian estimation
Bayes, Thomas 352, 355
Bayesian Estimation *354–358*
  Bayes Rule **355**, *355–357*
  Non-informative Priors 357
  Posterior Probability 355–357
  Prior Probability 355–357
Beliefs 4, **69**, 324–326
Bias **5**, 8–9, 30, 32, 53–57, 60, 81, 83, 93, 111,
  118–119, 123–124, 180, 227, 232, 336, 385
  See also Systematic Error
Binary Variable *see Dummy variable*
Binomial
  Distribution 120–121, 340, 342, 345, **368**
  Test **143**
Blake, William 355
Bonferroni's Test **155**, 158
Bortkiewicz, Ladislaus Josephovic 374
Box Plot **86**, *86–87*, 97
Box's M test **253**
Box-Whisker Diagram **86**
British Household Panel 35, **40**, 44
Brown & Forsythe Test 146, 159
Bunhill Fields 355
Burt Table 304, **375**

Canonical Correlation Analysis (CCA)
  208–216, 317, **375**
  Canonical Correlation 208, 253–254,
    257–258, **375**
  Canonical Coefficients **208–209**, 211
  Canonical Loadings **209**, 209–210,
    212, 216
  Canonical Redundancy Index **209**
  Canonical Roots **209**
  Canonical Scores **209**
  Canonical Variates **208**, 208–210
  In SAS 209–214
  In SPSS 214
  Nonlinear CCA **209**
  Partial Canonical Correlation Analysis
    209, 214
Cases *see Observations*
Case-wise Deletion *see List-wise Deletion*
Causality 174, 179, 184, 210, **318**, 320–321,
  383, 385

Census 64, 80, 105,122, **375**
Central Tendency Measures 85, 92, 100, 109,
  369–372, **370**
  See also Mean, Median, Mode
Characteristic Roots *see Eigenvalues*
Chartered Institute of Marketing 3
Charts 77, *85–91*, 99–100, 102
  *See also Bar Chart, Pie Chart, Line Chart,
    Histogram, Pareto Chart, Box Plot, Q-Q
    Plot, P-P Plot, Error Bar Plot,
    Box-Whisker Diagram, Scatter Diagram,
    Stacked Chart, Clustered Chart*
Chebychev Distance 266
Chi-square 143, 197, 199, 202–206, 214–215,
  301, *303–306*, 312, 322, 327–329, 343, 345
  Association Test *197–199*, 303–304, **376**
  Distribution 197, 202, 253, 376, 380
  Goodness-of-fit Test 143, 171, 202–206,
    253–254, 345–346, **376**
Choices
  Choice Experiments **339**
  Choice Models 337–339, *347–350*
  Choice Set 61, 347–350
  Experimental Design 140, 159, 163, 167,
    348–350
  *See also Discrete Choice Models, Conjoint
    Analysis*
City-block Distance 265–266
Classification Trees 261
Cluster analysis 52, 66, 219, 230, 244,
  247–249, 260, *263–282*, 285, 289–290,
  295, **376**
  Agglomeration Schedule **272**, 278
  Cluster Analysis and Outliers
    274–278, 281
  Cluster Analysis in SAS 276–279
  Cluster Analysis in SPSS 272–276
  Clustering algorithms *265–268*
  Hierarchical Clustering Methods
    266–271, 272, 276–278, 280, 282, **379**
  Non-hierarchical Clustering Methods
    266–271, 274, 276, 280, **382**
  Number of Clusters *See Clusters
    (Number of)*
  Two-step Cluster Analysis *270–271*, 272,
    274, 276
  Validation of Cluster Analysis 271,
    279–280
  *See also Single Linkage Method, Complete
    Linkage Method, Average Linkage Method,
    Ward's Method, k-means Clustering,
    Fuzzy Cluster Analysis*

Cluster Sampling 56, 83, *110–114*, **111**, 119, 123, 125–127

Clustered Chart **87**

Clusters (Number of) 265, 267, *268–270*, 271–282, 379, 382

　See also Arnold's Criterion, Cubic Clustering Criterion, k-nearest Neighbor, Pseudo F Statistic, Pseudo t2 Statistic, Agglomeration Schedule, Dendrogram, Icicle, Scree Diagram

Cochran Q Test 160, 166

Coefficient Ratio **329**

Coefficient of Determination *see R^2*

Coefficient of Variation 100, **372**

COICOP classification 11, 33, 35–36, **37–39**, 40

Collinearity *186*, 188, 190–191, 230, 320–321, **376**

　Condition Index **186**

　See also Regression

Common Factors 220, 222–224, 226–228

Communalities *223–224*, 226, *234–236*, 238, 243, 310, **376**

Comparative Fit Index (CFI) **322**, 330, 332

Comparative Scales **8**, 8–9, 287

　See also Constant Sum Scaling, Guttman Scale, Pair-wise Comparison, Q-Sorting, Rank Order Scale

Complete Linkage Method **267**

Compositional Methods *287–288*, 291, 303, **376**

Computer-assisted Interviews 58, 80, 106, 124

　Computer-assisted Personal Interviews (CAPI) 34, 36, 39–41, 43, *55–56*, 58, 74, **375**

　Computer-assisted Telephone Interviews (CATI) 31, 40, *53–54*, 58, 66–68, 74, **375**

　Computer-assisted Web Interviews (CAWI) *57–58*, **375**

Condition Index **186**

Confidence 136, 146, 306–307

　Confidence Interval(s) 86, 89–90, 103, 106, 115, 117, *130–136*, 137, 141, 146, 149, 156, 181–182, 205–206, 329, 343, 346, 355, 357, **376**, 380

　Confidence Level 115–117, 120–121, *131–133*, 134, 136, 146, 149, 175, 181, 184, 187, 209, 305, 322, 355, **376**, 380, 385

Confirmatory Factor Analysis 66, 317, *319–322*, 333, 376, **378**, 385

　See also Structural Equation Modeling

Conjoint Analysis 315, 337, 339, *347–350*, 351, 358, **376**, 378

　Adaptive Conjoint Analysis 348–349

　Choice-based Conjoint Analysis **349–350**

　Conjoint Analysis in SAS 349

　Conjoint Analysis in SPSS 349

　Experimental Design 140, 159, 163, 167, *348–350*

　Ideal Point Model *348*, **379**, 381

　Part-worth Model **348**

　Traditional Conjoint Analysis **348**

　Vector Linear Model **348**

　See also Utility, Choice

Consistency 9–10, 23, 54, 57–58, 68, *77–80*, **78**, 84, 97–100, 246, 384

　Statistical Consistency 109, **320**

Constant Sum Scaling **8**

Consumer Expenditure Survey (US) *39–40*, **122**

Consumer Behaviors 2, 35, 45, 47, **63**, 65, 76, 286, 315–316, 333, 338, 348, 363

Contingency

　Coefficient **197**

　Tables 173, 189, *196–199*, 200–201, 205, 214–215, 301, 303, 305–306, 312, 375, **376**, 379

　See also Association, Frequency

Continuous Rating Scale **9**

Contrasts (in ANOVA) 154, **376**

　Difference **154**

　Helmert **154**

　Linear **154**, 156, 169

　Polynomial **154**, 156

　Repeated **154**

　Reverse Helmert *see Difference Contrasts*

　Simple **154**

　See also Planned Comparisons

Convenience Sampling 105, *123–124*

Correlation *173–179*, **376**

　Bivariate Correlation Coefficient *see Pearson's Correlation Coefficient*

　Correlation in SAS *177–179*

　Correlation in SPSS *176–177*

　Correlation Matrix 224, 227, 232–234, 236, 304, **377**

　Multiple Correlation Coefficient 176, 182

　Part (Semi-partial) Correlation 172, 175, 186

　Partial Correlation Coefficient *175–179*, 186–188, **383**

　Pearson's Correlation Coefficient **174**, *174–179*, 183, 186, 193

Correlation (*continued*)
Zero-order Correlation *see Pearson's Correlation Coefficient*
See also Canonical Correlation
Correspondence Analysis 247, 285, 287, 300–313, **377**
Burt Table 304, **375**
Column Masses **303**, 304
Column Profile **303**, 306
Correspondence Table 302, 306–307, 310
In SAS 311
In SPSS 305–311
Inertia *304–310*, **380**
Multiple Correspondence Analysis 304, 310–313
Normalization Methods 304–305, **305**, 311
Row Masses **303**, 304
Row Profile **303**, 306
Covariance 183, *373–374*
Covariance Matrix 222–225, 231–233, 236, 242–243, 251, 253, 301, 321–322, 328–329, *373–374*, **374**, 376–377, 379
Cramer's V **197**, 199–200, 214
Cronbach's Alpha **10**, 68, 178, 225, 377
See also Reliability
Cross-tabulation *see Tabulation*
Cubic Clustering Criterion 270, 278

Data
Collection 1, 2, 5, 11, 23, 26, 28, 30–33, 35, 44, *46–76*, 97, 100, 287, 292–293, 315, 337–339, 347–348, 376, 378, 383, 385
Data Reduction Techniques *218–221*, 228, 230, 236, 242, 244–245, 265, 280, 292, 301, 378, 383
Editing 77–78, **78**, *78–80*, 97, 100
Post-editing 77–78, *81–84*, 96, 100
Preparation *78–84*
Quality 1, 28, 76, *80–84*, 93–98
Sources *see Secondary data*
Tabulation 33, 84–85, *91–94*, 99, 173, 196, 220, 238, 244, 278, 375
Warehousing **354**
Weights 78, 80, 82–84, **83**, 288
Data (Type of)
Primary data 1, 11, 23, 27–29, *46–77*, **383**
Secondary data 1, 11, 23, *27–45*, 47, 49, 77, **385**
Scan data 30, 35, *41–43*
Time series data 35, 184, 192
Data Mining 66, 315, *353–354*, 358
Data Problems *80–84*, 93–98

Consistency 9–10, 23, 54, 57–58, 68, *77–80*, **78**, 84, 97–100, 246, 384
Missing Data 22, 62, 77, 79, *80–83*, 84–85, 96–97, 100, 102, 310, 326, 380, **381**, 382
Outliers 1, 77, 80, 84–85, 87–88, *93–100*, 102, 146, 253, 266, 274, 276, 280–281, **382**
Data Reduction *218–221*, 228, 230, 236, 242, 244–245, 265, 280, 292, 301, 378, 383
See also Factor Analysis, Principal Component Analysis
Data Warehousing **354**
Data-set *78–85*
Coding 62, *78–81*
EFS Sample Data-set *11–18*, 87, 93, 144, 162
Trust Sample Data-set 11, 12, *18–21*, 23, 91–93, 97, 100, 134, 137, 140, 143–144, 176, 187, 191, 199, 208, 220, 235, 249, 301, 322–325, 342
Decompositional methods *287–289*, 348, 350, 376, **377**
Defoe, Daniel 355
Degeneracy (in MDS) *294–295*
See also Shepard's RNI, DeSarbo Intermixedness Index
Degrees of Freedom 54, 132, 138, 141, 145, 153, *321–322*, 327–329, **377**
Dendrogram **269**, 272, 278
DeSarbo's Intermixedness Index 295
Design Matrix 349
Discrete Choice Models 50, 66, 191–192, 248–249, 260, 262, 315, *337–347*, 349, 376, **377**, 379–384
Link Function 189, 192, *340–342*, 345–347, 349, 351, 379, **380**, 384
Logistic Regression 189, 191, 249, 251, *339–347*, 349, 379, **380**
Logit Model 66, 192, 340, 344–345, 379, **380–381**
Marginal effects 202, 204, 345, **347**
Multinomial Logit 249, *341–342*, 345, 349, 351, 376, 379, **381**
Multinomial Probit *341–342*, 358, 379, **381**
Odds Ratio 205, 217, 341–344, **382**
Ordered Logit 341, *345–346*, **382**
Ordered Probit 189, 341, 349, **382**
Probit Model 192, 341, 344, 347, 349, 379, **384**
See also Generalized Linear Model
Discriminant Analysis (DA) *247–260*, **377**
Cut-off Point 252, 256, 260
Discriminant Coefficients 251–252, 254

Discriminant Function *249–262*, **377**
Discriminant Loadings 254, 257
Discriminant Score *249–251*, 257, 260
Z Score **250**, 272
Fisher's Linear Discriminant Analysis 250–251
In SPSS 251–259
Leave-one-out Classification 259
Multiple Discriminant Analysis (MDA) 247, 249–251, 253, *254–260*, **377**
Quadratic Discriminant Function 251, 253, 261
Stepwise Discriminant Analysis 259
Structure Matrix 254, 257, 261
Dispersion 107, 152, 294, *371–372*
See also Variability
Distance Measure 7, 247, *265–266*, 272, 280, 289, 303
See also Mahalanobis Distance, City-block Distance, Euclidean Distance, Chebychev Distance, Minkowski Distance
Distribution (of probability) 148, *367–373*, **383**
Binomial 120–121, 340, 342, 345, **368**
Chi-square 197, 202, 253, 376, 380
Conditional 342, 355–357
F **145**, 153, 253
Gaussian *see Normal*
Joint **93**, 356
Multinomial 204, 215, 306
Normal 5–6, 87, 99, 106, 115, 120–121, *131–138*, 139, 141, 143–146, 148–149, 159, 174, 262, 321, 349, 357, 368–372, 386
Poisson 143, 148, 204, **368**
P-Value *133–138*, 141–142, 175, 179, 181, 184, 189, 253–254, 259, 322, 329, 343, **384**
Rectangular *see Uniform*
Studentized Range 155
Student-*t* 121, *132–133*, 137–138, 141, 148–149
Unconditional 342, 356–357
Uniform *120–121*, 368, 371
Dummy Variable 101, 160, 162, *191*, 199, 201, **377**
Trap 162, 169, **191**
Duncan's Procedure 155, 158
Dunnet Ç Test 155, 158
Dunn-Sidak Test 157

Effect size (of a test) 137, *155*, 168, **377**
See also Eta-squared Statistic, Analysis of Variance

Effects (in ANOVA)
Fixed Effects **159**, 168
Mixed Effects **159**
Random Effects **159**, 166, 168
Efficiency **109**, 111, 114, 118–119, 124, 185
Eigenvalues 186, 209, 231–232, 234–237, 253, 304, 306, *366*, **377–378**
Eigenvectors 231, 235, 241
Electoral Survey 30–31
Electronic Interviews 52–54, *57–58*, 375
Equamax Rotation **229**
Error
Coverage Error 36, **118**
Error Theory 6, 25
Experimentwise Error *see Family-wise Error*
Family-wise Error 151, **154**, 155, 168
Interviewer Error **48**
Measurement Error *4–6*, 10, 23, 32, 47, 76, 104, 107, 230, 289, 318, 321, 368–369
Non-response Error 48, 59, 62, 78, 105, *118–119*, 125, **382**
Non-sampling Error *47–48*, 53, 58, 61–62, 74, 103–104, 106, 118–119, 125, **382**
Random Error 1, 5, 10, 23, 47, 159, 197, 222, 317, 342, 359–360, **384**
Researcher Error **48**
Respondent Error **48**
Sampling Error 31, 47–48, 52, *103–107*, 115, 124–125, 135, 143, 151, 153, 155, 174, 271, **384**
Survey (Total) Error 45, *47–48*, **48**, 104, 106, 382, 384
Systematic Error 5–6, 23, 48, **385**
Type I Error *136–137*, 146, 151, 154–156, 163, 376, **385**
Type II Error *136–137*, 146, 155, 383, **385**
See also Bias
Error-Bar Plot *86–88*
Estimator 114–116, 118
Consistent 109, **320**
Efficient **109**, 111, 114, 118–119, 124, 185
Sample Estimators **52**, *106–109*, 114–116, 118–119, 126–127, 129, 131, 134
Unbiased 109, 117–118, 153
Eta-squared statistic 155
Euclidean Distance **265**, 272, 289, 305
EU-SILC **41**
Expected Value 109, 129, 192, 340, 342, 360, **370**
See also Mean
Expenditure and Food Survey 11, 25–26, 33, *36–39*, **122**

Experimental Auctions *see Choice Experiments*

Experimental Design 140, 159, 163, 167, *348–350*, **378**

Full Factorial Design 163–164, 348, 376, **378**

Fractional Factorial Design 376, **378**

See also Effects (in ANOVA)

Experimentwise Error *see Family-wise Error*

External Preference Mapping *291–292*, 299, **378**, 380

F-Distribution **145**, 153, 253

F-Test *145–146*, 155–156, 158, 161–162, 164, 181, 184, 187, 191, 270–271, 355, **378**

Face-to-face Interviews *see Personal Interviews*

Factor Analysis

Confirmatory 66, 221, 227, *316–333*, 376, **378**, 385

Exploratory *218–229*, **378**

Factor Interpretation 227–229

Factor Loadings 221, 223–229, 236, 238, 250, 291, 317, **378**

Factor Rotation 225–226, *227–230*, 237–238, 240, **382**

Factor Scores 222, 224–226, 230, 236, 238, 317, **374**

In SAS 239

In SPSS *234–239*

Indeterminacy Problem 225–226, 228

Number of Factors 220, 221, 224, *226–227*, 229–230, 232, 236, 239, 244, 246, 317, 319–320, 384

See also Communalities

Factorial ANOVA *see Multi-way ANOVA*

Factorial Design *see Experimental Design*

Factors *see Common Factors*

Family-wise Error 151, **154**, 155, 168

Field, Andy xii, xiv, 147, 154–155, 160, 168–169

Fisher, Sir Ronald Aylmer 251, 262, 355

Fixed Effects **159**, 168

Frequencies

Absolute 99, 101, 201, 205, 303, *367*, 370

Classes 83–84, 86–87, 91–92, 368, *370–371*, 378, *383*

Contingency Tables 173, 189, *196–199*, 200–201, 205, 214–215, 301, 303, 305–306, 312, 375, **376**, 379

Cumulative 92, 99, *367–371*

Distribution 86–87, 91–92, 197, *367–371*, 376, **378**

Relative 6, 93, 197, 303–304, *367–371*

See also Association, Contingency, Log-linear Analysis

Frequentist Paradigm *355–356*

See also Bayesian Paradigm

Friedman's test 160, 166

Function

Deterministic Function 230, **360**

Exponential Function **362**

Logarithmic Function **362–363**

Natural Logarithm **362**

Parameters 82–83, 160, 172, 179–181, 183, 193, 201, 205–207, 317, 321, 333, 338–340, 355–357, *360–364*

Stochastic Function 179, **360**

Fuzzy Cluster Analysis **281**

Gabriel's Test 155

Galton, Sir Francis 174, 221

Gamma Statistic *198–199*

Gauss, Carl Friedrich 3, 355

Gaussian curve *see Normal Curve*

Gaussian Distribution *see Normal Distribution*

General Household Survey 34

General Linear Model *160–161*, 165, 168, 173, 181, 189, 191–192, 199, 201, 214, 377, **378**, 380–382

General Log-linear Model (GLLM) *199–206*, 214, **379**, 381

Generalizability **10**

Generalized Linear Model 189, 192, 216, *344–345*, 347, 351, **379**, 380–382, 384

Link Function 189, 192, *340–342*, 345–347, 349, 351, 379, **380**, 384

See also Discrete Choice Models

Gfk 35, 41, 43, 45

GLM *see General Linear Model*

Goodman and Kruskal's Lambda **197**, 214

Goodness-of-fit *181–185*, 188, 205–206, 224, 227, 271, 288, **379**, 380, 385

In Discrete Choice Models *343–346*

In MDS *290*, 294

See also R²

Goodness-of-fit in SEM 320, 322, 328–330, 332–333

See also Minimum Sample Discrepancy (CMIN), Root Mean Square Residual (RMR), Goodness-of-fit Index (GFI), Normed Fit Index (NFI), Relative Fit Index (RFI), Incremental Fit Index (IFI), Tucker-Lewis Coefficient (TLI), Comparative Fit Index (CFI),

Non-centrality Parameter (NCP), Root Mean Square of Error Approximation (RMSEA), Test of Close Fit (PCLOSE), Hoelter's Critical N
Goodness-of-fit Index (GFI) **322**
Guttman scale **8**

Haphazard Sampling 29, **124**
Heteroskedasticity 180, 226, **379**
Hierarchical Clustering Methods *266–271*, 272, 276–278, 280, 282, **379**
  Agglomerative Hierarchical Methods 266, **379**
  Divisive Hierarchical Methods 267, **379**
Histogram 86, *86–88*, 91, 99–100, 368
Hochberg's GT2 Test **155**
Hoeffding's D Statistic **178**
Hoelter's Critical N **322**, 330–332
Homoskedasticity *180*, 195, 339, **379**
Hosmer and Lemeshow test **343**
Household Survey 34, 76
Household Budget (Expenditure) Survey 27, 30, **33**, 34, *36–40*, 64–68, 119, 122, 125
Hsu's MCB Test **155**
Hypothesis Testing 5, 93, 103, 106, 115, *130–147*, 151, 160, 174–175, 181, 193, 197, 317, 358, 369, **379**
  Alternative Hypothesis *135–137*, **375**
  Critical Values *130–133*, **377**
  Effect Size 137, *155*, 168, **377**
  In SPSS 142
  Non-parametric Tests 130, *142–145*, 146–147, 159–160, 167, 175–176, 178, 193, 270, **382**
  Null Hypothesis *130–136*, **382**
  One-tailed Tests **136**, 136–139, 145
  Parametric Tests *130–142*
  Power *136–137*, 146, 148, 155, 167, 205, **383**, 385
  P-value *133–138*, 141–142, 175, 179, 181, 184, 189, 253–254, 259, 322, 329, 343, **384**
  Significance Level 117, 129, 133, 135–138, 143, 145–146, 149, 151, 153, 157, 162, 168, 175, 185, 189, 199, 205, 217, 228, 236, 253, 329, *369*, 376, 382, **385**
  Tests on One Mean *137–139*
  Tests on Proportions *145–146*
  Tests on Two Means *139–142*
  Tests on Variances *145–146*
  Two-tailed tests *135–141*, 144–145

Type I error *136–137*, 146, 151, 154–156, 163, 376, **385**
Type II error *136–137*, 146, 155, 383, **385**

Icicle **272**
Ideal Point 285–287, *291–292*, 297, 298–299
  Ideal Point Model *348*, **379**, 381
Identification *321–323*, 329, 335, 377, 378, **379**
  Independence Model 322, 330
  Saturated Model *201–202*, 204, 322, 330–331
Image Factoring *224–225*, **225**
Incentives 57, 62, 67, 78
Incremental Fit Index (IFI) **322**
Independence **140**, 144, 160, 178, 186, 193, 197–198, 207, 214, 271
  Independence Model 322, 330
Indicator (in SEM) *see Manifest Variable(s)*
INDSCAL 290
Inertia *304–310*, **380**
Inference 5, 29, 50–51, 103–109, 112, 123–124, 133, 185, 347, *368–369*, **380**
Information Criteria 188, 195, **282**, 322, 330–331
Intention to Behave 3, 65, 69, 316, 324–326, 329
Internal Preference Mapping *291–293*, 295, 297, 378, **380**
Interquartile Range 97, 134–135, **371**
Interviewing Method *see Administration Method*
Interviews *see Administration Method*
Ipsos 41, 43, 45
IRI International 35, 41, 43, 47

Jonckheere-Terpstra Test **160**, 166
Jöreskog, Karl Gustav 225, 325, 335
Judgmental Sampling **124**

Kaiser's Rule *232–234*, 236, 243
Kehlmann, Daniel 26
Kendall's
  Tau-b Statistic 175–177, 198, 214, 294
  Tau-c Statistic 198–200, 214
  W test 160, 166
k-means Clustering *267–272*, 274, 276, 279, 281
*k*th Nearest Neighbor 270
Knowledge Discovery in Databases (KDD) *See Data Mining*
Kolmogorov-Smirnov Test **143**, 144
Kruskal-Wallis Test **160**, 166

Latent Constructs 8, 10, 220, 316, **318**, *318–321*, 325–326, 332–333, 336, 384
Latent Factors *see Latent Variables*
Latent Variables 222–223, 227–228, 317, 321, 324, *326–328*, 332, 336, 339–340, 342, 345, 349, 378, **380**, **385**
   See also Latent Constructs
Least Squares *180–182*, 192–193, 225–226, 320, 338
Levene's Test 141, **146,** 158
Lifestyle measurement 8, 11, 30, 34, 46, 208, 227, 333, 353
   Survey 33, **34,** 63–68, 74
   Brunso and Grunert Lifestyle Measurement 69–72, 76
Likelihood 106, 356, 369
   (Log-)Likelihood Function 343, 356–358, **380**
   Likelihood Ratio 202–206, 227, 232, **380**
   Maximum Likelihood 82, 224–227, 236, 238, 243, 321, 326, 328, 333, 340, **381**
Likert Scale 7, **9,** 23, 25, 120, 233, 286, 287
Line Chart **86,** *86–90*
Linear Composites *see Canonical Variates*
Linear Regression *see Regression*
Link Function 189, 192, *340–342*, 345–347, 349, 351, 379, **380**, 384
LISREL 21–22, **325**
List-wise Deletion 81, 255, 326, **380**
   See also Missing Data
Logistic Regression 189, 191, 249, 251, *339–347*, 349, 379, **380**
   Stepwise Logistic Regression 343
Logit Model 66, 192, 340, 344–345, 379, **380–381**
   Logit link function *340–342*, **342,** 380
   Multinomial Logit Model 249, *341–342*, 345, 349, 351, 376, 379, **381**
   Ordered Logit Model *345–346*, **382**
Log-linear Analysis 167, 179, 196, *199–207*, 214–217, 312–313, 322, 375, 379, **381**
   General Log-linear Model (GLLM) *199–206*, 214, **379**, 381
   Hierarchical *202–204*, 215, **379**
   k-way Effects *202–203*

Mahalanobis Distance 259, 266
Mail Interviews 52–53, **56,** *56–57*, 58, 61–62, 67, 74, 78
Main Effects 201, 204–205, 345
Mall-intercepts 52, *55–56*, 67, 114

Manhattan Distance 266
Manifest Variable *316–321*, 323, 325–326, 329, 333, 378, **381**, 385
Mann-Whitney U Test **144**, 160
Marginal Effects 202, 204, 345, **347**, see also Main Effects
Market Basket Analysis *201–207*, 345
Marketing Mix 46, *72–73*, **73,** 263
   See also New Product Development
Masses 303–304
   See also Correspondence Analysis
Matching 136, 140
Matrix *363–367*
Mean *369–370*, 92–93
   see also Expected Value
   Test on One Mean *137–139*
   Mean Comparison Tests 82, 93, *139–142*, 143, 151, 159–16, 167, 256
   See also Analysis of Variance
Means-End Chains 299
Measurement *1–10*, 23, 28, 32, 43, 47–48, 76, 286–288, **381**
   Error *4–6*, 10, 23, 32, 47, 76, 104, 107, 230, 289, 318, 321, 368–369
   Model 321, 329, 336, 381, **385**
   Scales 1–2, 4, *6–11*, 23, 25, 47, 59–62, 74, 232, 236, 372, **381**, 383
Median 86–87, 92, 100, 131, 143–144, 160, *369–371*, **370**
Meta Analysis 10–11, 30–31, **32**, 47, 49
Metric Scaling **289**
   See also Measurement Scales
Minimum Sample Discrepancy (CMIN) **322**, 329–332
Minkowski Distance 266
Missing Data 22, 62, 77, 79, *80–83*, 84–85, 96–97, 100, 102, 310, 326, 380, **381**, 382
   Missing at Random (MAR) *80–83*
   Statistical Imputation 78, 82, 84, 102, **380**
   List-wise (Case-wise) Deletion 81, 255, 326, **380**
   Pair-wise Deletion 81–82, **382**
Missing Values *See Missing Data*
Mixed Effects **159**
Mode 92, 100, 143, **371**
Moses Extreme Reaction Test 144
Multicollinearity *see Collinearity*
Multidimensional Scaling 247, *283–299*, 300, 303, 311, 348, 358, 376–377, 379, **381**, 383–384
   ALSCAL 290, 295, **299**
   Classical MDS **289**

Compositional methods (attribute-based) *287–288*, 291, 303, **376**
Decompositional methods (attribute-free) *287–289*, 348, 350, 376, **377**
Ideal Point 285–287, *291–292*, 297, 298–299
In SAS 296–297
In SPSS 293–296
INDSCAL 290
Metric MDS **289**
Non-metric MDS 289–290
PREFSCAL 293
Principal Co-ordinate Analysis *see Classical MDS*
STRESS Function *290*, 293–295, **385**
Unfolding 283, 291, *292–297*, 299, 379, **381**
Multinomial Distribution 204, 215, 306
Multinomial Logit 249, *341–342*, 345, 349, 351, 376, 379, **381**
Multinomial Probit *341–342*, 358, 379, **381**
Multiple Regression 161, 172, 182–183, *185–189*, 191, 193–194, 208, 378, **381**, 383
Multi-stage Sampling 18, **119**, 123, 125
Multivariate ANCOVA (MANCOVA) 165, 199, 378
Multivariate ANOVA (MANOVA) 161, *163*, 165, 167–168, 191, 199, 378–379, **381**
Multivariate Regression 161, *319–320*, **381**
Multi-way ANOVA *160–162*, 164–165, 167, 181, 191, 199, 378–379, **382**

National Diet and Nutrition Survey 34
Negatively Worded Questions 246, 332, 336
Neural Networks 261, 354
New Product Development 3, 47, 49–50, 73, 219, 247, 264, 283–287, 298–299, 337, 339, 347, 350
Neyman (optimal) allocation 110, *113*
Nominal Scale **7**, 300
Non-centrality Parameter (NCP) 322, 330, 332
Non-comparative Scaling **8**, *9*, 11, 23, 287
  See also Continuous Rating Scale, Likert Scale, Stapel Scale, Semantic Differential Scale
Non-hierarchical Clustering Methods *266–271*, 274, 276, 280, **382**
  See also k-means Clustering
Non-metric Scaling **289–290**
  See also Measurement Scales

Non-parametric Tests 130, *142–145*, 146–147, 159–160, 167, 175–176, 178, 193, 270, **382**
Non-probability Sampling 103–104, 106, 112, *123–125*, 128–129, 382, **384**
  See also Quota Sampling, Haphazard Sampling, Convenience Sampling, Snowball Sampling, Judgmental Sampling
Non-response 52, 54, 61, 67–68, 81, 83, 93, 100, 106, 113, 116, 124, 128–129, 381
Non-response Error 48, 59, 62, 78, 105, *118–119*, 125, **382**
  Not-at-home Error 48, 54, 78, 86
  Refusals 48, 61, **78**, 81, 86, 93, 105, 382
  See also Missing Values, Sensitive Questions
Non-sampling Error *47–48*, 53, 58, 61–62, 74, 103–104, 106, 118–119, 125, **382**
Normal Curve *5–6*, 132–133, 137, 372
  See also Error Theory
Normal Distribution 5–6, 87, 99, 106, 115, 120–121, *131–138*, 139, 141, 143–146, 148–149, 159, 174, 262, 321, 349, 357, 368–372, 386
Normed Fit Index (NFI) 322, 330, 332
Not-at-home 48, 54, 78, 86
NPD 41, 43, 45

Observations 80, *359–360*, **382**
Odds Ratio 205, 217, 341–344, **382**
OLAP cubes **99**
Omnibus Survey 34, 43
Ordinal Scaling *6–8*, 80, 164, 370
  See also Measurement Scales
Orthoblique Rotation **229**
Orthogonal 156–157, 225, 228–229, 235, 246, **367**
  Matrix 228, 246, 367
  Rotation 225, 228–229, 246, 380, **382**
  Transformation 367
  Vector *see Orthogonal Transformation*
Orthomax Rotation **229**
Outliers 1, 77, 80, 84–85, 87–88, *93–100*, 102, 146, 253, 266, 274, 276, 280–281, **382**
  Outlier treatment 96–97, 275–276

P-P Plot **99**
P-value *133–138*, 141–142, 175, 179, 181, 184, 189, 253–254, 259, 322, 329, 343, **384**
Paired Samples **140**, *142*, 144, 146

Pair-wise
  Comparison **8**, 26
  Deletion 81–82, **382**
  See also Missing Values
Panel Surveys 30, 33, *34–35*, 40–41, 43–44,
  45, 57, 66–67, 164
Pareto Chart *86–90*
Parsimax Rotation **229**
Part-worth Model **348**
Path Analysis *319–320*, 333, **383**, 385
Pearson, Karl 174
Pearson's
  Correlation Coefficient **174**, *174–179*, 183,
    186, 193
  Chi-square Association Test *see*
    Association, Chi-square Association Test
  Chi-square Goodness-of-fit Test *see*
    Chi-square Goodness-of-fit Test
Perceived Behavioral Control 69, 324–325,
  327–329
Percentile(s) 84, 92, 99, 357, **371**
Perceptions 7, 25, 66, 219, 247, 265, *283–287*,
  289–292, 297, 301, 376, 381, **383**
  Perceptual maps 219, 265, *285–287*, 290,
    292, 303, 376, 381, **383**
Personal Interviews 11, 33–34, 52–54, *55–56*,
  61, 63, 66–68, 74, 78, 119, 122, 375
Pie Chart **86**, *86–90*, 100
Planned Comparisons (in ANOVA) *154*,
  156–158, 162, 169, **376**, 383
  See also Contrasts
Poisson Distribution 143, 148, 204, **368**
Poisson, Siméon Denis 368
Population 4, 28–30, 48–52, 58, 65,
  74,*104–114*, 116–125, 128, 130–133,
  137–138, 140, 143–144, 160, 174–175,
  180–181, 355, 375, 380, **383**, 384
Post-hoc Testing (in ANOVA) *154–158*, 162,
  168, 376, **388**
  See also Scheffé's test, Bonferroni's test,
    Tukey's test, Dunnet C test, HSU test,
    Hochberg's GT2 test, Gabriel's test,
    Duncan's procedure, Studentized
    Newman-Keuls (SNK) procedure,
    REGWF procedure, REGWQ procedure
Post-stratification 32, 66, *118–119*
Power (of a Test) *136–137*, 146, 148, 155,
  167, 205, **383**, 385
Precision 4, 9, 51–52, 60, 86, 92, 103–104,
  107–113, *114–118*, 120–121, 125, 127,
  134, 148–149, 181, 183, 198, 375, **383**, 385
  See also Standard Error

Preference 247, 265, *283–292*, 297, 299–301,
  303, 311, 315
  External Preference Mapping *291–292*,
    299, **378**, 380
  Internal Preference Mapping *291–293*,
    295, 297, 378, **380**
  Preference Maps 283, 286, 291, 297, 377,
    381, **383**
  Revealed Preference Theory **338**, 351
  Stated Preference Theory *337–339*, **338**,
    347, 349, 351, 379
PREFSCAL 293
Principal Component Analysis 171,
  208–210, 218–221, *229–235*, 239–246,
  249–250, 253, 265–266, 272, 285, 297,
  299, 312, 376–377, 380, **382**, 384
  Component Loadings 231, *241–243*, 292,
    380, 382
  Component Rotation *235–237*, 243–244,
    246, 380, **382**
  Component Scores 230–231, *241–244*, 266,
    272, 282, 292, 301, **384**
  In SAS 241
  In SPSS 239–244
  Number of Components *232–234*, 243
  Qualitative PCA **313**
  See also Communalities, Correspondence
    Analysis
Principal Co-ordinate Analysis *see*
  Classical MDS
Principal Factoring 224, **225**
Probability *see Distribution of Probability*
Probability Sampling 104, 106, *108–114*, 125,
  383, **384**
  See also Simple Random Sampling,
    Systematic Sampling, Stratified Sampling,
    Cluster Sampling, Area Sampling,
    Multi-stage Sampling
Probit Model 192, 341, 344, 347, 349, 379, **384**
  Multinomial Probit Model *341–342*, 358,
    379, **381**
  Ordered Probit Model 189, 341, 349, **382**
Procrustes Rotation **229**
Profile *see Correspondence Analysis*
Promax Rotation **229**
Proximity matrix 266, 268, 289, 293, 378,
  380, **384**
Pseudo F Statistic 270, 278
Pseudo R^2 Statistics 345–346
Pseudo t^2 Statistic 270, 278
Psychographics 8, 11, 46, 59, *63-72*, 74, 125,
  see also Lifestyle

Q-Q Plot 86–87, 89, **99**
Q-sorting Scale 8, 287–288
Qualitative Research 4, 47, *49–50*, 59, 124
Qualitative Scale **7**, 86
Quantitative Scale **7**, 286
Quartiles 86–93, 97, 99–100, **371**
Quartimax Rotation **229**, 238
Questionnaire 51–53, *58–63*, 78–80, 385
   Coding 46, 79–80
   Design 53, *58–62*
   Layout 62
   Piloting 62–63
Question
   Structured 60
   Unstructured 60
   Multiple Choice 60–62, 79–80, 100
   Dichotomous 60
   Sensitive 30, 40–41, 50, 52, 55–56, 58,
      60–62, 67–68
   Wording 59–61, 246, 332, 336
Quota Sampling 29, 105–106, 112,
   123–125, 128

R^2 *181–183*, 185–186, 189, 217, 224, 271
   Adjusted R^2 *185*, 187–190, 375, **384**
   Pseudo R^2 345–346
Random Effects **159**, 166, 168,
Random Error 1, 5, 10, 23, 47, 159, 197, 222,
   317, 342, 359–360, **384**
Randomized Techniques *61*
   See also Sensitive Questions
Rank Order Scaling **8**, 287
Rao's V 259
Rating scales **9**
Rectangular Distribution *see Uniform*
   Distribution
Regression *179–193*, **380**
   (Ordinary) Least Squares (OLS) *180–182*,
      192–193, 225–226, 320, 338
   Assumptions 180–181, 185
   Bivariate Linear Regression *179–184*, 208,
      356–357
   Coefficients 179–181, 183–185, 193, 241,
      329, 333, 356, 385
   Diagnostics 181–184
   Forecasts 181, 185, 195
   Logistic Regression 189, 191, 249, 251,
      339–347, 349, 379, **380**
   Multiple Regression 161, 172, 182–183,
      185–189, 191, 193–194, 208, 378,
      381, 383
   Multivariate Regression 161, *319–320*, **381**
   Parameters *see Coefficients*

Residuals *180–182*, 187, 191–193,
   317, 320
   Spurious Regression **184**, 354
REGWF procedure 155
REGWQ procedure 155
Related Samples 136, **140**, 164, 166
   See also Non-parametric Tests
Relative Fit Index (RFI) 322, 330, 332
Reliability *9–11*, 225, 377, **384**
   See also Cronbach's Alpha
Representativeness 29, 33, 43, 52,
   58 64, 79, 83, 106, 110–113,
   123–124, **184**
Research Design *see Survey Design*
Retail Survey 33, **35**, 41, 43–44
Root Mean Square of Error Approximation
   (RMSEA) 322, 329–330, 332
Root Mean Square Residual (RMR) 322
R-square *see R^2*
Runs Test *142–144*

Samples *105–106*, **384**
   Independent Samples *see Unrelated*
     Samples
   Matched Samples 136, 140
   Paired Samples **140**, *142*, 144, 146
   Related Samples 136, **140**, 164, 166
   Sample Size 51–53, 67–68, 80, 81, 83, 97,
      104–107, 111, 113–115, *116–117*,
      120–121, 132–134, 137–138, 145,
      148–149
   Sample Statistics **52**, *106–109*, 114–116,
      118–119, 126–127, 129, 131, 134
   Selection Bias 57–58, 62, 74, 83, 114,
      123–124, 140, 180, 354
   Unrelated Samples *140–142*
Sampling *105–127*, **384**
   Accuracy 9–10, 103, 106, 106–9, *114–117*,
      125, 182, **375**
   Design 22, 33, 52, 63, 66, 83, 110–114, 119,
      122–123, 150
   Fraction **107**, 115
   Frame 31, *48–54*, 57–58, 65–66, 78, 106,
      110–112, 114, 116, 118–119, 122–124,
      128, **384**
   Precision 4, 9, 51–52, 60, 86, 92, 103–104,
      107–113, *114–118*, 120–121, 125, 127,
      134, 148–149, 181, 183, 198, 375,
      383, 385
   Reference Population *see Population*
   Sampling Error 31, 47–48, 52, *103–107*,
      115, 124–125, 135, 143, 151, 153, 155,
      174, 271, **384**

Sampling (*continued*)
  Space *108*, 129, 132
  Unit 18, 30, 34, 40, **51**, 52, 66, 105–106, 110, 112, 114, 118–119, 122, 124, 384
Sampling Methods
  Non-Probability Sampling 103–104, 106, 112, *123–125*, 128–129, 382, **384**
  Probability Sampling 104, 106, *108–114*, 125, 383, **384**
  See also Simple Random Sampling, Systematic Sampling, Stratified Sampling, Cluster Sampling, Area Sampling, Multi-stage Sampling, Two-stage Sampling, Quota Sampling, Haphazard Sampling, Convenience Sampling, Snowball Sampling, Judgmental Sampling
SAS 18, *21–22*, 25–26 *See also individual techniques in SAS*
Saturated Model *201–202*, 204, 322, 330–331
Scan Data 30, 35, *41–43*
Scatter Diagram *86–90*, 97–98, 100, 174, 257–258
Scatterplot *see Scatter Diagram*
Scheffé's Test 154–155, 158
Scree Diagram 227, 234, 236, 269–270, 272–273, 290, **384**
Segmentation 64, 66, 219, 247, *263–264*, 268–269, 283
Selection Bias 57–58, 62, 74, 83, 114, *123–124*, 140, 180, 354
Semantic Differential Scale **9**, 25, 287
Sensitive Questions 30, 40–41, 50, 52, 55–56, 58, *60–62*, 67–68
  See also Randomized Techniques
Shepard's Rough Nondegeneracy Index (RNI) 294–295
Significance Level 117, 129, 133, 135–138, 143, 145–146, 149, 151, 153, 157, 162, 168, 175, 185, 189, 199, 205, 217, 228, 236, 253, 329, *369*, 376, 382, **385**
  See also Type I Error, Confidence Level
Simple Random Sampling 31, 51, 66, 83, 105, **110**, *108–111*, 112, 114–120, 122–123, 125–127, 149
Simultaneous Equation Systems 22, *319–320*
  Reduced form **320**, 323
  Structural form **320**, 323
Single Linkage Method **267**
Smallest F ratio 259
Snowball Sampling **125**
Social Trend Survey 33–34
  see also Lifestyle Survey

Socrates 352–353, 358
Somer's d Statistic *198–200*, 214
Spearman, Charles Edward 221, 293
Spearman's Rho 175, 177–179, 193, 294
Sphericity **164**, 227, 236–237
  See also Bartlett's Sphericity Test
Stacked Chart **87**, 90–91
Standard Deviation 92–93, 107–108, 372, **385**
Standard Error 93, 107–108, *114–115*, 127, 131–133, 140–142, 148–149, **385**
Standardization (of Variables) 99, 137, 139, 174, 223–224, 232, 266, 272, 276, 282, *372–373*, **385**
Stapel Scale **9**, 25, 287
Statistical Software *18–22*
  AMOS 322, *325–333*
  Econometric Views 21, 347
  EQS 325
  LISREL 325
  LimDep 21–22, 347
  S-Plus 25
  SAS 18, *21–22*, 25–26 *see also individual techniques in SAS*
  SPSS 18, *21–22*, 25–26 *see also individual techniques in SPSS*
  Stata 21, 347
Stochastic Function 179, **360**
Stratified Sampling 35, 52, 66–68, 83, **110**, *112–113*, 118, 125–127
  See also Post-stratification
STRESS Function *290*, 293–295, **385**
Structural Equation Modeling (SEM) 66, 163, 175, *315–333*, 335, 339, 376–379, 381, 383, **385**
  Causality 174, 179, 184, 210, **318**, 320–321, 383, 385
  Coefficient Ratio **329**
  Confirmatory Factor Analysis 66, 317, *319–322*, 333, 376, **378**, 385
  EQS 325
  In SAS 325
  In SPSS AMOS 322, *325–333*
  Latent Constructs 8, 10, 220, 316, **318**, *318–321*, 325–326, 332–333, 336, 384
  Latent Variables 222–223, 227–228, 317, 321, 324, *326–328*, 332, 336, 339–340, 342, 345, 349, 378, **380**, **385**
  LISREL 325
  Manifest Variables *316–321*, 323, 325–326, 329, 333, 378, **381**, 385
  Measurement Model 321, 329, 336, 381, **385**

Path Diagram *320–321*, 325–326, 333, **383**, 385
Path Analysis *319–320*, 333, **383**, 385
Structural Model **321**, 326, 331, 379, 385
Studentized Newman-Keuls (SNK) procedure 155
Studentized Range Distribution 155
Student-t Distribution 121, *132–133*, 137–138, 141, 148–149
Subjective Norm 69, *324–325*, 329
Survey 28–32, *46–75*, **385**
　Design 32, 46–49, *51–53*, 59, 74, 82
　Error 45, *47–48*, **48**, 104, 106, 382, 384
　Incentives 57, 62, 67, 78
　See also Data, Secondary, Household Survey, Lifestyle Survey, Social Trend Survey, Panel Survey, Attitudinal Survey, Retail Survey, Omnibus Survey, National Diet & Nutrition Survey, British Household Panel, EU-SILC, General Household Survey, Consumer & Expenditure Survey (US), Expenditure & Food Survey, Trust Survey, Electoral Survey, Annual Population Survey
Systematic Error 5–6, 23, 48, **385**
　See also Bias
Systematic Sampling 41, 52, 56, 84, **110**, *111–112*, 114, 123, 125–127

t Distribution See Student-t distribution
t Test *140–142*, 155, 160–161, 181, 184, 189, 345, 378, **385**
Tabulation 33, 84–85, *91–94*, 99, 173, 196, 220, 238, 244, 278, 375
Taylor-Nelson Sofres 43
Telephone Interviews 41, *52–55*, 60, 67–68, 78–79, 375
　See also Computer-assisted Interviews
Test of Close Fit (PCLOSE) **322**, 330, 332
Tests *see Hypothesis Testing*
Theory of Planned Behavior 50, 59, 66, 69, 74, 316, *322–333*
　Attitudes 3–4, 8, 32–34, 35, 44, 50, 63–66, 69–70, 76, 174, 208, 219, 318, 324, 333
　Beliefs 4, **69**, 324–326
　Intention to Behave 3, 65, 69, 316, 324–326, 329
　Perceived Behavioral Control 69, 324–325, 327–329
　Subjective Norm 69, *324–325*, 329
Thurstone, Louis Leon 26, 221

Trust Survey 12, *18–21*, 23, 91–93, 97, 100, 134, 137, 140, 143–144, 176, 187, 191, 199, 208, 220, 235, 249, 301, 322–325, 342
Tucker-Lewis Coefficient (TLI) 322, 330, 332
Tukey's Test *154–155*
Two-stage Sampling *See Multi-stage Sampling*
Two-step Cluster Analysis *270–271*, 272, 274, 276
Type I error *136–137*, 146, 151, 154–156, 163, 376, **385**
　See also Significance Level, Confidence Level
Type II error *136–137*, 146, 155, 383, **385**
　See also Power

Unbiasedness 109, 117–118, 153
Uncertainty Coefficient 198
Unexplained Variance 259
Unfolding 283, 291, *292–297*, 299, 379, **381**
　See also Multidimensional Scaling
Uniform Distribution *120–121*, 368, 371
Unrelated Samples *140–142*
Utility 76, 347–348

Validity *9–10*, 26, 246, **386**
Variability 8, 10, 62, 77, 83, 90, *107–108*, 113–114, 120, 125, 131, 145–146, 148, 173, 180, 184, 193, 209–210, 219–221, 226–227, 229–231, 235, 243–244, 249, 289, 294–295, 306, 369, *371–374*, 385
　Coefficient of Variation 100, **372**
　Covariance Matrix 222–225, 231–233, 236, 242–243, 251, 253, 301, 321–322, 328–329, *373–374*, **374**, 376–377, 379
　Decomposition *151–153*, 181–183, 222–224
　Interquartile Range 97, 134–135, **371**
　Measures 85, 92, 100, 108–109, 126–127, 145, 172–173, 180–185, 232, *371–374*
　Standard Deviation 92–93, 107–108, *372*, **385**
　Variance (definition of) **372**
　See also Analysis of Variance
Variables *359–360*, **386**
　Categorical *see Non-metric Variables*
　Continuous 91, 93, 99, 192, 275, 340–341, 345, 371, 380, **386**
　Dependent 32, 82–83, 160–164, 172, 176, *179–187*, 191–193, 199, 201, 208–209, 249, 260, 315–317, 319–320, 337–349, 355, 360–362, **377**, 379–382, 384–385

Variables (*continued*)
 Discrete 85–87, 91–93, 191, 370, 380, **386**
 Dummy 101, 160, 162, *191*, 199, 201, **377**
 Endogenous *179*, 317, 319–322, 327, 333,
  378, 385
 Exogenous *179*, 317, 319–322, 327–328,
  333, 356, **378**, 385
 Explanatory 32, 52, 60, 82, 150, 159, 161,
  172, 179–182, *184–193*, 198–199,
  208–209, 226, 249–250, 259, 262,
  316–318, 332–333, 338–345, 355,
  360–362, 376–377, **378**, 379–381,
  383–385
 Independent *see Explanatory Variables*
 Metric 7, 84, 86, 91, 172, 179, 191, 193, 201,
  208, 249, 262, 275, 287–288, 311, 338,
  368, 371, 375–376, **381**, 384–386
 Nominal 1, 7, 24, 198, 214, 275, 293,
  310, **382**
 Non-metric 1, 7, 79, 91, 179, 191, 196, 199,
  208, 214, 285, 287–288, 301, 315, 337,
  339–340, 349, 375, 379–380, **382**, 386
 Omitted **187**
 Ordinal 1, 7–9, 24, 144, 175, 196–197, 247,
  262, 288–290, 300, 311, 339, 345, **382**
 Qualitative *See Non-metric Variables*
 Quantitative *see Metric Variables*

Random 107, 148, 223, 355–357, *359–360*,
  362, 370, 379, 383, **384**, 385
Redundant **187**
Scale *see Metric Variables*
Standardization 99, 137, 139, 174,
  223–224, 232, 266, 272, 276, 282,
  372–373, **385**
Variance *see Variability*
Variance Inflation Factor (VIF) **186**, 188, 190
  See also Collinearity
Variation *See Variability*
Varimax Rotation **229**, 237–240, 244–245
Vector *363–367*
Vector Linear Model **348**
von Humboldt, Alexander 3

Wald-Wolfowitz Test 144
Wang, Sam 31
Ward's Method **267**, 272–276
Welch Test 159
Wilcoxon
 Signed Rank Test 143–144
 Rank-sum Test *see Mann-Whitney U Test*
Wilks' Lambda 209, 253–256, 259

Zero-order Correlation *see Pearson's*
  Correlation Coefficient